T0296491

IRRESISTIBLE INTEGRALS

The problem of evaluating integrals is well known to every student who has
had a year of calculus. It was an especially important subject in nineteenth-
century analysis and it has now been revived with the appearance of symbolic
languages. In this book, the authors use the problem of exact evaluation of
definite integrals as a starting point for exploring many areas of mathematics.
The questions discussed here are as old as calculus itself.

In presenting the combination of methods required for the evaluation of
most integrals, the authors take the most interesting, rather than the shortest,
path to the results. Along the way, they illuminate connections with many subjects, including analysis, number theory, and algebra. This will be a guided
tour of exciting discovery for undergraduates and their teachers in mathematics, computer science, physics, and engineering.

George Boros was Assistant Professor of Mathematics at Xavier University
of Lousiana, New Orleans.

Victor Moll is Professor of Mathematics at Tulane University. He has published numerous articles in mathematical journals and is the co-author of
Elliptic Curves, also published by Cambridge University Press.

IRRESISTIBLE INTEGRALS

Symbolics, Analysis and Experiments in the
Evaluation of Integrals

GEORGE BOROS

Formerly of Xavier University of Lousiana

VICTOR MOLL

Tulane University

CAMBRIDGE
UNIVERSITY PRESS

CAMBRIDGE UNIVERSITY PRESS
Cambridge, New York, Melbourne, Madrid, Cape Town,
Singapore, São Paulo, Delhi, Mexico City

Cambridge University Press
The Edinburgh Building, Cambridge CB2 8RU, UK

Published in the United States of America by Cambridge University Press, New York

www.cambridge.org
Information on this title: www.cambridge.org/9780521796361

First published 2004
Reprinted 2006

A catalogue record for this publication is available from the British Library

Library of Congress Cataloguing in Publication Data
Boros George, 1947–
Irresistible integrals : symbolics, analysis and experiments in the evaluation of integrals /
George Boros, Victor H. Moll.
p. cm
Includes bibliographical references and index.
ISBN 0-521-79186-3 – ISBN 0-521-79636-9 (pbk)
1.Definite integrals. I. Moll, Victor. II. Title.
QA308.B67 2004
515'.43 – dc22 2003069574

ISBN 978-0-521-79186-1 Hardback
ISBN 978-0-521-79636-1 Paperback

To Marian

To Lisa, Alexander and Stefan

Contents

Preface

The idea of writing a book on all the areas of mathematics that appear in the evaluation of integrals occurred to us when we found many beautiful results scattered throughout the literature.

The original idea was naive: inspired by the paper "Integrals: An Introduction to Analytic Number Theory" by Ilan Vardi (1988) we decided to write a text in which we would prove every formula in *Table of Integrals, Series, and Products* by I. S. Gradshteyn and I. M. Rhyzik (1994) and its precursor by Bierens de Haan (1867). It took a short time to realize that this task was monumental.

In order to keep the book to a reasonable page limit, we have decided to keep the material at a level accesible to a junior/senior undergraduate student. We assume that the reader has a good knowledge of one-variable calculus and that he/she has had a class in which there has been some exposure to a rigorous proof. At Tulane University this is done in Discrete Mathematics, where the method of mathematical induction and the ideas behind recurrences are discussed in some detail, and in Real Analysis, where the student is exposed to the basic material of calculus, now with rigorous proofs. It is our experience that most students majoring in mathematics will have a class in linear algebra, but not all (we fear, few) study complex analysis. Therefore we have kept the use of these subjects to a minimum. In particular we have made an effort *not* to use complex analysis.

The goal of the book is to present to the reader the many facets involved in the evaluation of definite integrals. At the end, we decided to emphasize the connection with number theory. It is an unfortunate fact of undergraduate, and to some extent graduate, education that students tend to see mathematics as comprising distinct parts. We have tried to connect the discrete (prime numbers, binomial coefficients) with the continuous (integrals, special functions). The reader will tell if we have succeeded.

Many of the evaluations presented in this book involve parameters. These had to be restricted in order to make the resulting integrals convergent. We have decided not to write down these restrictions.

The symbolic language Mathematica™ is used throughout the book. We do not assume that the reader has much experience with this language, so we incorporate the commands employed by the authors in the text. We hope that the reader will be able to reproduce what we write. It has been our experience that the best way to learn a symbolic language is to learn the commands as you need them to attack your problem of interest. It is like learning a real language. You do not need to be fluent in Spanish in order to order *empanadas*, but more is required if you want to understand *Don Quixote*. This book is mostly at the empanada level.

Symbolic languages (like Mathematica) are in a constant state of improvement, thus the statement *this cannot be evaluated symbolically* should always be complemented with the phrase *at the time of writing this text*.

We have tried to motivate the results presented here, even to the point of *wasting* time. It is certainly shorter to present mathematics as facts followed by a proof, but that takes all the fun out of it.

Once the target audience was chosen we decided to write first about the elementary functions that the student encounters in the beginning sequence of courses. This constitutes the first seven chapters of the book. The last part of the book discusses different families of integrals. We begin with the study of a rational integral, and there we find a connection with the expansion of the double square root. The reader will find here a glimpse into the magic world of Ramanujan. The next three chapters contain the normal integral, the Eulerian integrals gamma and beta, and Euler's constant. The book concludes with a short study on the integrals that can be evaluated in terms of the famous Riemann zeta function and an introduction to logarithmic integrals; we finish with our master formula: a device that has produced many interesting evaluations.

We hope that the reader will see that with a good calculus background it is possible to enter the world of integrals and to experience some of its flavor. The more experienced reader will certainly know shorter proofs than the ones we have provided here. The beginning student should be able to read all the material presented here, accepting some results as given. Simply take these topics as a list of *things to learn later in life*.

As stated above, the main goal of the book is to evaluate integrals. We have tried to use this as *a springboard for many unexpected investigations and discoveries in mathematics* (quoted from an earlier review of this manuscript). We have tried to explore the many ramifications involved with a specific evaluation. We would be happy to hear about new ones.

The question of integrating certain functions produces many reactions. On page 580 of M. Spivak's calculus book (1980) we find

The impossibility of integrating certain functions in elementary terms is one of the most esoteric subjects in mathematics

and this should be compared with G. H. Hardy's famous remark

I could never resist an integral

and R. Askey's comment[1]

If things are nice there is probably a good reason why they are nice: and if you do not know at least one reason for this good fortune, then you still have work to do.

We have tried to keep these last two remarks in mind while writing.

The exercises are an essential part of the text. We have included alternative proofs and other connections with the material presented in the chapter. The level of the exercises is uneven, and we have provided hints for the ones we consider more difficult. The projects are exercises that we have not done in complete detail. We have provided some ideas on how to proceed, but for some of them we simply do not know where they will end nor how hard they could be. The author would like to hear from the reader on the solutions to these questions.

Finally the word *Experiments* in the subtitle requires an explanation. These are *computer experiments* in which the reader is required to guess a closed form expression for an analytic object (usually a definite integral) from enough data produced by a symbolic language. The final goal of the experiment is to provide a proof of the closed form. In turn, these proofs suggest new experiments.

The author would like to acknowledge many people who contributed to this book:

- First of all my special thanks to Dante Manna, who checked every formula in the book. He made sure that every $f_n^{(i+1)}$ was not a mistake for $f_{n-1}^{(i)}$. Naturally all the psosible errors are the author's responsibility.
- Bruce Berndt, Doron Zeilberger who always answered my emails.
- Michael Trott at Wolfram Research, Inc. who always answered my most trivial questions about the Mathematica language.
- Sage Briscoe, Frank Dang, Michael Joyce, Roopa Nalam, and Kirk Soodhalter worked on portions of the manuscript while they were undergraduates at Tulane.

---

[1] Quoted from a transparency by Doron Zeilberger.

- The students of SIMU 2000: Jenny Alvarez, Miguel Amadis, Encarnacion Gutierrez, Emilia Huerta, Aida Navarro, Lianette Passapera, Christian Roldan, Leobardo Rosales, Miguel Rosario, Maria Torres, David Uminsky, and Yvette Uresti and the teaching assistants: Dagan Karp and Jean Carlos Cortissoz.
- The students of SIMU 2002: Benjamin Aleman, Danielle Brooker, Sage Briscoe, Aaron Cardona, Angela Gallegos, Danielle Heckman, Laura Jimenez, Luis Medina, Jose Miranda, Sandra Moncada, Maria Osorio, and Juan Carlos Trujillo and the teaching assistants: Chris Duncan and Dante Manna.
- The organizers of SIMU: Ivelisse Rubio and Herbert Medina.
- The participants of a 1999 summer course on a preliminary version of this material given at Universidad Santa Maria, Valparaiso, Chile.

The second author acknowledges the partial support of NSF-DMS 0070567, Project Number 540623.

George Boros passed away during the final stages of this project. I have often expressed the professional influence he had on me, showing that integrals were interesting and fun. It is impossible to put in words what he meant as a person. *We miss him.*

— Victor Moll
New Orleans
January 2004

Notation

The notation used throughout the book is standard:

$\mathbb{N} = \{1,\ 2,\ 3,\ \dots\}$ are the natural numbers.

$\mathbb{N}_0 = \mathbb{N} \cup \{0\}$.

$\mathbb{Z} = \mathbb{N} \cup \{0\} \cup -\mathbb{N}$ are the integers.

$\mathbb{R}$ are the real numbers and $\mathbb{R}^+$ are the positive reals.

$\ln x$ is the natural logarithm.

$\lfloor x \rfloor$ is the integer part of $x \in \mathbb{R}$ and $\{x\}$ is the fractional part.

$n!$ is the factorial of $n \in \mathbb{N}$.

$\binom{n}{k}$ are the binomial coefficients.

C_m is the central binomial coefficients $\binom{2m}{m}$.

$(a)_k = a(a+1)(a+2) \cdots (a+k-1)$ is the ascending factorial or Pochhammer symbol.

1

Factorials and Binomial Coefficients

1.1. Introduction

In this chapter we discuss several properties of factorials and binomial coefficients. These functions will often appear as results of evaluations of definite integrals.

Definition 1.1.1. A function $f : \mathbb{N} \to \mathbb{N}$ is said to satisfy a **recurrence** if the value $f(n)$ is determined by the values $\{f(1), f(2), \ldots, f(n-1)\}$. The recurrence is of **order** k if $f(n)$ is determined by the values $\{f(n-1), f(n-2), \ldots, f(n-k)\}$, where k is a fixed positive integer. The notation f_n is sometimes used for $f(n)$.

For example, the **Fibonacci numbers** F_n satisfy the second-order recurrence

$$F_n = F_{n-1} + F_{n-2}. \tag{1.1.1}$$

Therefore, in order to compute F_n, one needs to know only F_1 and F_2. In this case $F_1 = 1$ and $F_2 = 1$. These values are called the **initial conditions** of the recurrence. The Mathematica command

```
F[n_] := If[n==0,1, If[n==1,1, F[n-1]+F[n-2]]]
```

gives the value of F_n. The modified command

```
F[n_] := F[n]= If[n==0,1, If[n==1,1, F[n-1]+F[n-2]]]
```

saves the previously computed values, so at every step there is a single sum to perform.

Exercise 1.1.1. Compare the times that it takes to evaluate

$$F_{30} = 832040 \tag{1.1.2}$$

using both versions of the function F.

1

A recurrence can also be used to *define* a sequence of numbers. For instance

$$D_{n+1} = n(D_n + D_{n-1}), \ n \geq 2 \tag{1.1.3}$$

with $D_1 = 0, \ D_2 = 1$ defines the **derangement numbers**. See Rosen (2003) for properties of this interesting sequence.

We now give a recursive definition of the factorials.

Definition 1.1.2. The **factorial** of $n \in \mathbb{N}$ is defined by

$$n! = n \cdot (n-1) \cdot (n-2) \cdots 3 \cdot 2 \cdot 1. \tag{1.1.4}$$

A recursive definition is given by

$$\begin{aligned} 1! &= 1 \\ n! &= n \times (n-1)!. \end{aligned} \tag{1.1.5}$$

The first exercise shows that the recursive definition characterizes $n!$. This technique will be used throughout the book: in order to prove some identity, you check that both sides satisfy the same recursion and that the initial conditions match.

Exercise 1.1.2. Prove that the factorial is the unique solution of the recursion

$$x_n = n \times x_{n-1} \tag{1.1.6}$$

satisfying the initial condition $x_1 = 1$. **Hint**. Let $y_n = x_n/n!$ and use (1.1.5) to produce a trivial recurrence for y_n.

Exercise 1.1.3. Establish the formula

$$D_n = n! \times \sum_{k=0}^{n} \frac{(-1)^k}{k!}. \tag{1.1.7}$$

Hint. Check that the right-hand side satisfies the same recurrence as D_n and then check the initial conditions.

The first values of the sequence $n!$ are

$$1! = 1, \quad 2! = 2, \quad 3! = 6, \quad 4! = 24, \tag{1.1.8}$$

and these grow very fast. For instance

50! = 30414093201713378043612608166064768844377641568960512000000000000

and 1000! has 2568 digits.

Mathematica 1.1.1. The Mathematica command for $n!$ is `Factorial [n]`. The reader should check the value 1000! stated above. The number of digits of an integer can be obtained with the Mathematica command `Length[IntegerDigits[n]]`.

The next exercise illustrates the fact that the extension of a function from $\mathbb{N}$ to $\mathbb{R}$ sometimes produces unexpected results.

Exercise 1.1.4. Use Mathematica to check that $\left(\dfrac{1}{2}\right)! = \dfrac{\sqrt{\pi}}{2}$.

The exercise is one of the instances in which the factorial is connected to π, the fundamental constant of trigonometry. Later we will see that the growth of $n!$ as $n \to \infty$ is related to e: the base of natural logarithms. These issues will be discussed in Chapters 5 and 6, respectively. To get a complete explanation for the appearance of π, the reader will have to wait until Chapter 10 where we introduce the **gamma function**.

1.2. Prime Numbers and the Factorization of *n*!

In this section we discuss the factorization of $n!$ into prime factors.

Definition 1.2.1. An integer $n \in \mathbb{N}$ is **prime** if its only divisors are 1 and itself.

The reader is refered to Hardy and Wright (1979) and Ribenboim (1989) for more information about prime numbers. In particular, Ribenboim's first chapter contains many proofs of the fact that there are infinitely many primes. Much more information about primes can be found at the site

http://www.utm.edu/research/primes/

The set of prime numbers can be used as building blocks for all integers. This is the content of the **Fundamental Theorem of Arithmetic** stated below.

Theorem 1.2.1. *Every positive integer can be written as a product of prime numbers. This factorization is unique up to the order of the prime factors.*

The proof of this result appears in every introductory book in number theory. For example, see Andrews (1994), page 26, for the standard argument.

Mathematica 1.2.1. The Mathematica command `FactorInteger[n]` gives the complete factorization of the integer n. For example `FactorInteger[1001]` gives the prime factorization $1001 = 7 \cdot 11 \cdot 13$. The concept of prime factorization can now be extended to rational numbers by allowing negative exponents. For example

$$\frac{1001}{1003} = 7 \cdot 11 \cdot 13 \cdot 17^{-1} \cdot 59^{-1}. \qquad (1.2.1)$$

The efficient complete factorization of a large integer n is one of the basic questions in computational number theory. The reader should be careful with requesting such a factorization from a symbolic language like Mathematica: the amount of time required can become very large. A safeguard is the command

```
FactorInteger[n, FactorComplete -> False]
```

which computes the small factors of n and leaves a part unfactored. The reader will find in Bressoud and Wagon (2000) more information about these issues.

Definition 1.2.2. Let p be prime and $r \in \mathbb{Q}^{+}$. Then there are unique integers a, b, not divisible by p, and $m \in \mathbb{Z}$ such that

$$r = \frac{a}{b} \times p^{m}. \qquad (1.2.2)$$

The p-**adic valuation** of r is defined by

$$\nu_p(r) = p^{-m}. \qquad (1.2.3)$$

The integer m in (1.2.2) will be called the **exponent of** p **in** m and will be denoted by $\mu_p(r)$, that is,

$$\nu_p(r) = p^{-\mu_p(r)}. \qquad (1.2.4)$$

Extra 1.2.1. The p-adic valuation of a rational number gives a new way of measuring its size. In this context, a number is small if it is divisible by a large power of p. This is the basic idea behind p-*adic Analysis*. Nice introductions to this topic can be found in Gouvea (1997) and Hardy and Wright (1979).

Exercise 1.2.1. Prove that the valuation ν_p satisfies

$$\nu_p(r_1 r_2) = \nu_p(r_1) \times \nu_p(r_2),$$
$$\nu_p(r_1/r_2) = \nu_p(r_1)/\nu_p(r_2),$$

and

$$v_p(r_1 + r_2) \leq \text{Max}\left(v_p(r_1), v_p(r_2)\right),$$

with equality unless $v_p(r_1) = v_p(r_2)$.

Extra 1.2.2. The p-adic numbers have many surprising properties. For instance, a series converges p-adically if and only if the general term converges to 0.

Definition 1.2.3. The **floor** of $x \in \mathbb{R}$, denoted by $\lfloor x \rfloor$, is the smallest integer less or equal than x. The Mathematica command is `Floor[x]`.

We now show that the factorization of $n!$ can be obtained *without* actually computing its value. This is useful considering that $n!$ grows very fast—for instance 10000! has 35660 digits.

Theorem 1.2.2. *Let p be prime and $n \in \mathbb{N}$. The exponent of p in $n!$ is given by*

$$\mu_p(n!) = \sum_{k=1}^{\infty} \left\lfloor \frac{n}{p^k} \right\rfloor. \tag{1.2.5}$$

Proof. In the product defining $n!$ one can divide out every multiple of p, and there are $\lfloor n/p \rfloor$ such numbers. The remaining factor might still be divisible by p and there are $\lfloor n/p^2 \rfloor$ such terms. Now continue with higher powers of p. $\qquad\qquad\square$

Note that the sum in (1.2.5) is finite, ending as soon as $p^k > n$. Also, this sum allows the fast factorization of $n!$. The next exercise illustrates how to do it.

Exercise 1.2.2. Count the number of divisions required to obtain

$$50! = 2^{47} \cdot 3^{22} \cdot 5^{12} \cdot 7^8 \cdot 11^4 \cdot 13^3 \cdot 17^2 \cdot 19^2 \cdot 23^2 \cdot 29 \cdot 31 \cdot 37 \cdot 41 \cdot 43 \cdot 47,$$

using (1.2.5).

Exercise 1.2.3. Prove that every prime $p \leq n$ appears in the prime factorization of $n!$ and that every prime $p > n/2$ appears to the first power.

There are many expressions for the function $\mu_p(n)$. We present a proof of one due to Legendre (1830). The result depends on the expansion of an integer in base p. The next exercise describes how to obtain such expansion.

Exercise 1.2.4. Let $n, p \in \mathbb{N}$. Prove that there are integers $n_0, n_1, \ldots, n_r$ such that

$$n = n_0 + n_1 p + n_2 p^2 + \cdots + n_r p^r \qquad (1.2.6)$$

where $0 \le n_i < p$ for $0 \le i \le r$. **Hint.** Recall the division algorithm: given $a, b \in \mathbb{N}$ there are integers q, r, with $0 \le r < b$ such that $a = qb + r$. To obtain the coefficients n_i first divide n by p.

Theorem 1.2.3. *The exponent of p in $n!$ is given by*

$$\mu_p(n!) = \frac{n - s_p(n)}{p - 1}, \qquad (1.2.7)$$

where $s_p(n) = n_0 + n_1 + \cdots + n_r$ is the sum of the base-p digits of n. In particular,

$$\mu_2(n!) = n - s_2(n). \qquad (1.2.8)$$

Proof. Write n in base p as in (1.2.6). Then

$$\mu_p(n!) = \sum_{k=1}^{\infty} \left\lfloor \frac{n}{p^k} \right\rfloor$$

$$= (n_1 + n_2 p + \cdots + n_r p^{r-1}) + (n_2 + n_3 p + \cdots + n_r p^{r-2})$$
$$+ \cdots + n_r,$$

so that

$$\mu_p(n!) = n_1 + n_2(1 + p) + n_3(1 + p + p^2) + \cdots + n_r(1 + p + \cdots + p^{r-1})$$

$$= \frac{1}{p-1}\left(n_1(p-1) + n_2(p^2 - 1) + \cdots + n_r(p^r - 1) \right)$$

$$= \frac{n - s_p(n)}{p - 1}. \qquad \square$$

Corollary 1.2.1. *The exponent of p in $n!$ satisfies*

$$\mu_p(n!) \le \frac{n - 1}{p - 1}, \qquad (1.2.9)$$

with equality if and only if n is a power of p.

Mathematica 1.2.2. The command `IntegerDigits[n,p]` gives the list of numbers n_i in Exercise 1.2.4.

Exercise 1.2.5. Define

$$A_1(m) = (2m + 1) \prod_{k=1}^{m}(4k - 1) - \prod_{k=1}^{m}(4k + 1). \qquad (1.2.10)$$

Prove that, for any prime $p \neq 2$,

$$\mu_p(A_1(m)) \geq \mu_p(m!). \qquad (1.2.11)$$

Hint. Let $a_m = \prod_{k=1}^{m}(4k - 1)$ and $b_m = \prod_{k=1}^{m}(4k + 1)$ so that a_m is the product of the least m positive integers congruent to 1 modulo 4. Observe that for $p \geq 3$ prime and $k \in \mathbb{N}$, exactly one of the first p^k positive integers congruent to 3 modulo 4 is divisible by p^k and the same is true for integers congruent to 1 modulo 4. Conclude that $A_1(m)$ is divisible by the odd part of $m!$. For instance,

$$\frac{A_1(30)}{30!} = \frac{359937762656357407018337533}{2^{24}}. \qquad (1.2.12)$$

The products in (1.2.10) will be considered in detail in Section 10.9.

1.3. The Role of Symbolic Languages

In this section we discuss how to use Mathematica to conjecture general closed form formulas. A simple example will illustrate the point.

Exercise 1.2.3 shows that $n!$ is divisible by a large number of consecutive prime numbers. We now turn this information around to empirically suggest closed-form formulas. Assume that in the middle of a calculation we have obtained the numbers

$$x_1 = 5356234211328000$$
$$x_2 = 1027936666719744000$$
$$x_3 = 2074369080655872000$$
$$x_4 = 43913881247588352000$$
$$x_5 = 9731608032706560000000,$$

and one hopes that these numbers obey a simple rule. The goal is to obtain a function $x : \mathbb{N} \to \mathbb{N}$ that interpolates the given values, that is, $x(i) = x_i$ for $1 \leq i \leq 5$. Naturally this question admits more than one solution, and we will use Mathematica to find one. The prime factorization of the data is

$$x_1 = 2^{23} \cdot 3^6 \cdot 5^3 \cdot 7^2 \cdot 11 \cdot 13$$
$$x_2 = 2^{15} \cdot 3^6 \cdot 5^3 \cdot 7^2 \cdot 11 \cdot 13 \cdot 17^2$$
$$x_3 = 2^{18} \cdot 3^{12} \cdot 5^3 \cdot 7^2 \cdot 11 \cdot 13 \cdot 17$$
$$x_4 = 2^{16} \cdot 3^8 \cdot 5^3 \cdot 7^2 \cdot 11 \cdot 13 \cdot 17 \cdot 19^3$$
$$x_5 = 2^{22} \cdot 3^8 \cdot 5^6 \cdot 7^2 \cdot 11 \cdot 13 \cdot 17 \cdot 19$$

and a moment of reflection reveals that x_i contains all primes less than $i + 15$. This is also true for $(i + 15)!$, leading to the consideration of $y_i = x_i / (i + 15)!$. We find that

$$y_1 = 256$$
$$y_2 = 289$$
$$y_3 = 324$$
$$y_4 = 361$$
$$y_5 = 400,$$

so that $y_i = (i + 15)^2$. Thus $x_i = (i + 15)^2 \times (i + 15)!$ is one of the possible rules for x_i. This can be then tested against more data, and if the rule still holds, we have produced the conjecture

$$z_i = i^2 \times i!, \tag{1.3.1}$$

where $z_i = x_{i+15}$.

Definition 1.3.1. Given a sequence of numbers $\{a_k : k \in \mathbb{N}\}$, the function

$$T(x) = \sum_{k=0}^{\infty} a_k x^k \tag{1.3.2}$$

is the **generating function** of the sequence. If the sequence is finite, then we obtain a **generating polynomial**

$$T_n(x) = \sum_{k=0}^{n} a_k x^k. \tag{1.3.3}$$

The generating function is one of the forms in which the sequence $\{a_k : 0 \leq k \leq n\}$ can be incorporated into an analytic object. Usually this makes it easier to perform calculations with them. Mathematica *knows* a large number of polynomials, so if $\{a_k\}$ is part of a known family, then a symbolic search will produce an expression for T_n.

Exercise 1.3.1. Obtain a closed-form for the generating function of the Fibonacci numbers. **Hint.** Let $f(x) = \sum_{n=0}^{\infty} F_n x^n$ be the generating function. Multiply the recurrence (1.1.1) by x^n and sum from $n = 1$ to ∞. In order to manipulate the resulting series observe that

$$\sum_{n=1}^{\infty} F_{n+1} x^n = \sum_{n=2}^{\infty} F_n x^{n-1}$$

$$= \frac{1}{x}(f(x) - F_0 - F_1 x).$$

The answer is $f(x) = x/(1 - x - x^2)$. The Mathematica command to generate the first n terms of this is

```
list[n_]:= CoefficientList
[Normal[Series[ x/(1-x-x^{2}), {x,0,n-1}]],x]
```

For example, `list[10]` gives $\{0, 1, 1, 2, 3, 5, 8, 13, 21, 34\}$.

It is often the case that the answer is expressed in terms of more complicated functions. For example, Mathematica evaluates the polynomial

$$G_n(x) = \sum_{k=0}^{n} k! x^k \tag{1.3.4}$$

as

$$G_n(x) = -\frac{e^{-1/x}}{x} \left\{ \Gamma(0, -\tfrac{1}{x}) + (-1)^n \Gamma(n+2)\Gamma(-1-n, -\tfrac{1}{x}) \right\}, \tag{1.3.5}$$

where e^u is the usual **exponential function**,

$$\Gamma(x) = \int_0^{\infty} t^{x-1} e^{-t}\, dt \tag{1.3.6}$$

is the **gamma function**, and

$$\Gamma(a, x) = \int_x^{\infty} t^{a-1} e^{-t}\, dt \tag{1.3.7}$$

is the **incomplete gamma function**. The exponential function will be discussed in Chapter 5, the gamma function in Chapter 10, and the study of $\Gamma(a, x)$ is postponed until Volume 2.

1.4. The Binomial Theorem

The goal of this section is to recall the binomial theorem and use it to find closed-form expressions for a class of sums involving binomial coefficients.

Definition 1.4.1. The **binomial coefficient** is

$$\binom{n}{k} := \frac{n!}{k!\,(n-k)!}, \quad 0 \le k \le n. \tag{1.4.1}$$

Theorem 1.4.1. *Let $a, b \in \mathbb{R}$ and $n \in \mathbb{N}$. Then*

$$(a+b)^n = \sum_{k=0}^{n} \binom{n}{k} a^{n-k} b^k. \tag{1.4.2}$$

Proof. We use induction. The identity $(a+b)^n = (a+b) \times (a+b)^{n-1}$ and the induction hypothesis yield

$$(a+b)^n = \sum_{k=0}^{n-1} \binom{n-1}{k} a^{n-k} b^k + \sum_{k=0}^{n-1} \binom{n-1}{k} a^{n-k-1} b^{k+1}$$

$$= a^n + \sum_{k=1}^{n-1} \left[\binom{n-1}{k} + \binom{n-1}{k-1} \right] a^{n-k} b^k + b^n.$$

The result now follows from the identity

$$\binom{n}{k} = \binom{n-1}{k} + \binom{n-1}{k-1}, \tag{1.4.3}$$

that admits a direct proof using (1.4.1). $\square$

Exercise 1.4.1. Check the details.

Note 1.4.1. The binomial theorem

$$(1+x)^n = \sum_{k=0}^{n} \binom{n}{k} x^k \tag{1.4.4}$$

shows that $(1+x)^n$ is the generating function of the binomial coefficients

$$\left\{ \binom{n}{k} : 0 \le k \le n \right\}.$$

In principle it is difficult to predict if a given sequence will have a *simple* generating function. Compare (1.3.5) with (1.4.4).

We now present a different proof of the binomial theorem in which we illustrate a general procedure that will be employed throughout the text. The goal is to find and prove an expression for $(a + b)^n$.

a) **Scaling**. The first step is to write

$$(a + b)^n = b^n(1 + x)^n \tag{1.4.5}$$

with $x = a/b$, so that it suffices to establish (1.4.2) for $a = 1$.

b) **Guessing the structure**. The second step is to formulate an educated guess on the form of $(1 + x)^n$. Expanding $(1 + x)^n$ (for any specific n) shows that it is a polynomial in x of degree n, with positive integer coefficients, that is,

$$(1 + x)^n = \sum_{k=0}^{n} b_{n,k} x^k \tag{1.4.6}$$

for some undetermined $b_{n,k} \in \mathbb{N}$. Observe that $x = 0$ yields $b_{n,0} = 1$.

c) The next step is to find a way to **understand the coefficients** $b_{n,k}$.

Exercise 1.4.2. Differentiate (1.4.6) to produce the recurrence

$$b_{n,k+1} = \frac{n}{k+1} b_{n-1,k} \quad 0 \le k \le n - 1. \tag{1.4.7}$$

Conclude that the numbers $b_{n,k}$ are determined from (1.4.7) and initial condition $b_{n,0} = 1$.

We now **guess the solution** to (1.4.7) by studying the list of coefficients

$$L[n] := \{b_{n,k} : 0 \le k \le n\}. \tag{1.4.8}$$

The list $L[n]$ can be generated symbolically by the command

```
term[n_,k_]:=If[n==0,1, If[ k==0, 1,
          n*term[n-1,k-1]/k]];
   L[n_]:= Table[ term[n,k], {k,0,n}];
```

that produces a list of the coefficients $b_{n,k}$ from (1.4.7). For instance,

$$L[0] = \{1\}$$
$$L[1] = \{1, 1\}$$
$$L[2] = \{1, 2, 1\}$$
$$L[3] = \{1, 3, 3, 1\}$$
$$L[4] = \{1, 4, 6, 4, 1\}$$
$$L[5] = \{1, 5, 10, 10, 5, 1\}$$
$$L[6] = \{1, 6, 15, 20, 15, 6, 1\}. \qquad (1.4.9)$$

The reader may now recognize the binomial coefficients (1.4.1) from the list (1.4.9) and conjecture the formula

$$b_{n,k} = \binom{n}{k} = \frac{n!}{k!\,(n-k)!} \qquad (1.4.10)$$

from this data. Naturally this requires *a priori* knowledge of the binomial coefficients. An alternative is to employ the procedure described in Section 1.3 to conjecture (1.4.10) from the data in the list $L[n]$.

The guessing of a closed-form formula from data is sometimes obscured by dealing with small numbers. Mathematica can be used to generate terms in the middle part of $L[100]$. The command

```
t:= Table[ L[100][[i]],{i,45,49}]
```

chooses the elements in positions 45 to 49 in $L[100]$:

$$L[100][[45]] = 49378235797073715747364762200$$
$$L[100][[46]] = 61448471214136179596720592960$$
$$L[100][[47]] = 73470998190814997343905056800$$
$$L[100][[48]] = 84413487283064039501507937600$$
$$L[100][[49]] = 93206558875049876949581681100, \qquad (1.4.11)$$

and, as before, we examine their prime factorizations to find a pattern. The prime factorization of $n = L[100][[45]]$ is

$$n = 2^3 \cdot 3^3 \cdot 5^2 \cdot 7 \cdot 19 \cdot 23 \cdot 29 \cdot 31 \cdot 47 \cdot 59 \cdot 61 \cdot 67 \cdot 71 \cdot 73 \cdot 79 \cdot 83 \cdot 89 \cdot 97,$$

suggesting the evaluation of $n/97!$. It turns out that this the reciprocal of an integer of 124 digits. Its factorization

$$\frac{97!}{n} = 2^{91} \cdot 3^{43} \cdot 5^{20} \cdot 7^{13} \cdot 11^8 \cdot 13^7 \cdot 17^5 \cdot 19^4 \cdot 23^3 \cdot 29^2 \cdot 31^2 \cdot 37^2 \cdot 41^2$$
$$\cdot 43^2 \cdot 47 \cdot 53$$

leads to the consideration of

$$\frac{97!}{n \times 53!} = 2^{42} \cdot 3^{20} \cdot 5^8 \cdot 7^5 \cdot 11^4 \cdot 13^3 \cdot 17^2 \cdot 19^2 \cdot 23 \cdot 29 \cdot 31 \cdot 37 \cdot 41 \cdot 43,$$

and then of

$$\frac{97!}{n \times 53! \times 43!} = \frac{2^3 \times 3 \times 11}{5 \times 7}. \tag{1.4.12}$$

The numbers 97, 53 and 43 now have to be slightly adjusted to produce

$$n = \frac{100!}{56! \times 44!} = \binom{100}{56}. \tag{1.4.13}$$

Repeating this procedure with the other elements in the list (1.4.11) leads to the conjecture (1.4.10).

Exercise 1.4.3. Use the method described above to suggest an analytic expression for

$t_1 = 334222131935032834453198400607001018901138886954416016636800,$
$t_2 = 478657831091859016375280557008832085120001822095490687560000,$
$t_3 = 632735060180453305105552747288270827687799251445377532080000,$
$t_4 = 772186537259697948007105490934041043000570796994194290795000.$

d) **Recurrences.** Finally, in order to prove that our guess is correct, define

$$a_{n,k} := \binom{n}{k}^{-1} b_{n,k} \tag{1.4.14}$$

and show that (1.4.7) becomes

$$a_{n,k+1} = a_{n-1,k}, \quad n \geq 1, \ 0 \leq k \leq n-1, \tag{1.4.15}$$

so that $a_{n,k} \equiv 1$.

Exercise 1.4.4. Check that $b_{n,k} = \binom{n}{k}$ by verifying that $\binom{n}{k}$ satisfies (1.4.7) and that this recurrence admits a unique solution with $b_{n,0} = 1$.

Note 1.4.2. The sequence of binomial coefficients has many interesting properties. We provide some of them in the next exercises. The reader will find much more in

$$\texttt{http://www.dms.umontreal.ca/\~{}andrew/Binomial/}$$
$$\texttt{index.htlm}$$

Exercise 1.4.5. Prove that the exponent of the **central binomial coefficients** $C_n = \binom{2n}{n}$ satisfies

$$\mu_p(C_n) = \frac{2s_p(n) - s_p(2n)}{p - 1}. \tag{1.4.16}$$

Hint. Let $n = a_0 + a_1 p + \cdots + a_r p^r$ be the expansion of n in base p. Define λ_j by $2a_j = \lambda_j p + v_j$, where $0 \le v_j \le p - 1$. Check that λ_j is either 0 or 1 and confirm the formula

$$\mu_p(C_n) = \sum_{j=0}^{r} \lambda_j. \tag{1.4.17}$$

In particular $\mu_p(C_n) \le r + 1$. Check that C_n is always even. When is $C_n/2$ odd?

The binomial theorem yields the evaluation of many finite sums involving binomial coefficients. The discussion on binomial sums presented in this book is nonsystematic; we see them as results of evaluations of some definite integrals. The reader will find in Koepf (1998) and Petkovsek et al. (1996). a more complete analysis of these ideas.

Exercise 1.4.6. Let $n \in \mathbb{N}$.
a) Establish the identities

$$\sum_{k=0}^{n} \binom{n}{k} = 2^n$$

$$\sum_{k=0}^{n} k \binom{n}{k} = 2^{n-1} n$$

$$\sum_{k=0}^{n} k^2 \binom{n}{k} = 2^{n-2} n(n + 1). \tag{1.4.18}$$

Hint. Apply the operator $x \frac{d}{dx}$ to the expansion of $(1 + x)^n$ and evaluate at $x = 1$. The operator $x \frac{d}{dx}$ will reappear in (4.1.12).

b) Determine formulas for the values of the alternating sums

$$\sum_{k=0}^{n}(-1)^k k^p \binom{n}{k}$$

for $p = 0$, 1 and 2. Make a general conjecture.

Mathematica 1.4.1. The sums in Exercise 1.4.6 can be evaluated directly by Mathematica by the command

```
s[n_,p_]:=Sum[k^p*Binomial[n,k],{k,0,n}]
```

For example,

$$\sum_{k=0}^{n} k^6 \binom{n}{k} = 2^{n-6}n(n + 1)(n^4 + 14n^3 + 31n^2 - 46n + 16). \quad (1.4.19)$$

The Appendix describes a technique developed by Wilf and Zeilberger that yields an automatic proof of identities like (1.4.19).

   The generalization of the sums in Exercise 1.4.6 is the subject of the next project.

Project 1.4.1. Consider the expression

$$Z_1(p, n) = \sum_{k=0}^{n} k^p \binom{n}{k}, \quad \text{for } n, \ p \in \mathbb{N}. \quad (1.4.20)$$

a) Use a symbolic language to observe that

$$Z_1(p, n) = 2^{n-p} T_p(n) \quad (1.4.21)$$

where $T_p(n)$ is a polynomial in n of degree p.
b) Explore properties of the coefficients of T_p.
c) What can you say about the factors of $T_p(n)$? The factorization of a polynomial can be accomplished by the Mathematica command `Factor`.
   The result of the next exercise will be employed in Section 7.5.

Exercise 1.4.7. Prove the identities

$$\sum_{k=0}^{n} \binom{2n+1}{2k} = 2^{2n}$$

$$\sum_{k=0}^{n} k \binom{2n+1}{2k} = (2n+1)2^{2n-2}$$

$$\sum_{k=0}^{n} k^2 \binom{2n+1}{2k} = (n+1)(2n+1)2^{2n-3}. \qquad (1.4.22)$$

Hint. Consider the polynomial $(1+x)^{2n+1} + (1-x)^{2n+1}$ and its derivatives at $x = 1$.

Project 1.4.2. Define the function

$$Z_2(p, n) = \sum_{k=0}^{n} k^p \binom{2n+1}{2k}. \qquad (1.4.23)$$

a) Observe that $2^{-2n} Z_2(p, n)$ is a polynomial in n with rational coefficients.
b) Make a prediction of the form of the denominators of the coefficients in $Z_2(p, n)$. **Hint.** First observe that these denominators are powers of 2. To obtain an exact formula for the exponents is slightly harder.
c) Study the factorization of $Z_2(p, n)$. Do you observe any patterns? Make a prediction on the signs of the coefficients of the polynomials in the factorization of $Z_2(p, n)$.

1.5. The Ascending Factorial Symbol

The binomial coefficient

$$\binom{n}{k} = \frac{n \cdot (n-1) \cdots (n-k+1)}{1 \cdot 2 \cdots k} \qquad (1.5.1)$$

contains products of consecutive integers. The **ascending factorial symbol**, also called the **Pochhammer symbol**, defined by

$$(a)_k := \begin{cases} 1 & \text{if } k = 0 \\ a(a+1)(a+2)\cdots(a+k-1) & \text{if } k > 0 \end{cases} \qquad (1.5.2)$$

generalizes this idea. In terms of the ascending factorial symbol we have

$$\binom{n}{k} = \frac{(n-k+1)_k}{(1)_k}. \tag{1.5.3}$$

Mathematica 1.5.1. The Mathematica command `Binomial[n,k]` gives $\binom{n}{k}$ and $(a)_k$ is given by `Pochhammer[a,k]`.

Exercise 1.5.1. Prove the following properties of the ascending factorial symbol:

a) Prove that $(1)_n = n!$.

b) Check that $(-x)_n = (-1)^n (x - n + 1)_n$.

c) Check the **dimidiation formulas**

$$(x)_{2n} = 2^{2n} \left(\frac{x}{2}\right)_n \left(\frac{1+x}{2}\right)_n,$$

$$(x)_{2n+1} = 2^{2n+1} \left(\frac{x}{2}\right)_{n+1} \left(\frac{1+x}{2}\right)_n. \tag{1.5.4}$$

A generalization of these formulas considered by Legendre is described in Section 10.4.

d) Establish the **duplication formulas**:

$$(2x)_n = \begin{cases} 2^n (x)_{\frac{n}{2}} (x + 1/2)_{\frac{n}{2}} & \text{for } n \text{ even} \\ 2^n (x)_{\frac{n+1}{2}} (x + 1/2)_{\frac{n-1}{2}} & \text{for } n \text{ odd.} \end{cases} \tag{1.5.5}$$

e) Prove that

$$\frac{(x)_n}{(x)_m} = \begin{cases} (x+m)_{n-m} & \text{if } n \geq m \\ (x+n)_{m-n}^{-1} & \text{if } n \leq m. \end{cases}$$

f) Prove that

$$\frac{d}{dx}(x)_n = (x)_n \sum_{j=0}^{n-1} \frac{1}{j+x}. \tag{1.5.6}$$

g) Find an expression for $(1/2)_n$.

The next exercise shows that the behavior of the ascending factorial with respect to addition is similar to the binomial theorem.

Exercise 1.5.2. Establish **Vandermonde's formula**

$$(x + y)_n = \sum_{j=0}^{n} \binom{n}{j} (x)_j (y)_{n-j}. \qquad (1.5.7)$$

Observe that (1.5.7) is obtained formally from the binomial theorem 1.4.2 by replacing x^j by $(x)_j$. **Hint.** Use induction.

The next comment requires some basic linear algebra. The set of polynomials $\{x, x^2, \cdots, x^n\}$ forms a basis for the vector space of polynomials of degree at most n, that vanish at $x = 0$. The same is true for the set $\{(x)_1, (x)_2, \cdots, (x)_n\}$. Therefore any polynomial in one of these sets can be written as a linear combination of the other one. The exercise below will prove this directly.

Exercise 1.5.3. a) Prove the existence of integers $c(n, k)$, $0 \le k \le n$, such that

$$(x)_n = \sum_{k=0}^{n} c(n, k) x^k. \qquad (1.5.8)$$

For instance

$$(x)_1 = x$$
$$(x)_2 = x(x + 1) = x^2 + x$$
$$(x)_3 = x(x + 1)(x + 2) = x^3 + 3x^2 + 2x.$$

b) The **(signed) Stirling numbers of the first kind** $S_n^{(k)}$ are defined by the generating function

$$x(x - 1)(x - 2) \cdots (x - n + 1) = \sum_{k=0}^{n} S_n^{(k)} x^k. \qquad (1.5.9)$$

Establish a relation between $c(n, k)$ and $S_n^{(k)}$. General information about these number appears in Weisstein (1999), page 1740.

c) Write a Mathematica command that generates a list of $c(n, k)$ for $0 \le k \le n$ as a function of n. Compare your list with the command `StirlingS1[n,k]`.

d) Check that $c(n, 0) = 0$ for $n \geq 1$.

e) Use $(x)_{n+1} = (x + n) \times (x)_n$ to obtain

$$c(n + 1, n + 1) = c(n, n)$$
$$c(n + 1, k) = c(n, k - 1) + nc(n, k) \quad \text{for } 2 \leq k \leq n$$
$$c(n + 1, 1) = nc(n, 1). \tag{1.5.10}$$

f) Prove that

$$\sum_{k=0}^{n} (-1)^k c(n, k) = 0 \quad \text{and} \quad \sum_{k=0}^{n} c(n, k) = n!. \tag{1.5.11}$$

1.6. The Integration of Polynomials

The Fundamental Theorem of Calculus relates the evaluation of the definite integral

$$I = \int_a^b f(x) \, dx \tag{1.6.1}$$

to the existence of a **primitive** function for f. This is a function $F(x)$ such that $F'(x) = f(x)$. In this case we obtain

$$\int_a^b f(x) \, dx = F(b) - F(a). \tag{1.6.2}$$

In theory every continuous function admits a primitive: the function defined by

$$F(x) := \int_a^x f(t) \, dt. \tag{1.6.3}$$

is a primitive for f and it is unique up to an additive constant. Many of the functions studied in elementary calculus appear in this form. For example, the **natural logarithm** and the **arctangent**, defined by

$$\ln x = \int_1^x \frac{dt}{t} \tag{1.6.4}$$

and

$$\tan^{-1} x = \int_0^x \frac{dt}{1 + t^2}, \tag{1.6.5}$$

appear in the basic calculus courses. These will be considered in Chapters 5 and 6 respectively.

The problem of primitives becomes more interesting if one considers a fixed class of functions. For example, if f is a **polynomial** of degree n, i.e.

$$f(x) = \sum_{k=0}^{n} p_k x^k,$$ (1.6.6)

then

$$F(x) = \sum_{k=0}^{n} \frac{p_k}{k+1} x^{k+1}$$ (1.6.7)

is a primitive for f. Therefore, the fundamental theorem of calculus yields the evaluation

$$\int_a^b f(x)\,dx = \sum_{k=0}^{n} p_k \frac{b^{k+1} - a^{k+1}}{k+1}.$$ (1.6.8)

The **coefficients** $p_0, p_1, \ldots, p_n$ in (1.6.6) can be considered as elements of a specific number system such as the **real numbers** $\mathbb{R}$ or the **integers** $\mathbb{Z}$, or they can be seen as **parameters,** that is, variables independent of x. This point of view allows us to perform analytic operations with respect to these coefficients. For example,

$$\frac{\partial f}{\partial p_k} = x^k,$$ (1.6.9)

so differentiating (1.6.8) with respect to p_k yields

$$\int_a^b x^k\,dx = \frac{b^{k+1} - a^{k+1}}{k+1},$$ (1.6.10)

and we recover a particular case of (1.6.8). Later chapters will show that differentiating a formula with respect to one of its parameters often yields new evaluations for which a direct proof is more difficult.

Note 1.6.1. The value of a definite integral can sometimes be obtained directly by Mathematica. By a **blind evaluation** we mean that one simply asks the machine to evaluate the integral without any intelligent input. For example, the command

$$\text{Integrate}[\ x\wedge n, \ \{x,a,b\}\ ]$$

provides a blind evaluation of (1.6.13). The command

$$\text{FullSimplify}[\ \text{Integrate}[x\wedge n, \ \{x,0,1\}],$$
$$\text{Element}[n,\text{Reals}]]$$

tells Mathematica to simplify the answer of the evaluation under the assumption that n is a real parameter.

Note 1.6.2. The justification of differentiation with respect to a parameter under an integral sign is given in Hijab (1997), page 189. Let $u(x;\lambda)$ be a function of the variable x and the parameter λ. Suppose u is differentiable in λ and $\dfrac{\partial u}{\partial \lambda}$ is a continuous function. Then differentiation with respect to the parameter λ,

$$\frac{\partial}{\partial \lambda}\int_a^b u(x;\lambda)\,dx = \int_a^b \frac{\partial}{\partial \lambda}u(x;\lambda)\,dx, \qquad (1.6.11)$$

holds.

Exercise 1.6.1. This exercise establishes the definite integral of the **power function** $x \mapsto x^n$.
a) Use the method of induction to check that

$$\frac{d}{dx}x^n = nx^{n-1} \qquad (1.6.12)$$

for $n \in \mathbb{N}$.
b) Extend (1.6.12) to $n = p/q \in \mathbb{Q}$, $q \neq 0$, by differentiating $y^q = x^p$ implicitly.
c) Establish the formula

$$\int_a^b x^n\,dx = \frac{b^{n+1}-a^{n+1}}{n+1}, \quad n \in \mathbb{Q},\ n \neq -1. \qquad (1.6.13)$$

In particular, we have

$$\int_0^1 x^n\,dx = \frac{1}{n+1}, \quad n \in \mathbb{Q},\ n \neq -1. \qquad (1.6.14)$$

Note 1.6.3. The extension of (1.6.14) to $n \in \mathbb{R}$, $n \neq -1$ presents analytic difficulties related to the *definition* of x^n for $n \notin \mathbb{Q}$. This is discussed in Section 5.6. For instance, one has to provide a meaning to the expression

$x^{\sqrt{2}}$. This difficulty is usually solved by introducing a sequence of rational numbers: given a sequence $a_n \in \mathbb{Q}$ such that $a_n \rightarrow \sqrt{2}$, we can define $x^{\sqrt{2}}$ as the limit of x^{a_n}. This is the subject of real analysis and can be found in many texts; Hijab (1997) and Stromberg (1981) are suggested for general information.

The reader is familiar with the method used to exchange the order of integration in a double integral. The next exercise provides a way to exchange double sums.

Exercise 1.6.2. Let $a_{k,j} : 0 \leq k,\ j \leq n$ be an array of numbers. Prove that

$$\sum_{k=0}^{n} \sum_{j=0}^{k} a_{k,j} = \sum_{j=0}^{n} \sum_{k=j}^{n} a_{k,j}. \qquad (1.6.15)$$

This identity is referred as *reversing the order of summation.*

Exercise 1.6.3. This exercise illustrates how to use a simple evaluation to obtain the value of some sums involving binomial coefficients.
a) Combine (1.6.13) with the change of variable $x = (b - a)t + a$ to produce

$$\sum_{j=0}^{n} \binom{n}{j} \frac{u^j}{j+1} = \frac{1}{a^n(n+1)} \times \frac{b^{n+1} - a^{n+1}}{b - a} \qquad (1.6.16)$$

with $u = b/a - 1$.
b) The special case $b = 2a$ yields

$$\sum_{j=0}^{n} \frac{1}{j+1} \binom{n}{j} = \frac{2^{n+1} - 1}{n+1}. \qquad (1.6.17)$$

Prove this directly by integrating the expansion

$$(1 + x)^n = \sum_{j=0}^{n} \binom{n}{j} x^j \qquad (1.6.18)$$

between 0 and 1.
c) Establish the identity

$$\frac{b^{n+1} - a^{n+1}}{b - a} = b^n + b^{n-1}a + \cdots + ba^{n-1} + a^n \qquad (1.6.19)$$

and use it in (1.6.16) to produce

$$\sum_{j=0}^{n} \binom{n}{j} \frac{u^j}{j+1} = \frac{1}{n+1} \sum_{k=0}^{n} (u+1)^k. \tag{1.6.20}$$

Expand the binomial $(u+1)^k$ and reverse the order of summation to obtain

$$\sum_{j=0}^{n} \binom{n}{j} \frac{u^j}{j+1} = \frac{1}{n+1} \sum_{j=0}^{n} \left[\sum_{k=j}^{n} \binom{k}{j} \right] u^j. \tag{1.6.21}$$

Conclude that

$$\sum_{k=j}^{n} \frac{\binom{k}{j}}{\binom{n}{j}} = \frac{n+1}{j+1}. \tag{1.6.22}$$

Hint. Both sides of (1.6.21) are polynomials in u so you can match coefficients of the same degree.

d) Use the ascending factorial symbol to write (1.6.22) as

$$\sum_{k=1}^{m} (k)_j = \frac{(m+j)!}{(m-1)!\,(j+1)}. \tag{1.6.23}$$

e) Replace m by $n+1$ and j by n in (1.6.23) to obtain

$$\sum_{k=0}^{n} \binom{n+k}{n} = \binom{2n+1}{n}. \tag{1.6.24}$$

Exercise 1.6.4. Use Mathematica to check that the sum appearing in (1.6.21) is given by

$$\sum_{k=j}^{n} \binom{k}{j} = \frac{\Gamma(2+n)}{\Gamma(2+j)\,\Gamma(1-j+n)}. \tag{1.6.25}$$

Similarly, the sum in (1.6.23) is given by

$$\sum_{k=1}^{m} (k)_j = \frac{\Gamma(1+j+m)}{(1+j)\,\Gamma(m)}. \tag{1.6.26}$$

Finally, check that a blind evaluation of (1.6.24) yields

$$\sum_{k=0}^{n} \binom{n+k}{n} = \frac{2^{2n+1}\,\Gamma(3/2+n)}{\sqrt{\pi}\,\Gamma(2+n)}. \tag{1.6.27}$$

Extra 1.6.1. The gamma function appearing in (1.6.25) will be studied in Chapter 10. In particular we show that

$$\Gamma(n) = (n-1)! \quad \text{and} \quad \Gamma(n+\tfrac{1}{2}) = \frac{\sqrt{\pi}}{2^{2n}} \cdot \frac{(2n)!}{n!}$$

so that (1.6.27) is consistent with (1.6.24).

2

The Method of Partial Fractions

2.1. Introduction

The method of partial fractions is normally introduced in calculus courses as a procedure to integrate rational functions $R(x) = P(x)/Q(x)$. The idea is very simple, and we illustrate it with an example.

The rational function

$$R(x) = \frac{x^2 - 2x - 5}{x^3 + 6x^2 + 11x + 6} \tag{2.1.1}$$

is to be integrated from 0 to 1. The basic idea of the method is to consider the integrand $R(x)$ as the *result* of a sum of rational functions with simpler denominators. Thus the method consists of two parts:

a) *The factorization of $Q(x)$.*
b) *The decomposition of $R(x)$ into simpler factors.*

In the example, $Q(x)$ factors as[1]

$$Q(x) = x^3 + 6x^2 + 11x + 6 = (x + 1)(x + 2)(x + 3), \tag{2.1.2}$$

so we seek a decomposition of $R(x)$ in the form

$$\frac{x^2 - 2x - 5}{(x + 1)(x + 2)(x + 3)} = \frac{a}{x + 1} + \frac{b}{x + 2} + \frac{c}{x + 3}, \tag{2.1.3}$$

for some constants a, b, c. It remains to find the coefficients a, b, c and to evaluate the simpler integrals

$$\int_0^1 \frac{x^2 - 2x - 5}{x^3 + 6x^2 + 11x + 6} dx = \int_0^1 \frac{a\,dx}{x + 1} + \int_0^1 \frac{b\,dx}{x + 2} + \int_0^1 \frac{c\,dx}{x + 3}. \tag{2.1.4}$$

---

[1] The reader now understands why we chose this $Q(x)$.

The solution to part b) is particularly easy when $Q(x)$ has simple real roots. The procedure is illustrated with (2.1.3). To obtain the value of a, multiply (2.1.3) by $x + 1$ to obtain

$$\frac{x^2 - 2x - 5}{(x+2)(x+3)} = a + \frac{b(x+1)}{x+2} + \frac{c(x+1)}{x+3}, \qquad (2.1.5)$$

and then let $x = -1$ to obtain $a = -1$. Similarly $b = -3$ and $c = 5$. Thus the integration of $R(x)$ is reduced to more elementary integrals. Each of the pieces in (2.1.4) can be evaluated as

$$\int_0^1 \frac{dx}{x+s} = \ln\left(\frac{1+s}{s}\right),$$

so that

$$\int_0^1 \frac{x^2 - 2x - 5}{x^3 + 6x^2 + 11x + 6}\, dx = -4\ln\left(\frac{9}{8}\right). \qquad (2.1.6)$$

Note 2.1.1. The reader has certainly encountered these evaluations in the basic calculus courses. Chapter 5 presents a discussion of the logarithm function.

Mathematica 2.1.1. The Mathematica command `Apart[R[x]]` gives the partial fraction decomposition of the rational function R, provided it can evaluate the roots of the denominator of R. For example,

$$\text{Apart}\left[\frac{x^2}{x^2 + 3x + 2}\right] = 1 + \frac{1}{1+x} - \frac{4}{2+x} \qquad (2.1.7)$$

and

$$\text{Apart}\left[\frac{x}{x^5 + 2x + 1}\right] = \frac{x}{x^5 + 2x + 1}. \qquad (2.1.8)$$

Mathematica 2.1.2. It is possible to ask Mathematica for a direct evaluation of the integral in (2.1.6) via the command

```
int:= Integrate[(x^2 - 2x - 5)/
(x^3 + 6x^2 + 11x + 6), x,0,1].
```

The answer given is

```
2 Log[2] - 8 Log[3] + 5 Log[4]
```

and we would like to simplify its result. The command `FullSim-plify` yields

$$6 \ \text{Log}[4] \ - \ 8 \ \text{Log}[3]$$

as the value, but it does not simplify automatically the expression `Log[4] = 2 Log[2]`. In order to reduce the arguments of logarithms to its minimal expression requires the introduction of a *Complexity Function*. The complete command is

```
penalyzeLogOfIntegerPowers[expr_] := (Plus @@ (
(Plus @@ (Last /@ FactorInteger[First[#]]) -
1)& /@ Cases[{expr}, Log[_Integer],
Infinity])) + (* avoid Log[rationalNumber]
too *) 10 Count[{expr}, Log[_Rational], Infinity]
```

simplifies the integral to the final result $4 (\ 3 \ \text{Log}[2] \ - \ 2 \ \text{Log}[3] \ )$ via the command

```
FullSimplify[ int, ComplexityFunction
    ->penalyzeLogOfIntegerPowers].
```

The optimal answer for the integral

$$I = \int_a^b R(x)\,dx \tag{2.1.9}$$

of a rational function

$$R(x) = \frac{P(x)}{Q(x)}, \tag{2.1.10}$$

with $P(x)$, $Q(x)$ polynomials in x, namely,

$$P(x) = p_m x^m + p_{m-1} x^{m-1} + \cdots + p_1 x + p_0$$
$$Q(x) = q_n x^n + q_{n-1} x^{n-1} + \cdots + q_1 x + q_0, \tag{2.1.11}$$

would be an explicit function of the parameters

$$\mathfrak{P} := \{a, b; \ m, n; \ p_m, \cdots, p_0; \ q_n, \cdots, q_0\}, \tag{2.1.12}$$

We begin the study of the evaluation of such an integral by normalizing the interval of integration into the half-line $[0, \infty)$.

Exercise 2.1.1. Let $R(x)$ be a rational function and $-\infty < a < b < \infty$. Use the change of variable $x \to y = (x - a)/(b - x)$ to show that we can always

28   *The Method of Partial Fractions*

assume $a = 0$ and $b = \infty$. Show that an alternate construction can be used in the case $a = -\infty$, $b = \infty$. Use this transformation to check that

$$\int_0^1 \frac{x\,dx}{x+1} = \int_0^\infty \frac{y\,dy}{(2y+1)(y+1)^2}. \qquad (2.1.13)$$

The example shows that the degree of R might increase in the normalization of the interval of integration.

Now that we have normalized the interval of integration to $[0, \infty)$ we show how to normalize some of the coefficients.

Exercise 2.1.2. Let I be the integral in (2.1.9).
a) Prove that if the integral I from 0 to ∞ is convergent, then q_n and q_0 must have the same sign.
b) Show that in the evaluation of (2.1.9), under the normalization $a = 0$ and $b = \infty$, we may assume $q_n = q_0 = 1$. **Hint.** Use a change of variable of the form $x \mapsto \lambda x$.
c) Compute the normalized form of the quartic integral

$$I_1 = \int_0^\infty \frac{dx}{bx^4 + 2ax^2 + c}. \qquad (2.1.14)$$

The method of partial fractions provides the value of the integral of a rational function in terms of the roots of its denominator. The next proposition gives the explicit formula in the case in which these roots are real and simple.

Proposition 2.1.1. *Let P and Q be polynomials, with* $\deg P \le \deg Q - 2$. *Assume that all the roots x_j of $Q(x) = 0$ are real, negative, and simple. Then*

$$\int_0^\infty \frac{P(x)}{Q(x)}dx = -\sum_{j=1}^m \frac{P(x_j)}{Q'(x_j)} \ln x_j. \qquad (2.1.15)$$

Proof. The constants α_j in the decomposition

$$\frac{P(x)}{Q(x)} = \sum_{j=1}^m \frac{\alpha_j}{x - x_j} \qquad (2.1.16)$$

can be evaluated by multiplying by $x - x_k$ to produce

$$P(x) \times \frac{x - x_k}{Q(x)} = \alpha_k + \sum_{j=1,\,j\neq k}^m \frac{\alpha_j(x - x_k)}{x - x_j};$$

letting $x \mapsto x_k$ we obtain

$$\alpha_k = \frac{P(x_k)}{Q'(x_k)}. \tag{2.1.17}$$

The fact that $Q'(x_k) \neq 0$ follows from the simplicity of the roots, see Exercise 2.4.2. Thus

$$\int_0^b \frac{P(x)}{Q(x)} dx = \sum_{j=1}^m \alpha_j \ln(1 - b/x_j). \tag{2.1.18}$$

Now observe that

$$\frac{P(1/s)}{Q(1/s)} = \sum_{j=1}^m \frac{\alpha_j s}{1 - s x_j},$$

so dividing both sides by s and letting $s \to 0$ we obtain

$$\sum_{j=1}^m \alpha_j = 0. \tag{2.1.19}$$

Computing the limit as $b \to \infty$ in (2.1.18) yields (2.1.15). □

Exercise 2.1.3. Check the details.

Throughout the text we will employ the normalization $[0, \infty)$ for the interval of integration. The next series of exercises presents an alternative normalization.

Exercise 2.1.4. Let R be a rational function. Prove that

$$R_1(x) = R(x) + \frac{1}{x^2} R(1/x) \tag{2.1.20}$$

is also a rational function, with the property

$$\int_0^\infty R(x) \, dx = \int_0^1 R_1(x) \, dx. \tag{2.1.21}$$

Conclude that we can always normalize the interval of integration to $[0, 1]$.

Exercise 2.1.5. A polynomial P is called **symmetric** if it satisfies

$$P(x) = x^{\deg(P)} P(1/x). \tag{2.1.22}$$

a) Describe this condition in terms of the coefficients of P.

b) Prove the identity

$$\int_0^\infty \frac{P(x)}{Q(x)}\,dx = \int_0^1 \frac{x^{m-2}P(1/x)Q(x) + x^m Q(1/x)P(x)}{x^m Q(x)Q(1/x)}\,dx \quad (2.1.23)$$

with $m = \deg(Q)$. **Hint.** Split the original integral at $x = 1$ and let $x \mapsto 1/x$ in the piece from $x = 1$ to $x = \infty$.

c) Check that the numerator and denominator of the rational function appearing on the right hand side of (2.1.23) are symmetric. Conclude that the integration of any rational function can be reduced to that of a symmetric one on $[0, 1]$.

d) Give the details in the case $P(x) = 1$ and $Q(x) = ax^2 + bx + c$.

The next project discusses some of the properties of the map that sends R to R_1. The reader with some background in linear algebra will see that this map is linear and has many eigenfunctions.

Project 2.1.1. The map $\mathfrak{T}(R(x)) = R_1(x)$, with $R_1(x)$ defined in (2.1.20) is a transformation on the space of rational functions. The goal of the project is to explore its properties.

a) Prove that for any rational function R, the image R_1 satisfies

$$\mathfrak{T}(R_1) = 2R_1. \quad (2.1.24)$$

Therefore the map $\mathfrak{T}$ has many eigenfunctions with eigenvalue 2.

b) Is it possible to characterize all other eigenvalues of $\mathfrak{T}$? These are solutions to $\mathfrak{T}(R(x)) = \lambda R(x)$.

c) Find all functions that are mapped to 0 under $\mathfrak{T}$.

d) Part a) shows that every function in the range of $\mathfrak{T}$ is an eigenfunction of $\mathfrak{T}$. Characterize this range.

2.2. An Elementary Example

In this section we consider the evaluation of

$$I_2(a, b) = \int_0^\infty \frac{dx}{x^2 + 2ax + b} \quad (2.2.1)$$

in terms of the parameters a and b. Completing the square we obtain

$$I_2(a, b) = \int_a^\infty \frac{dt}{t^2 - D}, \quad (2.2.2)$$

where $D = a^2 - b$ is (one quarter of) the **discriminant** of the quadratic $P_2(x) = x^2 + 2ax + b$. The evaluation of I_2 is discussed according to the sign of D.

2.2.1. Negative Discriminant

$a^2 < b$. In this case the evaluation employs the arctangent function defined in (1.6.5). Let $D = -c^2$, then

$$I_2(a, b) = \frac{1}{c} \int_{a/c}^{\infty} \frac{dt}{1 + t^2} = \frac{1}{c} \left[\frac{\pi}{2} - \tan^{-1} \left(\frac{a}{c} \right) \right]. \qquad (2.2.3)$$

Naturally

$$\int_0^{\infty} \frac{dt}{1 + t^2} = \frac{\pi}{2}. \qquad (2.2.4)$$

In the discussion of trigonometric functions given in Chapter 6, we actually use (2.2.4) to *define* π.

Exercise 2.2.1. Differentiate (2.2.3) to produce

$$\int_0^{\infty} \frac{dx}{(x^2 + 2ax + b)^2} = -\frac{a}{2b(b - a^2)} + \frac{\pi}{4(b - a^2)^{3/2}} - \frac{\tan^{-1}(a/\sqrt{b - a^2})}{2(b - a^2)^{3/2}}.$$
$$(2.2.5)$$

Obtain a similar formula for

$$\int_0^{\infty} \frac{dx}{(x^2 + 2ax + b)^3}. \qquad (2.2.6)$$

A generalization of this identity is discussed in Project 2.3.1.

2.2.2. Positive Discriminant

$a^2 > b$. In this case the quadratic $P_2(x)$ has two real roots $r_{\pm} = -a \pm \sqrt{D}$. The partial fraction decomposition of the integrand is

$$\frac{1}{x^2 + 2ax + b} = \frac{1}{2\sqrt{D}} \left(\frac{1}{x - r_+} - \frac{1}{x - r_-} \right). \qquad (2.2.7)$$

The integral is now expressed in terms of the logarithm function defined in (1.6.4).

Exercise 2.2.2. Prove that if $a^2 > b$, then the integral converges if and only if $b > 0$.

We now integrate (2.2.7) from 0 to ∞ to obtain

$$I_2(a, b) = \frac{1}{2\sqrt{D}} \ln \left(\frac{a + \sqrt{D}}{a - \sqrt{D}} \right). \tag{2.2.8}$$

We now summarize the discussion:

Theorem 2.2.1. *The integral*

$$I_2(a, b) = \int_0^\infty \frac{dx}{x^2 + 2ax + b} \tag{2.2.9}$$

converges precisely when $b > 0$. *Its value is*

$$I_2(a, b) = \begin{cases} \dfrac{1}{\sqrt{b - a^2}} \left[\dfrac{\pi}{2} - \tan^{-1} \left(\dfrac{a}{\sqrt{b - a^2}} \right) \right] & \text{if } b > a^2 \\[2ex] \dfrac{1}{2\sqrt{a^2 - b}} \ln \left(\dfrac{a + \sqrt{a^2 - b}}{a - \sqrt{a^2 - b}} \right) & \text{if } b < a^2 \\[2ex] \dfrac{1}{a} & \text{if } b = a^2. \end{cases}$$

2.3. Wallis' Formula

In this section we establish a formula of Wallis (1656) that is one the first exact evaluations of a definite integral. This example will reappear throughout the book.

The first proof uses the method of *introduction of a parameter*. In order to evaluate an integral, one consideres a more general problem obtained by changing numerical values by a parameter. In this proof we replace

$$\int_0^\infty \frac{dx}{(x^2 + 1)^{m+1}} \tag{2.3.1}$$

by

$$\int_0^\infty \frac{dx}{(x^2 + b)^{m+1}}. \tag{2.3.2}$$

The extra parameter gives more flexibility. For instance, we can differentiate with respect to b.

Theorem 2.3.1. *Let* $m \in \mathbb{N}$. *Then*

$$J_{2,m} := \int_0^\infty \frac{dx}{(x^2 + 1)^{m+1}} = \frac{\pi}{2^{2m+1}} \binom{2m}{m}. \tag{2.3.3}$$

Proof. Introduce the function

$$I_m(b) := \int_0^\infty \frac{dx}{(x^2+b)^{m+1}}, \qquad (2.3.4)$$

for $b > 0$. Thus $J_{2,m} = I_m(1)$.

Exercise 2.3.1. Check that $I_m(b)$ satisfies

$$I'_m(b) = -(m+1)I_{m+1}(b). \qquad (2.3.5)$$

The initial value

$$I_0(b) = \frac{\pi}{2\sqrt{b}} \qquad (2.3.6)$$

is a particular case of Theorem 2.2.1. Now use (2.3.5) to produce the first few values of $I_m(b)$:

$$I_1(b) = \frac{\pi}{4b^{3/2}}$$
$$I_2(b) = \frac{3\pi}{16b^{5/2}}.$$

This data suggests the definition

$$f_m(b) := b^{(2m+1)/2} I_m(b). \qquad (2.3.7)$$

Exercise 2.3.2. The goal of this exercise is to provide an analytic expression for $I_n(b)$.
a) Check that $f_m(b)$ satisfies

$$bf'_m(b) - \frac{2m+1}{2} f_m(b) = -(m+1)f_{m+1}(b). \qquad (2.3.8)$$

b) Use the value of $I_0(b)$ to check that $f_0(b)$ is independent of b. Now use induction and (2.3.8) to show that $f_m(b)$ is independent of b for all $m \in \mathbb{N}$. Conclude that f_m satisfies

$$f_{m+1} = \frac{2m+1}{2m+2} f_m. \qquad (2.3.9)$$

c) Prove that

$$f_m = \frac{\binom{2m}{m}}{2^{2m+1}}\pi. \qquad (2.3.10)$$

Hint. To guess the form of f_m from (2.3.9) compute

$$f_4 = \frac{7}{8}\cdot\frac{5}{6}\cdot\frac{3}{4}\cdot\frac{1}{2}\cdot\frac{\pi}{2} \qquad (2.3.11)$$

and insert in the numerator and denominator the missing even numbers to obtain

$$f_4 = \frac{8}{2 \cdot 4} \cdot \frac{7}{8} \cdot \frac{6}{2 \cdot 3} \cdot \frac{5}{6} \cdot \frac{4}{2 \cdot 2} \cdot \frac{3}{4} \cdot \frac{2}{2 \cdot 1} \cdot \frac{1}{2} \cdot \frac{\pi}{2}$$

$$= \frac{8!}{2^8 4!^2} \frac{\pi}{2}.$$

To prove the form of f_m define

$$g_m := \frac{f_m \cdot 2^{2m+1}}{\binom{2m}{m} \pi} \tag{2.3.12}$$

and check that $g_{m+1} = g_m$.

The proof of Wallis's formula is complete.   □

Project 2.3.1. The goal of this project is to discuss the structure of the integral

$$L_m(a) = \int_0^\infty \frac{dx}{(x^2 + 2ax + 1)^{m+1}} \tag{2.3.13}$$

as an explicit function of a and m. Properties of the indefinite version of $L_m(a)$ appear in Gradshteyn and Ryzhik (1994) [G & R], 2.171.3 and 2.171.4.

a) Describe the values of the parameter a for which the integral converges.

In the first part we assume $-1 < a < 1$, so that

$$L_0(a) = \frac{1}{\sqrt{1-a^2}} \left[\frac{\pi}{2} - \tan^{-1} \left(\frac{a}{\sqrt{1-a^2}} \right) \right]. \tag{2.3.14}$$

b) Prove that $L_m(a)$ satisfies the recurrence

$$L_m(a) = \frac{2m-1}{2m(1-a^2)} L_{m-1}(a) - \frac{a}{2m(1-a^2)}. \tag{2.3.15}$$

Hint. Write $1 = (x^2 + 2ax + 1) - \frac{x}{2}(2x + 2a) - ax$.

c) Use part b) to show that $L_m(a)$ can be written in the form

$$L_m(a) = \frac{R_m(a)}{(a^2 - 1)^m} + \frac{\pi}{\sqrt{1-a^2}} \cdot \frac{\alpha_m}{(a^2 - 1)^m} + \frac{\beta}{\sqrt{1-a^2}} \cdot \frac{\gamma_m}{(a^2 - 1)^m} \tag{2.3.16}$$

where

$$\beta = \tan^{-1} \left(\frac{a}{\sqrt{1-a^2}} \right), \tag{2.3.17}$$

α_m, γ_m are constants and $R_m(a)$ is a function of a. These will be determined in the remainder of the exercise.

d) Use (2.3.15) to show that α_m and γ_m satisfy the recurrence

$$2mx_m = -(2m-1)x_{m-1}. \tag{2.3.18}$$

To solve this recurrence, obtain a new recurrence for $y_m := \binom{2m}{m}x_m$.

e) Prove that the function $R_m(a)$ satisfies the recurrence

$$R_m(a) = -\frac{2m-1}{2m}R_{m-1}(a) + \frac{a}{2m}(a^2-1)^{m-1}, \tag{2.3.19}$$

with initial condition $R_0(a) = 0$. Conclude that $R_m(a)$ is a polynomial. The precise closed form for its coefficients seems to be difficult to obtain.

f) Prove that $R_m(a)$ is an odd polynomial (only odd powers appear), of degree $2m-1$ and that its coefficients alternate sign starting with a positive leadind term.

g) Prove, or at least provide convincing symbolic evidence, that the least common denominator t_m of the coefficients of $R_m(a)$ satisfies

$$\frac{t_m}{2mt_{m-1}} = \begin{cases} 1 & \text{if } 2m-1 \text{ is prime} \\ \dfrac{1}{\dfrac{2m}{p}-1} & \text{if } 2m-1 \text{ is not prime but not a prime power} \\ \dfrac{\dfrac{2m}{p}}{2m-1} & \text{if } 2m-1 = p^n \text{ for some prime } p \text{ and } 1 < n \in \mathbb{N}. \end{cases}$$

h) Obtain similar results for the case $a^2 > 1$, in which case the integral is expressed in terms of logarithms.

i) Evaluate $L_m(1)$.

j) Discuss the evaluation of $L_m(a)$ by considering the derivatives of

$$h(c) := \int_0^\infty \frac{dx}{x^2 + 2ax + 1 + c} \tag{2.3.20}$$

with respect to the parameter c at $c = 0$.

Mathematica 2.3.1. The command

```
Integrate[1/(x^2+2*a*x+1), {x, 0 , Infinity},
          Assumptions -> a > 0]
```

gives the value of $L_0(a)$ in (2.3.14).

Note 2.3.1. The quartic analog integral

$$g(c) := \int_0^\infty \frac{dx}{x^4 + 2ax^2 + 1 + c} \tag{2.3.21}$$

appears in Section 7.7 and plays a crucial role in the evaluation of integrals with denominators of degree 4 and in the Taylor expansion of the function $h(c) = \sqrt{a + \sqrt{1+c}}$.

2.4. The Solution of Polynomial Equations

The zeros of the polynomial

$$Q(x) = x^n + q_{n-1}x^{n-1} + q_{n-2}x^{n-2} + \cdots + q_1 x + q_0 \qquad (2.4.1)$$

with $q_i \in \mathbb{R}$, form an essential part of the decomposition of the rational function $R(x) = P(x)/Q(x)$ into partial fractions. In this section we discuss the question of how to produce formulas for the solutions of $Q(x) = 0$ in terms of the coefficients $\{q_0, q_1, \cdots, q_{n-1}\}$. We will provide details for polynomials of small degree and use these formulas to give explicit closed forms for a class of integrals of rational functions.

Exercise 2.4.1. Let x_0 be a real root of $Q(x) = 0$. Prove that there exists a polynomial $Q_1(x)$ with real coefficients such that $Q(x) = (x - x_0)Q_1(x)$. **Hint**. Use (1.6.19) to factor $Q(x) - Q(x_0)$.

Exercise 2.4.2. Let x_0 be a *double* root of $Q(x) = 0$, that is, $Q(x) = (x - x_0)^2 Q_1(x)$ for some polynomial Q_1 with $Q_1(x_0) \neq 0$. Prove that this is equivalent to $Q(x_0) = Q'(x_0) = 0$ and $Q''(x_0) \neq 0$. Check that a root is double if it appears exactly twice in the list of all roots of $Q(x) = 0$. Generalize to roots of higher multiplicity (this being the number of times a root appears in the list).

Extra 2.4.1. The reader is familiar with the fact that a real polynomial might not have any real roots, for example $P(x) = x^2 + 1$. The solution to this question was one of the motivation for the creation of complex numbers:

$$\mathbb{C} = \{a + bi : a, b \in \mathbb{R}, \; i^2 = -1\}. \qquad (2.4.2)$$

The operations in $\mathbb{C}$ are defined in a natural way: treat the *number i* as a variable and simplify the expressions using $i^2 = -1$. The complex numbers come with an extra operation: **conjugation**. The conjugate of $z = a + bi$ is $\bar{z} = a - bi$.

Exercise 2.4.3. Prove that if $a + bi$ is a root of a polynomial with real coefficients, then so is its conjugate. **Hint**: Do it first for polynomials of small degree to get the general idea.

We conclude from the previous exercise that the polynomial $Q(x)$ can be factored in the form

$$Q(x) = (x - x_1)^{n_1} (x - x_2)^{n_2} \cdots (x - x_k)^{n_k} \qquad (2.4.3)$$

where x_j are the roots of $Q(x) = 0$, some of which might be complex. A real form of this factorization is obtained by combining the complex conjugate pairs into a form

$$Q(x) = (x - r_1)^{n_1} (x - r_2)^{n_2} \cdots (x - r_j)^{n_j}$$
$$\times (x^2 - 2a_1 x + a_1^2 + b_1^2)^{m_1} \cdots (x^2 - 2a_k x + a_k^2 + b_k^2)^{m_k}. \quad (2.4.4)$$

The next exercise produces a relation between the roots of a polynomial and its coefficients. In Chapter 11, Exercise 11.3.1, we will use an extension of this result to give Euler's proof of the identity

$$\sum_{n=1}^{\infty} \frac{1}{n^2} = \frac{\pi^2}{6}. \qquad (2.4.5)$$

Exercise 2.4.4. a) Prove that the polynomial Q can be written as

$$Q(x) = C x^r \prod_{j=1}^{m} \left(1 - \frac{x}{x_j} \right) \qquad (2.4.6)$$

where the product runs over all the nonzero roots of $Q(x) = 0$.
b) Check that

$$C = (-1)^m \prod_{j=1}^{m} x_j. \qquad (2.4.7)$$

c) Assume that $Q(0) \neq 0$. Prove that

$$\sum_{j=1}^{m} \frac{1}{x_j} = -C q_1. \qquad (2.4.8)$$

Extra 2.4.2. Once the factorization (2.4.4) is given the rational function $P(x)/Q(x)$ can be written as

$$\frac{P(x)}{Q(x)} = \frac{w_1(x)}{(x - r_1)^{n_1}} + \cdots + \frac{w_j(x)}{(x - r_j)^{n_j}}$$
$$+ \frac{z_1(x)}{(x^2 - 2a_1 x + a_1^2 + b_1^2)^{m_1}} + \cdots + \frac{z_k(x)}{(x^2 - 2a_k x + a_k^2 + b_k^2)^{m_k}},$$

for some polynomials $w_1, \ldots, w_j, z_1, \ldots, z_k$. This is the partial fraction decomposition of R.

Thus if the roots of the polynomial Q are assumed to be known, the integration of a rational function is reduced to integrals of the form

$$\int_0^\infty \frac{x^j \, dx}{(x-r)^n}$$

and

$$\int_0^\infty \frac{x^j \, dx}{(x^2 - 2ax + a^2 + b^2)^n},$$

corresponding to the real and complex roots of $Q(x) = 0$, respectively.

2.4.1. Quadratics

The question of finding the roots of a polynomial equation starts with the familiar quadratic formula, which expresses the roots of

$$x^2 + q_1 x + q_0 = 0 \tag{2.4.9}$$

as

$$x_\pm = \frac{1}{2}\left(-q_1 \pm \sqrt{q_1^2 - 4q_0}\right). \tag{2.4.10}$$

Exercise 2.4.5. Determine the restrictions on the parameters q_0, q_1, q_2 so that

$$\int_0^\infty \frac{P(x)\,dx}{(q_2 x^2 + q_1 x + q_0)^{m+1}}$$

converges.

2.4.2. Cubics

There are similar formulas that express the solution of

$$x^3 + q_2 x^2 + q_1 x + q_0 = 0 \tag{2.4.11}$$

in terms of the coefficients $\{q_0, q_1, q_2\}$. The next exercise describes how to find them. The first step is to eliminate the coefficient q_2.

Exercise 2.4.6. Prove that the transformation $y = x + q_2/3$ transforms (2.4.11) into

$$y^3 + q_1^* y + q_0^* = 0 \tag{2.4.12}$$

with

$$q_1^* = q_1 - \frac{1}{3}q_2^2 \quad \text{and} \quad q_0^* = \frac{2}{27}q_2^3 - \frac{1}{3}q_2q_1 + q_0.$$

We now present a method due to Cardano (1545/1968) to solve

$$x^3 + 3q_1x + 2q_0 = 0. \tag{2.4.13}$$

The scaling factors 3 and 2 are introduced so the final formulas have a simpler form.

Define u and v by

$$u + v = x \tag{2.4.14}$$
$$uv = -q_1.$$

Substituting in (2.4.13) gives

$$u^3 + v^3 = -2q_0. \tag{2.4.15}$$

We thus have a symmetric system

$$U + V = -2q_0 \tag{2.4.16}$$
$$UV = -q_1^3$$

where $U = u^3$ and $V = v^3$. It is easy to check that U and V are roots of the quadratic

$$X^2 + 2q_0X - q_1^3 = 0. \tag{2.4.17}$$

The discriminant of this quadratic is

$$D = q_0^2 + q_1^3 \tag{2.4.18}$$

and D is also known as the **discriminant of the cubic** $x^3 + 3q_1x + 2q_0$. The solutions of (2.4.13) can now be expressed as

$$x_1 = \sqrt[3]{-q_0 + \sqrt{D}} + \sqrt[3]{-q_0 - \sqrt{D}} \tag{2.4.19}$$
$$x_2 = \rho \sqrt[3]{-q_0 + \sqrt{D}} + \rho^2 \sqrt[3]{-q_0 - \sqrt{D}}$$
$$x_3 = \rho^2 \sqrt[3]{-q_0 + \sqrt{D}} + \rho \sqrt[3]{-q_0 - \sqrt{D}}$$

where $\rho = (-1 + i\sqrt{3})/2$ is a primitive cube root of 1, that is, a cube root of 1 not equal to 1.

Note 2.4.1. Hellman (1958, 1959) presents an interesting discussion of the solution of the cubic and quartic equation.

Exercise 2.4.7. Use Cardano's method to solve the cubic $x^3 - 6x^2 + 11x - 6 = 0$.

Exercise 2.4.8. Check that the roots of the cubic $x^3 + 3x^2 + 2x + 1 = 0$ can be given in terms of

$$\alpha = \left(108 + 12\sqrt{69}\right)^{1/3} \qquad (2.4.20)$$

as

$$x_1 = -\left(\frac{\alpha}{6} + \frac{2}{\alpha} + 1\right)$$

$$x_{2,3} = \left(\frac{\alpha}{12} + \frac{1}{\alpha} - 1\right) \pm \frac{\sqrt{-3}}{2}\left(\frac{2}{\alpha} - \frac{\alpha}{6}\right).$$

This specific cubic will appear in Section 3.8.

Mathematica 2.4.1. The Mathematica command to find the roots of $x^3 + 3x^2 + 2x + 1 = 0$ is

```
Solve[ x^3 + 3*x^2 + 2*x + 1 == 0,x ]
```

and this yields the three roots in the form

$$\left\{ x \rightarrow -1 - \left(\frac{2}{3(9 - \sqrt{69})}\right)^{1/3} - \frac{\left(\frac{1}{2}(9 - \sqrt{69})\right)^{1/3}}{3^{2/3}} \right\}, \qquad (2.4.21)$$

and with a similar expression for the complex roots. An attempt to simplify this root produces

```
Root[1 + 2 #1  + 3#1^2 + #1^3 &, 1]
```

that simply identifies the number as the first root of the original equation.

Project 2.4.1. The goal of this project is to determine the region in the (a, b) plane on which the integral

$$I(a, b) = \int_0^\infty \frac{dx}{x^3 + ax^2 + bx + 1} \qquad (2.4.22)$$

converges. Observe first that there are no problems near infinity, so the question of convergence of the integral is controlled by the location of the zeros of the denominator.

Define

$$\Lambda := \{(a, b) \in \mathbb{R}^2 : I(a, b) < \infty\} \tag{2.4.23}$$

and $P_3(x) = x^3 + ax^2 + bx + 1$.

a) Prove that $I(a, b)$ converges if and only if the equation $P_3(x) = 0$ has no real positive roots. **Hint.** Consider first the case of three real roots r_1, r_2, r_3, and then the case of a single real root and a pair of complex conjugate roots. The first case should be divided into three subcases: 1) all roots distinct, 2) $r_1 = r_2 \neq r_3$ and 3) $r_1 = r_2 = r_3$.

b) Suppose $a^2 < 3b$. Then $(a, b) \in \Lambda$. **Hint.** This corresponds to the case in which $P_3'(x) > 0$ for all $x \in \mathbb{R}$, so $P_3(x)$ has one real root and two complex (not real) roots.

c) Assume $a^2 > 3b$. Let $t_+ = (-a + \sqrt{a^2 - 3b})/3$ be the largest of the critical points of P. Prove that $t_+ < 0$ is equivalent to $a, b > 0$. Confirm that if $t_+ < 0$ then $(a, b) \in \Lambda$.

d) Suppose $a^2 > 3b$ and $t_+ > 0$. Prove that $(a, b) \in \Lambda$ if and only if $P(t_+) > 0$.

e) Suppose again that $a^2 > 3b$ and $t_+ > 0$. Show that if $27 + 2a^3 - 9ab < 0$, then $(a, b) \notin \Lambda$.

f) The **discriminant curve**

$$R(a, b) = 4a^3 + 4b^3 - 18ab - a^2b^2 + 27 = 0 \tag{2.4.24}$$

consists of two separate branches $R_{\pm}(a, b)$. Let $R_-(a, b)$ be the branch containing $(-1, -1)$. Prove that

$$(a, b) \in \Lambda \Leftrightarrow R_-(a, b) > 0. \tag{2.4.25}$$

g) Check that the point $(3, 3)$ is on the discriminant curve. Compute the Taylor series of $R(a, b) = 0$ at $(3, 3)$ and confirm that this is a cusp. **Hint.** Consider the Taylor series up to third order, let $u = a + 3$, $v = b + 3$ to translate the cusp from $(3, 3)$ to the origin. Now write everything in the new coordinates $x = (u - v)/2$ and $y = (u + v)/2$.

Extra 2.4.3. The reader will find that region Λ is related to the dynamical system

$$a_{n+1} = \frac{a_n b_n + 5a_n + 5b_n + 9}{(a_n + b_n + 2)^{4/3}}$$

$$b_{n+1} = \frac{a_n + b_n + 6}{(a_n + b_n + 2)^{2/3}}.$$

These expressions comes from the fact that the integral

$$U_6(a, b; c, d, e) = \int_0^\infty \frac{cx^4 + dx^2 + e}{x^6 + ax^4 + bx^2 + 1}\, dx \qquad (2.4.26)$$

remains the same if (a, b) is replaced by (a_1, b_1) and (c, d, e) are changed according to similar rules. More information about these results is given by Moll (2002). See also Extra 5.4.2 for the original invariant integral.

Exercise 2.4.9. Use the method of partial fractions to evaluate the integrals

$$C_{0,0}(a, b) = \int_0^\infty \frac{dx}{x^3 + ax^2 + bx + 1} \qquad (2.4.27)$$

and

$$C_{1,0}(a, b) = \int_0^\infty \frac{x\, dx}{x^3 + ax^2 + bx + 1} \qquad (2.4.28)$$

in terms of the roots of the cubic equation $x^3 + ax^2 + bx + 1$. Describe how to obtain the values of

$$C_{0,1}(a, b) = \int_0^\infty \frac{dx}{(x^3 + ax^2 + bx + 1)^2} \qquad (2.4.29)$$

and

$$C_{1,1}(a, b) = \int_0^\infty \frac{x\, dx}{(x^3 + ax^2 + bx + 1)^2} \qquad (2.4.30)$$

by differentiation with respect to the parameters a and b.

These are special cases of the family

$$C_{j,m}(a, b) = \int_0^\infty \frac{x^j\, dx}{(x^3 + ax^2 + bx + 1)^{m+1}}. \qquad (2.4.31)$$

2.4.3. Quartics

A simple procedure can now be used to describe the roots of the quartic

$$x^4 + q_3 x^3 + q_2 x^2 + q_1 x + q_0 = 0. \qquad (2.4.32)$$

The details are left as an exercise.

Exercise 2.4.10. We present here methods developed by Cardano and Descartes to solve the general equation of degree 4.

a) Use a translation of the variable x to show that the general quartic can be written as

$$x^4 + q_2^* x^2 + q_1^* x + q_0^* = 0. \qquad (2.4.33)$$

b) Show that one can choose the parameter γ in $z = x^2 + \gamma$ to reduce (2.4.33) to the form

$$(z^2 + \gamma)^2 = (Az + B)^2. \qquad (2.4.34)$$

This involves the solution of a cubic equation. The case $\gamma = q_2/2$ requires special treatment. The equation (2.4.34) can be solved directly.

c) An alternative method due to Descartes is based on the factorization

$$x^4 + q_2^* x^2 + q_1^* x + q_0^* = (x^2 + ax + b)(x^2 + cx + d).$$

Show that a^2 satisfies a cubic equation, so it can be solved by Cardano's formulas, and that b, c, d are rational functions of a.

d) Use these methods to solve the quartic $x^4 + 10x^3 + 35x^2 + 50x + 24 = 0$.

Extra 2.4.4. In theory, the formula to solve the general cubic and quartic equations provides analytic expressions for the integral of any rational function such that every term in its partial fraction decomposition has denominators of degree at most 4. In particular they may be used to evaluate the integral

$$\int_0^\infty \frac{P(x)\,dx}{(x^4 + q_3 x^3 + q_2 x^2 + q_1 x + q_0)^{m+1}}$$

for a polynomial $P(x)$ of degree at most $4m + 2$ in terms of the roots of the quartic.

The fact that the roots of the general equation of degree 5, $Q_5(x) = 0$, can not be expressed in terms of the coefficients of Q_5 using only radicals was indicated by Ruffini (1799/1950) and proved by Abel (1826) and Galois (1831) at the beginning of the 19th century; see McKean and Moll (1997), Chapter 4 for a discussion of this classical problem. The reader will find in Ayoub (1982) the proof of the nonsolvability of the general polynomial equation.

The existence of formulas for the solution of the quintic is full of beautiful connections: it involves the study of symmetries of the icosahedron; see Shurman (1997) for an introduction to these ideas. In Chapter 6 we show how to solve the general cubic and quartic using trigonometric functions; this idea extends to degrees 5 and 6. It is possible to express the roots of a general quintic in terms of *doubly-periodic functions*. Details can be found in

McKean and Moll (1997). It turns out that the solutions to a polynomial equation can be given (explicitly?) in terms of *theta functions* of a hyperelliptic curve. This is quite advanced, the details appear in Umemura (1983).

In his classical treatise (1958), page 9, Hardy states

The solution of the problem[2] in the case of rational functions may therefore be said to be complete; for the difficulty with regard to the explicit solution of algebraic equations is one not of inadequate knowledge but of proved impossibility.

2.5. The Integration of a Biquadratic

The method of partial fractions is now used to evaluate the integral

$$N_{0,4}(a; m) = \int_0^\infty \frac{dx}{(x^4 + 2ax^2 + 1)^{m+1}}. \tag{2.5.1}$$

As the notation indicates, this integral is part of a family that will be discussed in Chapter 7.

The quartic denominator factors as

$$x^4 + 2ax^2 + 1 = (x^2 + r_1^2)(x^2 + r_2^2), \tag{2.5.2}$$

where

$$r_1^2 = a + \sqrt{a^2 - 1} \quad \text{and} \quad r_2^2 = a - \sqrt{a^2 - 1}. \tag{2.5.3}$$

Exercise 2.5.1. Prove that the integral converges if $a > -1$. **Hint.** Consider the cases $a \geq 1$ and $-1 < a < 1$ separately.

In this evaluation we assume that $a > 1$ and $r_1 > 1 > r_2$.

Exercise 2.5.2. Prove that

$$r_1 = \frac{1}{\sqrt{2}} \left(\sqrt{a+1} + \sqrt{a-1} \right) \quad \text{and} \quad r_2 = \frac{1}{\sqrt{2}} \left(\sqrt{a+1} - \sqrt{a-1} \right). \tag{2.5.4}$$

Now write $t = x^2$ and consider the partial fraction decomposition of the integrand

$$h_m(t) = \frac{1}{(t + t_1)^{m+1} (t + t_2)^{m+1}}, \tag{2.5.5}$$

---

[2] Hardy is discussing the problem of integrating a function in elementary terms.

where we have written $t_1 = r_1^2$ and $t_2 = r_2^2$. The Mathematica command

$$f[m\_,t\_] := \text{Apart}[\ h[m,t]\ ]$$

yields

$$h_0(t) = \frac{s}{t + t_1} - \frac{s}{t + t_2} \qquad (2.5.6)$$

$$h_1(t) = \left(\frac{s}{(t + t_1)^2} - \frac{2s^3}{t + t_1} \right) + \left(\frac{s^2}{(t + t_2)^2} + \frac{2s^3}{t + t_2} \right)$$

and

$$h_2(t) = \left(\frac{s^3}{(t + t_1)^3} - \frac{3s^4}{(t + t_1)^2} + \frac{6s^5}{t + t_1} \right) - \left(\frac{s^3}{(t + t_2)^3} + \frac{3s^4}{(t + t_2)^2} + \frac{6s^5}{t + t_2} \right).$$

$$(2.5.7)$$

with $s = 1/(t_2 - t_1)$. These examples suggest the conjecture

$$h_m(t) = \sum_{j=0}^{m} \frac{(-1)^j s^{m+1+j}}{(t + t_1)^{m+1-j}} B_{m,j} + (-1)^{m+1} \sum_{j=0}^{m} \frac{s^{m+1+j}}{(t + t_2)^{m+1-j}} \qquad (2.5.8)$$

for some coefficients $B_{m,j} : 0 \le j \le m$.

The first few coefficients are

$$\begin{array}{cccc} 1 & & & \\ 1 & 2 & & \\ 1 & 3 & 6 & \\ 1 & 4 & 10 & 20 \end{array}$$

and we recognize the binomial coefficients

$$B_{m,j} = \binom{m + j}{j} \qquad (2.5.9)$$

from the previous data.

Exercise 2.5.3. Use Mathematica to generate the partial fraction decomposition of $h_m(t)$ for $3 \le m \le 10$ and use the guessing method described in Chapter 1 to produce a formula for $B_{m,j}$.

An alternative procedure is to access Neil Sloane's web site *The On-Line Encyclopedia of Integer Sequences* at http://www.research.att.com/~njas/sequences to find out (2.5.9).

Thus the conjecture (2.5.8) becomes

$$h_m(t) = (-1)^{m+1} \sum_{j=0}^{m} \binom{m+j}{m} \left(\frac{s^{m+1+j}}{(t+t_2)^{m+1-j}} + \frac{(-s)^{m+1+j}}{(t+t_1)^{m+1-j}} \right).$$

Exercise 2.5.4. Prove this conjecture by induction on m. **Hint.** Expand the identity

$$1 = \left(\frac{t+t_2}{t_2 - t_1} - \frac{t+t_1}{t_2 - t_1} \right)^{m+1} \tag{2.5.10}$$

by the binomial theorem, and repeatedly use

$$1 = \frac{t+t_2}{t_2 - t_1} - \frac{t+t_1}{t_2 - t_1} \tag{2.5.11}$$

in order to show

$$1 = (-1)^{m+1} \sum_{j=0}^{m} \binom{m+j}{j}$$
$$\times \left[\frac{(t+t_1)^{m+1}(t+t_2)^j}{(t_2-t_1)^{m+j+1}} - \frac{(-1)^{m+j+1}(t+t_1)^j(t+t_2)^{m+1}}{(t_2-t_1)^{m+j+1}} \right]. \tag{2.5.12}$$

Exercise 2.5.5. Use Wallis's formula given in Theorem 2.3.1 and the expression (2.5.8) to evaluate $N_{0,4}$ as

$$N_{0,4}(a;m) := \int_0^\infty \frac{dx}{(x^4 + 2ax^2 + 1)^{m+1}} = (-1)^{m+1} \pi \sum_{j=0}^{m} \frac{\binom{m+j}{j}\binom{2m-2j}{m-j}}{2^{4m-2j+5/2}}$$
$$\times \left(\frac{(P-N)^{2m+1-2j} + (-1)^{m+1+j}(P+N)^{2m+1-2j}}{N^{m+1+j} P^{m+1+j}} \right) \tag{2.5.13}$$

where $P = \sqrt{a+1}$ and $N = \sqrt{a-1}$.

Project 2.5.1. Chapter 7 describe proofs that the function

$$P_m(a) := \frac{2^{m+3/2}}{\pi} (a+1)^{m+1/2} N_{0,4}(a;m) \tag{2.5.14}$$

is a polynomial in a, of degree m, with positive rational coefficients. The goal of this project is to establish this result directly from Exercise 2.5.5. Observe that the expression in this exercise contains the term $(a-1)^{m+1/2}$

in the denominator. In view of (2.5.14) the first step in the project is to show
that these denominators must cancel.

Obtain a closed form expression for the coefficients $d_l(m)$ in

$$P_m(a) = \sum_{l=0}^{m} d_l(m)a^l. \tag{2.5.15}$$

Exercise 2.5.6. Use the method of partial fractions described above to check
the evaluation

$$\int_0^\infty \frac{dx}{(x^4 + 2ax^2 + 1)^2} = \frac{(2a + 3)\,\pi}{2^{7/2}\,(a + 1)^{3/2}}. \tag{2.5.16}$$

Conclude that $P_1(a) = (2a + 3)/2$. Repeat the procedure to check that the
next polynomial is $P_2(a) = \frac{3}{8}(4a^2 + 10a + 7)$.

3

A Simple Rational Function

3.1. Introduction

The method of partial fractions described in Chapter 2 shows that the complexity of an integral increases with the number of poles of Q. In this chapter we evaluate the definite integral of a rational function R that has a single pole of multiplicity $m + 1$; that is, we consider the integral

$$\int_0^\infty \frac{P(x)}{(q_1 x + q_0)^{m+1}} \, dx, \quad m \in \mathbb{N}, \tag{3.1.1}$$

where $P(x)$ is a polynomial of degree at most $m - 1$. The goal is to describe the integral in (3.1.1) in terms of the parameters

$$\mathfrak{P}_1 = \{m; \, q_0, \, q_1\} \cup \{ \text{ coefficients of } P\}. \tag{3.1.2}$$

We will show that

$$\int_0^\infty \frac{x^n \, dx}{(q_1 x + q_0)^{m+1}} = \frac{1}{q_0^{m-n} q_1^{n+1}} \sum_{j=0}^{n} \frac{(-1)^{n-j}}{m - j} \binom{n}{j}. \tag{3.1.3}$$

The next question in this evaluation is whether one accepts a finite sum of binomial coefficients as an *admissible closed form*. In this case we show that this sum can be reduced to a much simpler form.

The identity (3.1.3) is of the form

$$\int = \sum \tag{3.1.4}$$

where a sum is equal to an integral. The expression (3.1.4) is a variation of a colloquium title given by Doron Zeilberger at Tulane University on April 9, 1999. Many more similar expressions will appear throughout this book.

3.2. Rational Functions with a Single Multiple Pole

The integral in (3.1.1) is a linear combination of

$$I(m, n) = \int_0^\infty \frac{x^n \, dx}{(q_1 x + q_0)^{m+1}},$$ (3.2.1)

for $0 \leq n \leq m - 1$, so it suffices to give a closed form of these.

The next exercise establishes a link between $I(m, n)$ and a finite sum.

Exercise 3.2.1. Define

$$S(m, n) = \sum_{j=0}^n \frac{(-1)^{n-j}}{m - j} \binom{n}{j}, \quad n < m.$$ (3.2.2)

Prove that

$$I(m, n) = q_1^{-n-1} q_0^{-m+n} S(m, n).$$ (3.2.3)

Hint. Use the change of variable $u = q_1 x + q_0$ and expand the integrand by the binomial theorem.

Section 3.4 presents an evaluation of $I(m, n)$ based on a recursion. The initial data for this recursion is established in the next exercise.

Exercise 3.2.2. Use the result of Exercise 3.2.1 to obtain the value

$$I(m, 0) = \frac{1}{mq_1 q_0^m}.$$ (3.2.4)

We have obtained the value of $I(m, n)$ in terms of a finite sum. This is an explicit formula that can be used to evaluate $I(m, n)$ in specific cases. Note that the sum $S(m, n)$ becomes undefined if $n \geq m$, which is a reflection of the condition $n < m$ for convergence of the integral.

3.3. An Empirical Derivation

The goal of this section is to use the empirical method described in Chapter 1 to guess the value of the sum

$$S(m, n) = \sum_{j=0}^n \frac{(-1)^{n-j}}{m - j} \binom{n}{j}, \quad n < m$$ (3.3.1)

introduced in Section 3.2. An alternative way to obtain a **simple formula** for $S(m, n)$ exists is discussed in the Appendix.

The first step is to use Mathematica to produce a large number of evaluations of the sum $S(m, n)$ via the command

```
S[m_,n_] := Sum[ (-1)^{n-j}*Binomial[n,j]/
                 (m-j),{j,0,n} ].
```

The first few evaluations

$$S(3, 2) = \frac{1}{3}$$

$$S(5, 2) = \frac{1}{30}$$

$$S(10, 5) = \frac{1}{1260}$$

suggest that $S(m, n)$ is the reciprocal of a positive integer. The example

$$S(100, 50) = \frac{1}{504456722727820966674062486280 0} \qquad (3.3.2)$$

should convince the reader of the validity of this statement.

The next step is to study the factorization of the numbers

$$S_1(m, n) = \frac{1}{S(m, n)} \qquad (3.3.3)$$

conjectured to be integers.

Exercise 3.3.1. Determine the factorization of $y = S_1(500, 300)$. Check that y has 148 digits and its prime factorization contains all primes from 307 to 499. Moreover these primes appear to the first power.

The presence of consecutive primes suggests that 500! might be linked to $S_1(500, 300)$. We therefore compute $S_1(500, 300)/500!$ in order to eliminate these prime factors. The resulting rational number contains in the denominator all primes between 211 and 293. We therefore compute $S_1(500, 300) \times 300!/500!$. We continue this process and multiply by 200! to obtain

$$S_1(500, 300) \times \frac{300! \cdot 200!}{500!} = 200. \qquad (3.3.4)$$

At this point one has the strong belief that this cancellation is not accidental. We therefore compute

$$S_1(650, 301) \times \frac{301! \cdot 349!}{650!} = 349, \qquad (3.3.5)$$

which leads to the conjecture

$$S_1(m, n) = \frac{m!}{n! \cdot (m-n)!} \times (m-n) = \binom{m}{n} \times (m-n). \quad (3.3.6)$$

This will be proved in the next section.

Exercise 3.3.2. Use the procedure described above to evaluate $S_1(1000, 500)$ and check that (3.3.6) holds for these values.

Exercise 3.3.3. The four numbers

$$x_1 = 32118821490799144825027127893311699600$$
$$x_2 = 72088910457126969496171998160544036880$$
$$x_3 = 156715022732884716296026082957704428000$$
$$x_4 = 330101856394799721559714515166228476000$$

are known to be consecutive terms of a sequence. Use the technique described in this section to develop a reasonable conjecture for an analytic formula that yields these values. (After finishing the problem, the reader should look back at Section 1.4).

3.4. Scaling and a Recursion

The goal of this section is to provide a proof of the identity (3.3.6). The first part of the proof consists in scaling the integral $I(m, n)$, so as to describe the role of the parameters q_0, q_1.

Exercise 3.4.1. Prove that

$$I(m, n) = \frac{J(m, n)}{q_1^{n+1} q_0^{m-n}}, \quad (3.4.1)$$

where

$$J(m, n) := \int_0^\infty \frac{t^n \, dt}{(t+1)^{m+1}}. \quad (3.4.2)$$

Conclude that $S(m, n) = J(m, n)$. **Hint.** Consider a change of variable of the form $x \mapsto \lambda x$ with an appropriate λ.

The change of variable in the previous exercise is a natural one for this problem. It allows us to factor both parameters from the integrand so that

52 A Simple Rational Function

the dependence of $I(m, n)$ upon q_0, q_1 is explicit and we are left with the evaluation of a simpler integral.

The next exercise produces a closed-form formula for $J(m, n)$.

Exercise 3.4.2. Integrate (3.4.2) by parts to produce the recurrence

$$J(m, n) = \frac{n}{m} J(m - 1, n - 1).$$ (3.4.3)

Iterate this to obtain

$$J(m, n) = \frac{n(n - 1) \cdots (n - j + 1)}{m(m - 1) \cdots (m - j + 1)} J(m - j, n - j).$$

Now use $J(m, 0) = 1/m$ to obtain

$$J(m, n) = \frac{n! \, (m - n)!}{m! \, (m - n)} = \left[\binom{m}{n} (m - n) \right]^{-1}.$$ (3.4.4)

This completes the evaluation of the original integral. We summarize the previous discussion in a theorem.

Theorem 3.4.1. *Let* $m, n \in \mathbb{N}$ *and* $n < m$. *Then*

$$\int_0^\infty \frac{x^n \, dx}{(q_1 x + q_0)^{m+1}} = \frac{1}{q_1^{n+1} q_0^{m-n}} \left[\binom{m}{n} (m - n) \right]^{-1}.$$ (3.4.5)

The evaluation of integrals is a subject full of *unintended consequences*. The first one presented here is the evaluation of a finite sum involving binomial coefficients.

Corollary 3.4.1. *Let* $m, n \in \mathbb{N}$, *with* $n > m$. *Then*

$$\sum_{j=0}^{n} \frac{(-1)^{n-j}}{m - j} \binom{n}{j} = \left[\binom{m}{n} (m - n) \right]^{-1}.$$ (3.4.6)

The question of how to evaluate the sum (3.4.6) directly is discussed the Appendix. Observe that the sum is positive. The reader should try to give a direct proof of this, that is, without evaluating the sum.

The next exercise describes a certain symmetry of $J(m, n)$.

Exercise 3.4.3. Prove that the integral $J(m, n)$ satisfies the relation

$$J(m, n) = J(m, m - n - 1). \tag{3.4.7}$$

Hint. The change of variable $x \mapsto 1/x$ does it.

In terms of the sum $S(m, n)$ (3.4.7) states

$$\sum_{j=0}^{n} \frac{(-1)^j}{m - j} \binom{n}{j} = \sum_{j=0}^{m-n-1} \frac{(-1)^{m-1-j}}{m - j} \binom{m - n - 1}{j}. \tag{3.4.8}$$

The fact that two finite sums are equal

$$\sum_{k=0}^{n} a_k = \sum_{k=0}^{n} b_k \tag{3.4.9}$$

sometimes indicates that the sets of numbers $\{a_k\}$ and $\{b_k\}$ are identical. For instance, for any $m, n \in \mathbb{N}$, $0 \le m \le n$, we have

$$\sum_{k=0}^{m} \binom{n}{k} = \sum_{k=0}^{m} \left\{ \binom{n - 1}{k} + \binom{n - 1}{k - 1} \right\}, \tag{3.4.10}$$

in view of a basic identity for binomial coefficients. The sums in (3.4.8) are not of this type: the individual terms do not agree, only their sums do. In fact, it is clear that the limits of the sums may not agree.

3.5. A Symbolic Evaluation

In this section we describe the results of a blind evaluation of the sum $S(m, n)$ using Mathematica.

A direct symbolic evaluation of the sum $S(m, n)$ can be achieved by the Mathematica command

```
FullSimplify[Sum[(-1)^(n-j)*Binomial[n,j]/
           (m-j),{j,0,n}]].
```

Mathematica yields the answer

$$S(m, n) = \frac{(-1)^{n+1} n \Gamma(-m) \Gamma(n)}{\Gamma(1 - m + n)}, \tag{3.5.1}$$

where $\Gamma(x)$ is the gamma function defined in (1.3.6).

A symbolic evaluation of the primitive of the integrand in $I(m, n)$ yields

$$\int \frac{x^n \, dx}{(q_1 x + q_0)^{m+1}} = \frac{x^{n+1}}{q_0^{m+1}(n+1)} {}_2F_1 \left[1+n, \, 1+m, \, 2+n, \, -\frac{q_1 x}{q_0} \right],$$

(3.5.2)

and

$$I(m, n) = \frac{1}{q_1^{n+1} q_0^{m-n}} \frac{\Gamma(m-n) \, \Gamma(n+1)}{\Gamma(m+1)}$$

(3.5.3)

for the definite integral.

The **hypergeometric function** appearing in (3.5.2) is defined by the series

$$_2F_1 \, [a, \, b, \, c \, ; x] := \sum_{k=0}^{\infty} \frac{(a)_k \, (b)_k}{(c)_k} \frac{x^k}{k!},$$

(3.5.4)

where $(a)_k$ is the ascending factorial symbol. This is a special case of the function

$$_pF_q \left[\{a_1, a_2, \cdots, a_p\}, \{b_1, b_2, \cdots, b_q\}; x \right] := \sum_{k=0}^{\infty} \frac{(a_1)_k \, (a_2)_k \cdots (a_p)_k}{(b_1)_k \, (b_2)_k \cdots (b_q)_k} \frac{x^k}{k!}.$$

(3.5.5)

The reader is referred to Andrews et al. (1999) for more information about this function. Most of the functions that appear in this book can be expressed as special cases of this hypergeometric series. See Exercises 5.2.6 and 6.2.8 for details.

Exercise 3.5.1. The goal of this exercise is to provide an alternative way to evaluate the integral (3.5.2).
a) Prove that

$$\int \frac{x^n \, dx}{(q_1 x + q_0)^{m+1}} = q_0^{m-n} q_1^{-n-1} \int \frac{t^n \, dt}{(t+1)^{m+1}}.$$

(3.5.6)

b) Let $u = t + 1$ to obtain a closed form for the integral.

Extra 3.5.1. The formula for the sum $S(m, n)$ obtained from (3.5.3) by using (3.2.3) is now compared to the one in (3.5.1) to produce the identity

$$\frac{(-1)^{n+1} n \, \Gamma(-m) \, \Gamma(n)}{\Gamma(1-m+n)} = \frac{\Gamma(m-n) \, \Gamma(n+1)}{\Gamma(m+1)},$$

(3.5.7)

which will be established in Chapter 10, Exercise 10.1.6. Similarly (3.4.8) becomes

$$\frac{(-1)^{m+1} \Gamma(m-n)}{\Gamma(-n)} = \frac{\Gamma(n+1)}{\Gamma(1-m+n)}.$$

(3.5.8)

The point to be made here is that different evaluations of the same object sometimes lead to interesting identities.

3.6. A Search in Gradshteyn and Ryzhik

In view of the existence of very complete tables of integrals, throughout the text we encourage the reader to search for a given integral in these tables. For instance, in an attempt to find $I(m, n)$, we look at the index of Gradshteyn and Ryzhik (1994) [G & R] and find under Sections 2.19–2.23 the title: *Combinations of powers of x and powers of binomials of the form* $(\alpha + \beta x)$. In this section, we find[1] 3.194.4

$$\int_0^\infty \frac{x^{\mu-1}\,dx}{(1+\beta x)^{m+1}} = (-1)^m \frac{\pi}{\beta^\mu}\binom{\mu-1}{m}\frac{1}{\sin(\mu\pi)}, \qquad (3.6.1)$$

for $\beta \in \mathbb{R}^-$ and $0 < \mu < m + 1$. The formula in [G & R] refers to the *Bateman Manuscript Project* (Bateman, 1953). This is large table of formulas of special functions compiled by a group of mathematicians known as the Staff of the Bateman Manuscript Project. The formula (3.6.1) appears in Volume I, chapter 5 (*Mellin transforms*), section 5.2 (*Algebraic functions and powers with arbitrary index*), formula 6. The Bateman compendium contains no proofs and the reader is referred to Doetsch (1950) and Titchmarsh (1948) for details on the Mellin transform.

Observe that the denominator of (3.6.1) vanishes when μ becomes an integer. Therefore, the evaluation of $J(m, n)$ requires the limit of (3.6.1) as $\mu \to n + 1 \in \mathbb{N}$. This can now be computed directly. Indeed, we have

$$(-1)^m \frac{\pi}{\beta^\mu}\binom{\mu-1}{m}\frac{1}{\sin\mu\pi} = \frac{(-1)^m\pi(\mu-1)(\mu-2)\cdots(\mu-m)}{\beta^\mu m!\sin(\mu\pi)}$$

and now isolate from the numerator the factor that vanishes in the limit to get

$$(-1)^m \frac{\pi}{\beta^\mu}\binom{\mu-1}{m}\frac{1}{\sin\mu\pi} = \frac{(-1)^m\pi}{\beta^\mu m!}\prod_{j=1}^n(\mu-j)\times\frac{\mu-n-1}{\sin(\mu\pi)}$$

$$\times\prod_{j=n+2}^m(\mu-j).$$

Passing to the limit as $\mu \to n + 1$ yields (3.4.5). The required limit

$$\lim_{\mu\to n+1}\frac{\mu-n-1}{\sin(\mu\pi)} = \frac{(-1)^{n+1}}{\pi} \qquad (3.6.2)$$

---

[1] We have changed n to m to be consistent with our notation.

is based on elementary properties of trigonometric functions. These are described in Chapter 6.

3.7. Some Consequences of the Evaluation

In this section we discuss the evaluation of some series that follow from the integral evaluated in Section 3.4. This evaluation can be written as

$$\int_0^\infty \frac{x^n\, dx}{(q_1 x + q_0)^{r+n+1}} = \frac{1}{q_1^{n+1} q_0^r}\left[r \binom{r+n}{n}\right]^{-1} \tag{3.7.1}$$

for $r, n \in \mathbb{N}$. In the special case $q_0 = q_1 = 1$ we have

$$\int_0^\infty \frac{x^n\, dx}{(x+1)^{r+n+1}} = \left[r \binom{r+n}{n}\right]^{-1}. \tag{3.7.2}$$

The result (3.7.2) is next employed to obtain the sum of a series involving binomial coefficients. The technique illustrated here will be used throughout the book: once we succeed in evaluating an integral that contains parameters, summing over a certain range sometimes yields new results. The question of convergence of the series require to understand the asymptotic behavior of $n!$. This will be discussed in Chapter 5.

Exercise 3.7.1. Prove that

$$\sum_{r=1}^\infty \frac{1}{r \binom{r+n}{n}} = \frac{1}{n}. \tag{3.7.3}$$

Hint. Sum (3.7.2) from $r = 1$ to $r = \infty$ and recognize the resulting integral as $J(n, n-1)$. Check the result with Mathematica.

Some interesting series can be produced from (3.7.2) by choosing r as an appropriate function of n before summing. The next exercise illustrates this point.

Exercise 3.7.2. Prove that

$$\sum_{n=1}^\infty \frac{1}{n \binom{2n}{n}} = \frac{\pi}{3\sqrt{3}}. \tag{3.7.4}$$

Hint. Let $r = n$ in (3.7.2) and then sum over $n \in \mathbb{N}$. The resulting integral can be evaluated directly. A slick proof follows from the normalization described in Exercise 2.1.4. A different proof appears in Exercise 6.6.1.

Exercise 3.7.3. Prove that for $n \in \mathbb{N}$,

$$\sum_{r=1}^{\infty} \frac{(-1)^{r+1}}{r \binom{r+n}{n}} = \int_0^{\infty} \frac{x^n \, dx}{(x+1)^{n+1} (x+2)}. \tag{3.7.5}$$

Conclude that

$$\int_0^{\infty} \frac{x^n \, dx}{(x+1)^{n+1} (x+2)} = 2^n \ln 2 - \sum_{j=0}^{n-1} \binom{n}{j} \frac{(-1)^{n-j-1}}{n-j}(2^n - 2^j). \tag{3.7.6}$$

Hint. Multiply (3.7.2) by $(-1)^r$ and then sum over $r \in \mathbb{N}$. The change of variable $x \mapsto x/(x+1)$ is useful in the resulting integral.

Project 3.7.1. This project generalizes the integral in (3.7.6):
a) For $y < 0$ define

$$f_n(y) = \int_0^{\infty} \frac{x^n \, dx}{(x+1)^{n+1} (x-y)} \tag{3.7.7}$$

and

$$p_n(y) = n! \left\{ (y+1)^{n+1} f_n(y) + y^n \ln(-y) \right\}. \tag{3.7.8}$$

Prove that f_n satisfies

$$f_n'(y) = -\frac{n+1}{y} f_{n+1}(y) + \frac{n}{y} f_n(y) \tag{3.7.9}$$

and use it to check that $p_n(y)$ satisfies the recurrence

$$p_{n+1}(y) = n(2y+1)p_n(y) - y(y+1)p_n'(y) + n! y^n(y+1), \tag{3.7.10}$$

with $p_0(y) = 0$. Conclude that p_n is a polynomial of degree n.
b) Confirm the values

$$p_1(y) = y + 1,$$
$$p_2(y) = 2y^2 + 3y + 1,$$
$$p_3(y) = 6y^3 + 11y^2 + 7y + 2.$$

c) Derive a recurrence for the coefficients of $p_n(y)$ and conclude that they are positive integers.
d) Can you find a closed-form expressions for the coefficients?

3.8. A Complicated Integral

In this section we illustrate the method of partial fractions to evaluate a definite integral.

The identity (3.7.2) is

$$\int_0^\infty \frac{x^n \, dx}{(x+1)^{r+n+1}} = \left[r \binom{r+n}{n} \right]^{-1},$$

and for $r = 2n$ yields

$$\int_0^\infty \frac{x^n \, dx}{(x+1)^{3n+1}} = \left[2n \binom{3n}{n} \right]^{-1}.$$

Summing from $n = 1$ to infinity yields

$$\sum_{n=1}^\infty \frac{1}{n \binom{3n}{n}} = \int_0^\infty \frac{2x \, dx}{(x+1)(x^3 + 3x^2 + 2x + 1)}. \qquad (3.8.1)$$

The method of partial fractions requires the roots $\{x_1, x_2, x_3\}$ of the cubic equation $x^3 + 3x^2 + 2x + 1 = 0$. These are given in Exercise 2.4.8. Now let $a := -x_1$ and introduce the quadratic factor

$$x^2 + bx + c = (x - x_2)(x - x_3) \qquad (3.8.2)$$

so that

$$x^3 + 3x^2 + 2x + 1 = (x + a)(x^2 + bx + c) \qquad (3.8.3)$$

with $b = -(x_2 + x_3)$ and $c = x_2 x_3$. The reader will observe that the co-efficients b and c are real even though the roots x_2 and x_3 are not. The integrand in (3.8.1) can be expanded in partial fractions in the form

$$\frac{2x}{(x+1)(x^3 + 3x^2 + 2x + 1)} = \frac{a_1}{x+1} + \frac{b_1}{x+a} + \frac{c_1 + d_1 x}{x^2 + bx + c}. \qquad (3.8.4)$$

An elementary calculation, similar as the one in (2.1.3), yields the values

$$a_1 = \frac{2}{(1-a)(1-b+c)}$$

$$b_1 = \frac{2a}{(a-1)(a^2 - ab + c)}.$$

To evaluate the remaining two constants, it suffices to give x two specific values in (3.8.4), say $x = 0$ and $x = 1$, to produce a linear system that yields

$$c_1 = \frac{2c(1 + a - b)}{(1 - b + c)(a^2 - ab + c)}$$

$$d_1 = \frac{2(a - c)}{(1 - b + c)(a^2 - ab + c)}.$$

We now write (3.8.4) in the form

$$\frac{2x}{(x + 1)(x^3 + 3x^2 + 2x + 1)} = \frac{a_1}{x + 1} + \frac{b_1}{x + a} + \frac{d_1}{2} \frac{2x + b}{x^2 + bx + c}$$
$$+ \frac{1}{2} \frac{(2c_1 - bd_1)}{x^2 + bx + c}$$

and integrate term by term to produce

$$\int_0^\infty \frac{2x \, dx}{x^3 + 3x^2 + 2x + 1} = -b_1 \ln a - \frac{d_1}{2} \ln c + \frac{(2c_1 - bd_1)}{\sqrt{4c - b^2}}$$
$$\times \left(\frac{\pi}{2} - \tan^{-1} \left(\frac{b}{\sqrt{4c - b^2}} \right) \right).$$

This answer illustrates the fact that an explicit answer to the integral of a rational function cannot always be expressed in simple algebraic form. The real root x_1, obtained by Cardano's formula described in Section 2.4, is

$$x_1 := -1 - \sqrt[3]{\frac{2}{3(9 - \sqrt{69})}} - \frac{1}{\sqrt[3]{9}} \sqrt[3]{\frac{9 - \sqrt{69}}{2}}.$$

3.8.1. Warning

A symbolic evaluation of the integral in (3.8.1) yields the answer ∞. This is clearly incorrect. Mathematica performs a partial fraction decomposition of the integrand to obtain

$$\frac{2x}{(x + 1)(x^3 + 3x^2 + 2x + 1)} = -\frac{2}{1 + x} + \frac{2(x + 1)^2}{x^3 + 3x^2 + 2x + 1},$$

and the divergence of the integrals of the parts leads to the incorrect evaluation.

Note 3.8.1. A symbolic evaluation of

$$f(k) := \sum_{n=1}^\infty \frac{1}{n \binom{kn}{n}} \tag{3.8.5}$$

A Simple Rational Function

using Mathematica yields the answer

$$f(3) = \frac{1}{3} \, {}_3F_2 \left[\{1, 1, \tfrac{3}{2}\}, \{\tfrac{4}{3}, \tfrac{5}{3}\}, \tfrac{2^2}{3^3}\right], \tag{3.8.6}$$

$$f(4) = \frac{1}{4} \, {}_4F_3 \left[\{1, 1, \tfrac{4}{3}, \tfrac{5}{3}\}, \{\tfrac{5}{4}, \tfrac{6}{4}, \tfrac{7}{4}\}, \tfrac{3^3}{4^4}\right], \tag{3.8.7}$$

and

$$f(5) = \frac{1}{5} \, {}_5F_4 \left[\{1, 1, \tfrac{5}{4}, \tfrac{6}{4}, \tfrac{7}{4}\}, \{\tfrac{6}{5}, \tfrac{7}{5}, \tfrac{8}{5}, \tfrac{9}{5}\}, \tfrac{4^4}{5^5}\right], \tag{3.8.8}$$

The pattern is now clear. The simplicity of $f(2)$ is due to the fact that

$$f(2) = \tfrac{1}{2} \, {}_2F_1 \left[\{1, 1\}, \{\tfrac{3}{2}\}; \tfrac{1}{4}\right] \tag{3.8.9}$$

can be expressed in terms of simpler numbers, namely $\pi/3\sqrt{3}$.

The question of special values of the hypergeometric function can be expressed in simple terms is a difficult one. We will touch upon this in the simpler case of the trigonometric functions in Extra 6.6.1.

4

A Review of Power Series

4.1. Introduction

The class of polynomial functions

$$P(x) = p_0 + p_1 x + p_2 x^2 + \cdots + p_n x^n \qquad (4.1.1)$$

is the most elementary class considered in calculus. The goal of this chapter is to give a brief overview of **power series**. These are representations of a function, similar to (4.1.1), in which the degree n is allowed to become infinite.

Definition 4.1.1. A **power series** centered at $x = a$ is a sum of the form

$$f(x) = \sum_{k=0}^{\infty} c_k (x - a)^k \qquad (4.1.2)$$

where the coefficients $c_k \in \mathbb{R}$.

Most of the functions considered in this book have a power series representation. This includes the *advanced functions* that appear from blind Mathematica evaluations. For example, the hypergeometric function is defined in (3.5.4) by its power series.

Given a value of x, the sum in (4.1.2) becomes a sum of real numbers and as such it may or may not converge. A simple argument shows that the set of points x for which the series converges is an interval of the form $(a - R, a + R)$, called the **interval of convergence** of the series. The number R is the **radius of convergence**. The expression (4.1.2) defines a function for $x \in (a - R, a + R)$.

Example 4.1.1. The **geometric series**

$$\sum_{k=0}^{\infty} x^k = \frac{1}{1 - x} \qquad (4.1.3)$$

has radius of convergence 1.

Exercise 4.1.1. Establish (4.1.3) and check that

$$\frac{1}{1+x^2} = \sum_{k=0}^{\infty} (-1)^k x^{2k}. \tag{4.1.4}$$

Hint. Use (1.6.19) and then pass to the limit.

Note 4.1.1. Observe that the function $f(x) = 1/(1-x)$ is well defined on $\mathbb{R}$ with the single exception of $x = 1$, but its power series representation centered at $x = 0$ converges only on $(-1, 1)$. The incorrect use of f outside its radius of convergence is at the center of false identities like

$$1 + 2 + 4 + 8 + 16 + \cdots = -1. \tag{4.1.5}$$

The presence of a discontinuity at $x = 1$ is reflected in the interval of convergence of the series representation for f. This phenomena is sometimes harder to visualize: the function in the left-hand side of (4.1.4) is well defined for all $x \in \mathbb{R}$ but the series converges only for $x \in (-1, 1)$. In this case the singularity at $x = i \in \mathbb{C}$ is what prevents the series from converging in a larger region.

Note 4.1.2. The radius of convergence of a series is given by

$$R = \left(\lim_{n \to \infty} \frac{|c_{n+1}|}{|c_n|} \right)^{-1} \tag{4.1.6}$$

with the conventions $1/0 = \infty$ and $1/\infty = 0$. An alternative expression is given by

$$R = \left(\lim_{n \to \infty} |c_n|^{1/n} \right)^{-1}. \tag{4.1.7}$$

These formulas are obtained by applying to (4.1.2) the ratio and root test respectively.

Exercise 4.1.2. Compute the radius of convergence of the hypergeometric series as a function of the parameters p and q.

It is a surprising fact that functions defined by power series with coefficients given by simple formulas are sometimes not elementary. For instance, the **dilogarithm function** defined by

$$\text{DiLog}(x) = \sum_{k=1}^{\infty} \frac{x^k}{k^2} \tag{4.1.8}$$

was considered by Euler and after a period of silence it has reappeared in many aspects of modern mathematics. The reader will find in Lewin (1981) a fascinating collection of results including the evaluation

$$\int_0^{(\sqrt{5}-1)/2} \frac{\ln(1-x)}{x}\,dx = \ln^2\left(\frac{\sqrt{5}-1}{2}\right) - \frac{\pi^2}{10} \qquad (4.1.9)$$

that is connected to the dilogarithm. The proceedings (Lewin, 1991) present much advanced material on this function. On the other hand, the **tangent function** that is one of the elementary functions of calculus has the power series

$$\tan x = \sum_{k=1}^{\infty} \frac{2^{2k}(2^{2k}-1)}{(2k)!}|B_{2k}|x^{2k-1} \qquad (4.1.10)$$

where B_{2k} are the **Bernoulli numbers**. These numbers will be considered in Chapter 5 and the Taylor series for the tangent is established in Exercise 6.9.5.

Note 4.1.3. In this text we will manipulate power series as if they were finite sums. The justification of differentiation and integration rules are given in the next two theorems. The details of the proofs can be found in Hijab (1997), pages 89 and 107 respectively.

Theorem 4.1.1. *Let* $f(x) = \sum_{n=0}^{\infty} a_n x^n$ *be a power series with radius of convergence* $R > 0$. *Then*

$$\sum_{n=1}^{\infty} n a_n x^{n-1} = a_1 + 2a_2 x^2 + 3a_3 x^2 + \cdots \qquad (4.1.11)$$

has radius of convergence R, f *is differentiable on* $(-R, R)$, *and* $f'(x)$ *equals (4.1.11) for all* x *in* $(-R, R)$.

Similarly R is the radius of convergence of $a_0 x + a_1 x^2/2 + a_2 x^3/3 + \cdots$ and

$$\int_0^x f(t)\,dt = \sum_{k=0}^{\infty} \frac{a_n}{n+1}x^{n+1} \quad \text{on } (-R, R).$$

The formula (4.1.11) can be written as

$$x\frac{d}{dx}\sum_{n=0}^{\infty} a_n x^n = \sum_{n=0}^{\infty} n a_n x^n \qquad (4.1.12)$$

so, in the context of power series, the operator $\theta = x\frac{d}{dx}$ is more natural than differentiation.

The next exercise introduces a new family of polynomials studied originally by Euler.

Exercise 4.1.3. Define

$$A_n(x) := (1-x)^{n+1} \left(x\frac{d}{dx} \right)^n \frac{1}{1-x}. \tag{4.1.13}$$

a) Prove that $A_n(x)$ is a polynomial in x. These are the **Eulerian polynomials**.
b) Check that

$$A_0(x) = 1,$$
$$A_1(x) = x,$$
$$A_2(x) = x + x^2,$$
$$A_3(x) = x + 4x^2 + x^3,$$
$$A_4(x) = x + 11x^2 + 11x^3 + x^4.$$

c) Establish the recurrence

$$A_n(x) = nxA_{n-1}(x) + x(1-x)A'_{n-1}(x) \tag{4.1.14}$$

and derive from it a recurrence for the coefficients of $A_n(x)$:

$$A_n(x) = \sum_{k=0}^{n} A_{n,k} x^k. \tag{4.1.15}$$

Exercise 4.1.4. Evaluate the sums

$$S(n) = \sum_{k=0}^{\infty} k^n x^k \tag{4.1.16}$$

and

$$S_1(n) = \sum_{k=0}^{\infty} (k)_n x^k. \tag{4.1.17}$$

Use the evaluations to establish the identity

$$\sum_{j=0}^{n} c(n,j)(1-x)^{n-j} A_j(x) = n!\,x, \tag{4.1.18}$$

where $A_j(x)$ are the Eulerian polynomials defined in Exercise 4.1.3 and $c(n,k)$ appear in Exercise 1.5.3.

4.2. Taylor Series

The behavior of a function $f(x)$ near a point $x = a$ can be determined by expanding f in powers of $x - a$. For example, the binomial theorem yields

$$x^n = [(x - a) + a]^n$$

$$= \sum_{j=0}^{n} \binom{n}{j} a^{n-j} (x - a)^j.$$

The coefficients in this expansion are

$$\binom{n}{j} a^{n-j} = \frac{1}{j!} \left(\frac{d}{dx} \right)^j x^n \bigg|_{x=a}.$$

Definition 4.2.1. The **Taylor series** of the function $f(x)$ centered at $x = a$ is the sum

$$\sum_{j=0}^{\infty} \frac{f^{(j)}(a)}{j!} (x - a)^j. \tag{4.2.1}$$

A function is called **analytic** if, for x inside the interval of convergence of the Taylor series, this series agrees with the function f.

Exercise 4.2.1. Use the Taylor series of $f(x) = x^{n+1}$ and (1.6.19) to establish the identity

$$\sum_{j=0}^{n} (-1)^j \binom{n+k+1}{j+k+1} \binom{j+k}{k} = 1, \tag{4.2.2}$$

for all $n, k \in \mathbb{N}$. Can you provide a direct proof?

Exercise 4.2.2. In this exercise we consider the extension of the binomial theorem

$$(1 + x)^n = \sum_{k=0}^{n} \binom{n}{k} x^k \tag{4.2.3}$$

to noninteger exponents.
    Define the **extended binomial coefficients** by

$$\binom{a}{m} := \frac{(a - m + 1)_m}{m!}, \quad a \in \mathbb{R}, \; m \in \mathbb{N}, \tag{4.2.4}$$

where $(a - m + 1)_m$ is the ascending factorial defined in Section 1.5.

a) Check that for $a \in \mathbb{N}$ and $0 \le m \le a$ this definition yields the usual binomial coefficients.

b) Compute the Taylor series of the function $f(x) = (1 + x)^a$ and establish the binomial theorem

$$(1 + x)^a = \sum_{k=0}^{\infty} \binom{a}{k} x^k. \tag{4.2.5}$$

Ignore the issues of convergence.

c) Compute $\binom{-1}{k}$ and confirm that (4.1.3) and (4.2.5) are consistent.

Many special cases of the binomial theorem produce closed-form expressions for some classes of binomial coefficients. The next exercise illustrates the point.

Exercise 4.2.3. Let $C_k := \binom{2k}{k}$ for $k \in \mathbb{N}$ be the central binomial coefficient considered in Exercise 1.4.5.

a) Use the binomial theorem established in Exercise 4.2.2 to prove that the generating function of the central binomial coefficients C_k is given by

$$\sum_{n=0}^{\infty} \binom{2n}{n} x^n = \frac{1}{\sqrt{1 - 4x}}. \tag{4.2.6}$$

b) Use the ratio test (4.1.6) to check that the series converges for $|x| < \frac{1}{4}$. This is consistent with the singularity of the radical in (4.2.6).

c) Establish an asymptotic formula for C_k using (4.1.7).

In the next exercise we establish a formula due to Cauchy to multiply power series.

Exercise 4.2.4. Prove Cauchy's formula:

$$\left(\sum_{n=0}^{\infty} a_n x^n \right) \left(\sum_{n=0}^{\infty} b_n x^n \right) = \sum_{n=0}^{\infty} c_n x^n, \tag{4.2.7}$$

where

$$c_n = \sum_{j=0}^{n} a_j b_{n-j}. \tag{4.2.8}$$

The reader will observe that the expression for c_n can be written as a sum is over all possible solutions of the equation $i + j = n$, where $i, \, j \in \mathbb{N}$:

$$c_n = \sum_{i+j=n} a_i b_j. \tag{4.2.9}$$

The next exercise will be used in Chapter 5 to check that two different approaches to the exponential function are consistent.

Exercise 4.2.5. Prove that

$$\left(\sum_{k=0}^{\infty} \frac{1}{k!}\right) \times \left(\sum_{j=0}^{\infty} \frac{(-1)^j}{j!}\right) = 1. \qquad (4.2.10)$$

The reader will recognize this identity as $e \times \frac{1}{e} = 1$.

Exercise 4.2.6. Square the identity (4.2.6) to obtain the evaluation

$$\sum_{j=0}^{n} \binom{2j}{j} \binom{2n-2j}{n-j} = 2^{2n}. \qquad (4.2.11)$$

The formula of Cauchy shows how to multiply two power series. The next exercise outlines the basic properties of division of power series. The word *formal proof* simply means that convergence issues are to be ignored.

Exercise 4.2.7. Give a formal proof that the reciprocal of the power series

$$f(x) = \sum_{n=0}^{\infty} a_n x^n \qquad (4.2.12)$$

exists if and only if $a_0 \neq 0$. **Hint.** Consider the identity

$$\sum_{n=0}^{\infty} a_n x^n \times \sum_{n=0}^{\infty} b_n x^n = 1 \qquad (4.2.13)$$

and show that one can solve for the unknown b_n provided $a_0 \neq 0$. Compute b_n for $0 \leq n \leq 4$ in terms of the coefficients of f.

4.3. Taylor Series of Rational Functions

In this section we describe properties of the Taylor series of a rational function

$$R(x) = \frac{P(x)}{Q(x)}$$

where

$$P(x) = p_n x^n + p_{n-1} x^{n-1} + \cdots + p_1 x + p_0$$
$$Q(x) = q_m x^m + q_{m-1} x^{m-1} + \cdots + q_1 x + q_0. \qquad (4.3.1)$$

We can reduce to the case $n < m$ by dividing P by Q. The algorithm is similar to the division algorithm of elementary school: it produces a *quotient* quot(x) and a *remainder* rem(x) with degree smaller than that of Q such that

$$\frac{P(x)}{Q(x)} = \text{quot}(x) + \frac{\text{rem}(x)}{Q(x)}. \tag{4.3.2}$$

For example,

$$\frac{x^4}{x^4 + 2ax^2 + 1} = 1 + \frac{-2ax^2 - 1}{x^4 + 2ax^2 + 1}. \tag{4.3.3}$$

Theorem 4.3.1. The division algorithm for polynomials. *Let P and Q as above, with p_i, $q_i \in \mathbb{R}$ and assume $n \geq m$. Then there exists unique polynomials* quot(x) *and* rem(x)*, such that*

$$P(x) = \text{quot}(x)Q(x) + \text{rem}(x), \tag{4.3.4}$$

and either rem(x) $= 0$ *or*

$$\deg(\text{rem}) < \deg(Q) = m.$$

Proof. We proceed by induction on the degree of P to show the existence of the quotient and remainder. Observe that

$$P_1(x) = P(x) - \frac{p_n}{q_m} x^{n-m} Q(x) \tag{4.3.5}$$

is a polynomial of degree strictly less than P. This reduction of degree can be iterated as long as the new polynomials have degree larger than $\deg(Q)$, completing the induction.

In order to prove uniqueness assume that

$$\text{quot}(x)Q(x) + \text{rem}(x) = \text{quot}_1(x)Q(x) + \text{rem}_1(x) \tag{4.3.6}$$

Then

$$(\text{quot}(x) - \text{quot}_1(x))\, Q(x) = \text{rem}_1(x) - \text{rem}(x). \tag{4.3.7}$$

The right-hand side is a multiple of Q of degree smaller than $\deg(Q)$, so it must vanish. This shows that rem(x) $\equiv$ rem$_1(x)$ and (4.3.7) yields quot(x) $\equiv$ quot$_1(x)$. $\qquad\square$

Extra 4.3.1. The division algorithm can be used to find the greatest common divisor of the polynomials P and Q. After (4.3.4) has been obtained one replaces the pair (P, Q) by (Q, rem). In this process the degree of the second component decreases so eventually becomes 0. The last nonzero remainder is the greatest common divisor. The example below will reappear in Section 7.6.

$$P(x) = x^4 + 2ax^2 + 1 \quad \text{and} \quad Q(x) = P'(x) = 4x^3 + 4ax. \quad (4.3.8)$$

Dividing P by Q yields

$$x^4 + 2ax^2 + 1 = \frac{x}{4}\left(4x^3 + 4ax\right) + (ax^2 + 1).$$

In the next step we divide $4x^3 + 4ax$ by $ax^2 + 1$ to obtain

$$4x^3 + 4ax = \frac{4x}{a}\left(ax^2 + 1\right) + 4(a - 1/a)x.$$

In the final step we obtain

$$ax^2 + 1 = \frac{a^2 x}{4(a^2 - 1)} \cdot 4\,(a - 1/a)\,x + 1.$$

The reader can check that the next step will leave a zero remainder. Therefore the greatest common divisor of P and Q is 1.

Exercise 4.3.1. Use the calculations described above to find polynomials α, β such that

$$\alpha(x)P(x) + \beta(x)P'(x) = 1. \quad (4.3.9)$$

Hint. The last step yields a polynomial $\gamma(x)$ such that

$$1 = (ax^2 + 1) + \gamma(x) \cdot 4(a - 1/a)x.$$

Solve for $4(a - 1/a)x$ in the second step and express 1 as a combination of $4x^3 + 4ax$ and $ax^2 + 1$. Repeat one more time to finish the calculation.

The subject of Taylor coefficients of a rational function or, equivalently, the sequences with rational generating function is very beautiful. For instance, a remarkable result of Lech, Mahler and Skolem states that the coefficients of the Taylor series of a rational function that vanish must be contained in a finite number of arithmetic progressions; see Myerson and van der Poorten (1995) and van der Poorten (1984) for details. The reader can find more information about this topic in Stanley (1999) Chapter 4 and in Everest et al. (2003).

The next theorem states that the sequence of coefficients of a rational function satisfies a linear recurrence. Before presenting the proof we illustrate the result with the example

$$R(x) = \frac{1+x}{1 - 2x - x^2}.$$ (4.3.10)

Let

$$R(x) = \sum_{k=0}^{\infty} r_k x^k$$ (4.3.11)

be the Taylor series of R. Then

$$1 + x = \sum_{k=0}^{\infty} r_k x^k - 2 \sum_{k=0}^{\infty} r_k x^{k+1} - \sum_{k=0}^{\infty} r_k x^{k+2}$$

$$= r_0 + (r_1 - 2r_0)x + \sum_{k=0}^{\infty} (r_{k+2} - 2r_{k+1} - r_k i) x^{k+2}.$$

Matching equal powers we obtain $r_0 = 1$, $r_1 = 3$ and for $k \geq 2$,

$$r_k = 2r_{k-1} + r_{k-2}.$$

This is a linear recursion, of order 2 (the degree of the denominator of R).

Theorem 4.3.2. *Let $R(x) = P(x)/Q(x)$ be a rational function, with P and Q given in (4.3.1). Then the coefficients r_k of the Taylor series of R satisfy the linear recurrence*

$$q_0 r_{k+m} + q_1 r_{k+m-1} + q_2 r_{k+m-2} + \cdots + q_m r_k = 0.$$ (4.3.12)

Proof. Let

$$R(x) = \sum_{k=0}^{\infty} r_k x^k$$ (4.3.13)

be the Taylor expansion centered at $x = 0$. The identity $R(x)Q(x) = P(x)$ now yields

$$\left(\sum_{j=0}^{\infty} r_j x^j \right) \left(\sum_{i=0}^{m} q_i x^i \right) = \sum_{l=0}^{n} p_l x^l.$$ (4.3.14)

Define $q_i = 0$ for $i > m$, so that (4.3.14) yields

$$\sum_{k=0}^{\infty} \left(\sum_{v=0}^{k} r_{k-v} q_v \right) x^k = \sum_{l=0}^{n} p_l x^l.$$

The series on the left is now split at $k = m$ to produce

$$\sum_{k=0}^{m} \left(\sum_{v=0}^{k} r_{k-v} q_v \right) x^k + \sum_{k=m+1}^{\infty} \left(\sum_{v=0}^{m} r_{k-v} q_v \right) x^k = \sum_{l=0}^{n} p_l x^l.$$

For $0 \le i \le n$ we obtain

$$\sum_{v=0}^{i} q_{i-v} r_v = p_i, \qquad (4.3.15)$$

for $n + 1 \le i \le m$,

$$\sum_{v=0}^{i} q_{i-v} r_v = 0, \qquad (4.3.16)$$

and for $i > m$,

$$\sum_{v=i-m}^{i} q_{i-v} r_v = 0. \qquad (4.3.17)$$

This is a recurrence of the type stated. The leading order term is $q_0 r_i$ with $q_0 = Q(0) \ne 0$, so the recurrence always be solved for r_i. $\qquad \square$

Example 4.3.1. Consider the function

$$R(x) = \frac{1}{1 - x - x^2}. \qquad (4.3.18)$$

Then $n = 0$ and $m = 2$, and the only nonzero coefficients are $p_0 = 1$, $q_0 = 1$, $q_1 = -1$ and $q_2 = -1$. Then (4.3.15) gives $r_0 = 1$, (4.3.16) then gives $r_1 = 1$, $r_2 = 2$. Finally, (4.3.17) produces

$$r_i = r_{i-1} + r_{i-2}, \quad \text{for } i > 2. \qquad (4.3.19)$$

We conclude that r_i is the Fibonacci number F_{i+1} and we recognize $R(x)$ as the generating function of this sequence.

Extra 4.3.2. It is possible to get a closed form solution for the recurrence

$$F_i = F_{i-1} + F_{i-2} \qquad (4.3.20)$$

satisfied by the Fibonacci sequence with initial conditions $F_1 = F_2 = 1$. The method consists in looking for a solution of (4.3.20) of the form $F_i = \alpha^i$. Replacing in (4.3.20) yields the values $\alpha_{\pm} = (1 \pm \sqrt{5})/2$. The theory of difference equations now states that

$$F_i = A\alpha_+^i + B\alpha_-^i \qquad (4.3.21)$$

for some constants A and B. See Rosen (2003) for details. The initial conditions determine these constants. The final result is

$$F_i = \frac{1}{\sqrt{5}}\left(\frac{1+\sqrt{5}}{2}\right)^i - \frac{1}{\sqrt{5}}\left(\frac{1-\sqrt{5}}{2}\right)^i. \qquad (4.3.22)$$

Extra 4.3.3. The result of Theorem 4.3.2 actually is equivalent to the fact that the sequence $\{r_k\}$ has a rational generating function. See Stanley (1999) for details.

Exercise 4.3.2. Determine the recurrence satisfied by the coefficients of

$$R(x) = \frac{x(1-x)}{x^2 + 3x + 1}.$$

Can you find a closed form for the coefficients of the Taylor expansion of R? **Hint.** Let $R(x) = r_0 + r_1 x + r_2 x^2 + \cdots$ satisfy $r_0 = 0$, $r_1 = 1$, $r_2 = -4$ and $r_{i-2} + 3r_{i-1} + r_i = 0$ for $i \geq 3$. Use the method described in Extra 4.3.2 to obtain

$$r_i = (-1)^{i+1}\left(\frac{1-\sqrt{5}}{2}\right)^{2i-1} + (-1)^{i+1}\left(\frac{1+\sqrt{5}}{2}\right)^{2i-1}, \quad i \geq 1. \quad (4.3.23)$$

Exercise 4.3.3. Repeat the previous problem with the function

$$R(x) = \frac{1}{(x^2 - 1)(x^3 - 1)}. \qquad (4.3.24)$$

Hint. Use the factorization $x^3 - 1 = (x-1)(x-\rho)(x-\rho^2)$, where $\rho = (-1 + i\sqrt{3})/2$. To simplify the expressions for the Taylor coefficients of R, keep in mind that $1 + \rho + \rho^2 = 0$.

5

The Exponential and Logarithm Functions

5.1. Introduction

The rule of integration

$$\int_1^x t^n dt = \frac{x^{n+1} - 1}{n + 1} \qquad (5.1.1)$$

has been discussed in Chapter 1 for $n \in \mathbb{Q}$, $n \neq -1$. The evaluation of (5.1.1) is elementary since the power function $f(x) = x^n$ admits a primitive that is also a power. In order to complete the integration of powers, we need to discuss the case $n = -1$. The primitive of $f(x) = 1/x$ is called the **logarithm**, that is,

$$\ln x := \int_1^x \frac{dt}{t}. \qquad (5.1.2)$$

The logarithm function is often introduced in an informal and unmotivated manner in which the student is simply made aware of its properties as a form of definition. In this chapter we develop many of the same elementary properties of $\ln x$ from its integral representation. This approach presents a pedagogical problem: the student has the feeling that one is proving properties that are already known.

We also introduce one of the basic constants of analysis, the **Euler number**[1] e, and discuss some of its arithmetical properties.

The technique employed in (5.1.2) of defining a function as the primitive of a simpler one will be repeated throughout this text. This will allow us to increase the list of **known functions**. The reader should be aware of the possibility that the new functions defined might be already known. Theorem 5.2.3 states that there is no rational function $R(x)$ that satisfies $R'(x) = 1/x$. The Mathematica notation for the logarithm is $\text{Log}[x]$.

---

[1] This is not be confused with the **Euler constant** defined in Chapter 9.

5.2. The Logarithm

The **logarithm function** defined by (5.1.2) is the unique primitive of $f(x) = 1/x$ that satisfies $\ln 1 = 0$:

$$\frac{d}{dx}\ln x = \frac{1}{x} \text{ for } x \in \mathbb{R}_+.$$

This definition completes the family given in (5.1.1). It remains to establish some properties of $\ln x$ directly from (5.1.2). This discussion is given in some detail because it will serve as a model for more complicated functions. We have tried to provide proofs that employ techniques from integration.

The first result describes functional properties of $\ln x$.

Theorem 5.2.1. *The logarithm function satisfies*
a) $\ln(xy) = \ln x + \ln y, \quad for\ x, y \in \mathbb{R}_+.$
b) $\ln(x^a) = a \ln x, \quad for\ x, a \in \mathbb{R}_+.$

Proof. To prove a) observe that

$$\ln xy = \int_1^{xy} \frac{dt}{t} = \int_1^x \frac{dt}{t} + \int_x^{xy} \frac{dt}{t}.$$

Then *a*) follows by the change of variable $t \mapsto xt$ in the second integral. Property *b*) appears from the change of variable $t \mapsto t^a$ in the integral that defines $\ln(x^a)$. □

Note 5.2.1. Implicit in this argument is the validity of the formula

$$\frac{d}{dx} x^a = ax^{a-1} \tag{5.2.1}$$

for any $a \in \mathbb{R}$. This is justified in Section 5.6.

Exercise 5.2.1. Conclude from the form of the integrand in (5.1.2) that $\ln x > 0$ for $x > 1$ and $\ln x < 0$ for $0 < x < 1$.

Exercise 5.2.2. Prove that $\ln\left(\frac{x}{y}\right) = \ln x - \ln y$.

Exercise 5.2.3. Verify that $\ln x$ is increasing and concave down.

Exercise 5.2.4. Prove that $\ln x < x - 1$ for $x > 1$. **Hint.** Determine a bound for the integrand. Derive from this Napier's inequality

$$\frac{1}{b} < \frac{\ln b - \ln a}{b - a} < \frac{1}{a} \tag{5.2.2}$$

for $b > a$. **Hint.** Let $t = b/a$.

Exercise 5.2.5. Prove the arithmetic–logarithmic–geometric mean inequality

$$\frac{a+b}{2} > \frac{b-a}{\ln b - \ln a} > \sqrt{ab} \tag{5.2.3}$$

for $b > a$. **Hint.** Let $t = b/a$ again. Compare derivatives of each side with respect to t.

We now establish the Taylor series of $\ln(1+x)$.

Theorem 5.2.2. *The Taylor series expansion of* $\ln(1+x)$ *at* $x = 0$ *is given by*

$$\ln(1+x) = \sum_{n=1}^{\infty} \frac{(-1)^{n-1}}{n} x^n \tag{5.2.4}$$

which is valid for $|x| < 1$.

Proof. This follows by expanding the integrand in

$$\ln(1+x) = \int_0^x \frac{dt}{1+t}$$

in a geometric series. ☐

Extra 5.2.1. The evaluation of the power series (5.2.4) at $x = 1$ gives

$$\ln 2 = \sum_{n=1}^{\infty} \frac{(-1)^{n-1}}{n}. \tag{5.2.5}$$

This is justified by *Abel's limit theorem:* Assume that we have

$$f(x) = \sum_{n=0}^{\infty} a_n x^n, \quad \text{if } -r < x < r.$$

If the series converges at $x = r$, then the limit $\lim\limits_{x \to r^-} f(x)$ exists and we have

$$\lim_{x \to r^-} f(x) = \sum_{n=0}^{\infty} a_n r^n.$$

See Apostol (1957), page 421 for details.

Exercise 5.2.6. Check that the logarithm function can be expressed in terms of the hypergeometric series (3.5.4) as

$$\ln(1+x) = x \cdot {}_2F_1\left[\{1, 1\}, \{2\}; -x\right]. \tag{5.2.6}$$

Exercise 5.2.7. This exercise uses the expansion of $\ln(1 + x)$ to produce the expansion of other functions.
a) Check that

$$\ln(1 - x) = -\sum_{n=1}^{\infty} \frac{x^n}{n}. \tag{5.2.7}$$

b) Given the expansion

$$f(x) = \sum_{n=1}^{\infty} a_n x^n, \tag{5.2.8}$$

determine the expansion of $f(x)/(1 - x)$. **Hint.** Use Cauchy's formula to multiply power series given in Exercise 4.2.4.
c) Use part b) to establish

$$\frac{-\ln(1 - x)}{1 - x} = \sum_{n=1}^{\infty} H_n x^n, \tag{5.2.9}$$

where

$$H_n := 1 + \frac{1}{2} + \cdots + \frac{1}{n} \tag{5.2.10}$$

are the **harmonic numbers**.
d) Integrate (5.2.9) to obtain

$$\ln^2(1 - x) = 2 \sum_{n=1}^{\infty} \frac{H_n}{n + 1} x^{n+1}. \tag{5.2.11}$$

This appears in [G & R] 1.516.1.

Project 5.2.1. Express the power series for $\ln^m(1 - x)$ in terms of the harmonic numbers. For instance,

$$\ln^3(1 - x) = -6 \sum_{n=1}^{\infty} \frac{x^{n+2}}{n + 2} \sum_{k=1}^{n} \frac{H_k}{k + 1}. \tag{5.2.12}$$

This appears in [G & R] 1.516.2.

Exercise 5.2.8. Verify the Taylor series expansion

$$\ln(1 + x) \ln(1 - x) = \sum_{n=1}^{\infty} \frac{x^{2n}}{n} \sum_{k=1}^{2n-1} \frac{(-1)^k}{k} \tag{5.2.13}$$

$$= \sum_{n=1}^{\infty} \frac{1}{n} \left(H_n - H_{2n} - \frac{1}{2n} \right) x^{2n}.$$

This appears in [G & R] 1.516.3.

Exercise 5.2.9. Prove that the harmonic numbers H_n, $n \geq 2$ are not integers.
Hint from Graham et al. (1989). Let 2^k be the largest power of 2 less or
equal than n. Consider the number $2^{k-1} H_n - \frac{1}{2}$. **Hint for a second proof**:
use induction and the inequality

$$\nu_2(H_n) \leq \text{Max}\{\nu_2(H_{n-1}), \ \nu_2(1/n)\}, \tag{5.2.14}$$

with equality unless $\nu_2(H_{n-1}) = \nu_2(1/n)$, to prove $\nu_2(H_n) > 1$. Separate into
two cases according to whether $\nu_2(H_{n-1}) = \nu_2(1/n)$ or not.

Extra 5.2.2. An apparently simpler solution of Exercise 5.2.9 starts by writing

$$H_n = \frac{2 \cdot 3 \cdots n + 1 \cdot 3 \cdots n + 1 \cdot 2 \cdot 4 \cdots n + \cdots + 1 \cdot 2 \cdots (n-1)}{n!}$$

and let p be the largest prime less than or equal to n. Then p divides $n!$
and every term in the numerator is divisible by p, with the single exception
of $1 \cdot 2 \cdots (p-1) \cdot (p+1) \cdots n$, provided $2p > n$. This is a fact of prime
numbers called **Bertrand's postulate**. The reader will find the proof in Hardy
and Wright (1979), page 343 quite readable.

The harmonic number also appear in the remarkable inequality

$$\sum_{d|n} d \leq H_n + \exp(H_n) \ln(H_n) \tag{5.2.15}$$

where the sum on the left is over all the divisors of n. Lagarias proved that this
bound is equivalent to **the Riemann hypothesis**, one of the most important
unsolved problems in mathematics. See Chapter 11 for more comments and
Lagarias (2002) for the details.

We now discuss some basic analytic properties of the logarithm.

Lemma 5.2.1. *The function $f(x) = \ln x$ is increasing and satisfies*

$$\lim_{x \to \infty} \ln x = \infty \tag{5.2.16}$$

Proof. The only part that needs to be proven is (5.2.16). The upper and lower
Riemann sums for $1/x$ yield

$$H_n - 1 < \ln n < H_n - \frac{1}{n}, \tag{5.2.17}$$

where H_n is the harmonic number. Therefore the fact that $\ln x \to \infty$ as $x \to$
∞ is equivalent to the divergence of the harmonic series. The standard proof
of this second fact is presented is the next exercise. $\qquad \square$

Exercise 5.2.10. The harmonic series $\sum\limits_{n=1}^{\infty} 1/n$ diverges. **Hint**. Prove that the tails of partial sums

$$S_n := \sum_{j=2^n+1}^{2^{n+1}} \frac{1}{j} \qquad (5.2.18)$$

satisfy $S_n > 1/2$.

Exercise 5.2.11. Evaluate

$$\lim_{n\to\infty} \sum_{k=n}^{2n} \frac{1}{k}.$$

This appears in Goode (1956).

Exercise 5.2.12. Prove that

$$\sum_{n=1}^{\infty} \frac{H_n}{n^2} = 2 \sum_{n=1}^{\infty} \frac{1}{n^3}. \qquad (5.2.19)$$

This appears in Klamkin (1951, 1952). Prove also the result of Klamkin (1955):

$$\sum_{n=1}^{\infty} \frac{H_n}{n^3} = \frac{\pi^4}{72}. \qquad (5.2.20)$$

Extra 5.2.3. Boas and Wrench (1971) examined the partial sums of the harmonic series. For a given $a \in \mathbb{R}$ they established a relation between

$$n_a := \min\{n \in \mathbb{N} : 1 + \tfrac{1}{2} + \cdots + \tfrac{1}{n} \geq a\}$$

and the Euler constant $\gamma = \lim\limits_{n\to\infty} 1 + \tfrac{1}{2} + \cdots + \tfrac{1}{n} - \ln n$. This constant is studied in Chapter 9.

Exercise 5.2.13. This exercise establishes the divergence of the harmonic series given that $\ln x \to \infty$ as $x \to \infty$. The proof is due to D. Bradley (2000). **Hint**. Observe that

$$\sum_{k=1}^{n-1} \ln(1 + 1/k) = \ln n \qquad (5.2.21)$$

and now use Exercise 5.2.4.

Note 5.2.2. The fact that $\ln x \to \infty$ as $x \to \infty$ was implicitly used in the proof of Proposition 2.1.1.

The growth of $\ln x$ at infinity is slower than any power of x. A clever proof of this was presented by Greenstein (1965):

Lemma 5.2.2. *Let* $a > 0$. *Then*

$$\lim_{x \to \infty} \frac{\ln x}{x^a} = 0.$$

Proof. Exercise 5.2.4 yields, for $x > 1$,

$$0 < \frac{\ln x}{x} = \frac{2 \ln \sqrt{x}}{x} < \frac{2\sqrt{x}}{x} = \frac{2}{\sqrt{x}},$$

so that $\lim_{x \to \infty} x^{-1} \ln x = 0$. The result follows by replacing x by x^a. $\qquad\square$

We now follow Hamming (1970) to prove that $\ln x$ is not a rational function.

Theorem 5.2.3. *The function* $\ln x$ *is not rational.*

Proof. Suppose $\ln x = N(x)/D(x)$, with N and D polynomials without a common factor. Differentiate to produce

$$D^2 = x \left(DN' - D'N \right), \qquad (5.2.22)$$

so x divides $D(x)$, say $D(x) = x^k D_1(x)$, with $D_1(0) \neq 0$ and $k \geq 1$. Replacing in (5.2.22) we obtain

$$x^k D_1^2 = x D_1 N' - k N D_1 - x N D_1',$$

and letting $x = 0$ shows that $N(0) = 0$, so x must also divide N. This is a contradiction that completes the proof. $\qquad\square$

Note 5.2.3. Observe that the previous proof is purely algebraic. The only property of $\ln x$ that is used is the fact that its derivative is $1/x$.

Exercise 5.2.14. Give a direct analytic proof of Theorem 5.2.3 by using Lemma 5.2.2.

Extra 5.2.4. An **algebraic function** $y = f(x)$ is one that satisfies a polynomial equation, where the coefficients are polynomials in x. Thus y is algebraic if there exist polynomials $\{a_j(x) : 0 \le j \le n\}$, with $a_n(x) \not\equiv 0$, such that

$$\sum_{k=0}^{n} a_k(x)y^k = 0. \tag{5.2.23}$$

For example every rational function is algebraic and so is the **double square root** function

$$y = \sqrt{a + \sqrt{1+x}} \tag{5.2.24}$$

since it satisfies $y^4 - 2ay^2 - x + a^2 - 1 = 0$. This function will make a mysterious appearance in Section 7.7.

This is a generalization of the notion of **algebraic number**: these are numbers x that are solutions of a polynomial equation

$$\sum_{k=0}^{n} a_k x^k = 0. \tag{5.2.25}$$

with integers a_k. For example $x = \sqrt{3}$ solves $x^3 - 1 = 0$ and $x = \sqrt{5 + \sqrt{5}}$ solves $x^4 - 10x^2 + 20 = 0$.

Not every algebraic number can be expressed in terms of radicals and this issue is connected with the fundamental questions of algebra of the 19th century. It was one of the motivating forces to in the development of the subject.

The set of algebraic numbers is a field, in the sense that if x, y are algebraic, so is $x + y$, xy and $1/x$ if $x \ne 0$.

Exercise 5.2.15. This exercise outlines a proof by Hamming (1970) of the fact that $y = \ln x$ is not an algebraic function.

Suppose $y = \ln x$ satisfies (5.2.23).
a) Prove that there is an equation of minimal degree, and this is unique up to scaling.
b) Differentiate the minimal equation

$$\ln^n x + \frac{a_{n-1}(x)}{a_n(x)} \ln^{n-1} x + \cdots + \frac{a_0(x)}{a_n(x)} = 0 \tag{5.2.26}$$

to produce

$$n \ln^{n-1} x + x \frac{d}{dx}\left(\frac{a_{n-1}(x)}{a_n(x)}\right) \ln^{n-1} x + \cdots = 0. \tag{5.2.27}$$

If all the coefficients of $\ln^j x$ do not vanish identically, we get an equation for

$\ln x$ of degree lower than n. Otherwise we obtain

$$\frac{n}{x} + \frac{d}{dx}\frac{a_{n-1}(x)}{a_n(x)} = 0.$$

Integrate to conclude that $\ln x$ is a rational function. This is a contradiction.

5.3. Some Logarithmic Integrals

This section describes some indefinite integrals involving the logarithm functions. This is an introduction to the material described in Chapter 12. The implicit constant of integration is omitted throughout.

Exercise 5.3.1. Consider the class of functions

$$\mathfrak{L} := \left\{ f(x) = \sum_{i=0}^{n}\sum_{j=0}^{m} a_{i,j}x^i \ln^j x : a_{i,j} \in \mathbb{R}; \ n, \ m \in \mathbb{N} \right\}. \quad (5.3.1)$$

For example $\mathfrak{L}$ contains the function $3x \ln x + x^7 \ln^2 x$. The goal of the exercise is to establish that $\mathfrak{L}$ is **closed under primitives**, that is, any function in $\mathfrak{L}$ admits a primitive in $\mathfrak{L}$.
a) Prove that

$$I(m, n) = \int x^n \ln^m x \, dx \quad (5.3.2)$$

satisfies the recurrence

$$I(m, n) = \frac{1}{n+1}x^{n+1} \ln^m x - \frac{m}{n+1}I(m-1, n). \quad (5.3.3)$$

b) Establish the formula

$$\int x^n \ln x \, dx = \frac{x^{n+1} \ln x}{n+1} - \frac{x^{n+1}}{(n+1)^2} \quad (5.3.4)$$

for $n \in \mathbb{N}$. The special case

$$\int \ln x \, dx = x \ln x - x \quad (5.3.5)$$

appears in [G & R]: 2.711.
c) Prove the identity

$$\int \ln^n x \, dx = (-1)^n x \sum_{k=0}^{n} \frac{n!}{k!}(-\ln x)^k. \quad (5.3.6)$$

This is also part of [G & R] 2.711.

d) Derive [G & R] 2.722

$$I(m, n) = \frac{x^{n+1}}{m+1} \sum_{k=0}^{m} \frac{(-1)^k (m+1-k)_k \ln^{m-k} x}{(n+1)^{k+1}}.$$

e) Conclude that $\mathfrak{L}$ is closed under primitives.

Exercise 5.3.2. Introduce the polynomial

$$\mathrm{Ex}(x, n) := \sum_{k=0}^{n} \frac{x^k}{k!}, \tag{5.3.7}$$

so (5.3.6) can be written as

$$\int \ln^n x \, dx = (-1)^n n! x \, \mathrm{Ex}(-\ln x, n). \tag{5.3.8}$$

Use a symbolic language to check that

$$\mathrm{Ex}(x, n) = \frac{e^x \, \Gamma(n+1, x)}{\Gamma(n+1)}, \tag{5.3.9}$$

where $\Gamma(a, x)$ is the incomplete gamma function defined in (1.3.7).

Project 5.3.1. This project deals with the iterated integral of the function $\ln x$. We thank T. Amdeberhan for this example.
a) Define $f_n(x)$ inductively by $f_0(x) = \ln x$ and

$$f_n(x) = \int f_{n-1}(x) \, dx, \quad \text{for } n \geq 1. \tag{5.3.10}$$

Prove that there exist coefficients a_n and b_n such that

$$f_n(x) = a_n x^n \ln x - b_n x^n. \tag{5.3.11}$$

Describe the choices made on the constants of integration to obtain this form.
b) Find a recurrence for a_n and conclude that $a_n = 1/n!$.
c) Check that b_n satisfies

$$b_n = \frac{1}{n} (a_n + b_{n-1}). \tag{5.3.12}$$

Prove that $\alpha_m := m! b_m$ satisfies $\alpha_m = \alpha_{m-1} + \frac{1}{m}$ and obtain

$$b_n = \frac{H_n}{n!}. \tag{5.3.13}$$

Project 5.3.2. This project deals with the iterated integral of the function $\ln(1 + x)$.

a) Check that

$$\int \ln(1+x)\,dx = (1+x)\ln(1+x) - x$$

and

$$\int [(1+x)\ln(1+x) - x]\,dx = \frac{(x+1)^2}{2}\ln(1+x) - \frac{x(3x+2)}{4}.$$

b) Define $f_n(x)$ inductively by $f_0(x) = \ln(1+x)$ and

$$f_n(x) = \int f_{n-1}(x)\,dx. \tag{5.3.14}$$

Prove that there exist polynomials $a_n(x)$ and $b_n(x)$ such that

$$f_n(x) = a_n(x) + b_n(x)\ln(1+x), \tag{5.3.15}$$

where the implicit constant of integration is chosen so that $a_n(0) = 0$.

c) Prove that $b_n(x) = (1+x)^n/n!$. **Hint**. Use (5.3.14) to derive a recurrence for b_n.

d) Prove that $a_n(x) = -x\,c_n(x)$, where $c_n(x)$ is a polynomial with positive rational coefficients.

e) Establish the recurrence

$$a'_{n+1}(x) = a_n(x) - \frac{(1+x)^n}{(n+1)!}. \tag{5.3.16}$$

f) Write $c_n(x) = d_n(x)/r_n$, where d_n is a polynomial with integer coefficients and r_n is the least common multiple of the denominators in $c_n(x)$. Check that

$$d_1(x) = 1$$
$$d_2(x) = 3x + 2$$
$$d_3(x) = 11x^2 + 15x + 6$$
$$d_4(x) = 25x^3 + 52x^2 + 42x + 12.$$

g) Define $s_n = r_n/(n\,r_{n-1})$. Provide convincing symbolic evidence that

$$s_n = \begin{cases} 1 & \text{if } n \text{ is divisible by two disctinct primes} \\ p & \text{if } n = p^k \text{ for some } k \in \mathbb{N}. \end{cases}$$

Thus

$$s_n = e^{\Lambda(n)}, \tag{5.3.17}$$

where $\Lambda(n)$ is the classical **von Mangoldt function**

$$\Lambda(n) = \begin{cases} 0 & \text{if } n \text{ is divisible by two distinct primes} \\ \ln p & \text{if } n = p^k \text{ for some } k \in \mathbb{N}. \end{cases}$$

Conclude that

$$r_n = n! \times e^{\sum_{k=0}^{n} \Lambda(k)} \tag{5.3.18}$$

where e is the well-known base of the natural logarithm. This constant is discussed in the next section. The reader will find in Weisstein (1999), page 1135, more information about the von Mangoldt function.

h) The more general problem

$$f_0(x) = \ln(a + x)$$

$$f_n(x) = \int_0^x f_{n-1}(t)\,dt,$$

where $a > 0$, is described in Underwood (1924). Prove that

$$n!\,f_n(x) = (x + a)^n \left[\ln(x + a) - H_n\right] + \ln a \left[x^n - (x + a)^n\right] \tag{5.3.19}$$

$$+ \sum_{r=1}^{n} H_r \binom{n}{r} a^r x^{n-r},$$

where H_n are the harmonic numbers. Discuss the connection between the Mangoldt function and the harmonic numbers H_n that is derived from this identity.

i) Conclude that

$$a_n(x) = \frac{1}{n!} \left(x \left[\sum_{r=1}^{n-1} H_r \binom{n}{r} x^{n-r-1}\right] + H_n \left[1 - (x + 1)^n\right] \right). \tag{5.3.20}$$

5.4. The Number e

There are many real numbers that are important in several areas of mathematics. This prominent role is reflected in that they have been given special symbols. In this section we introduce the first of these numbers. We define the **Euler number** e by the relation

$$\ln e = 1,$$

that is,

$$\int_1^e \frac{dt}{t} = 1. \tag{5.4.1}$$

Note 5.4.1. The reader will find in Coolidge (1950), Finch (2003) and Maor (1998) interesting information about e.

The function $\ln x$ increases from $\ln 1 = 0$ to $\ln \infty = \infty$, so the number e is uniquely defined by (5.4.1).

There are many alternative ways to introduce this constant. In the rest of this section, we prove that e can be expressed as a limit

$$e = \lim_{n \to \infty} \left(1 + \frac{1}{n}\right)^n. \tag{5.4.2}$$

or as a series

$$e = \sum_{k=0}^{\infty} \frac{1}{k!}. \tag{5.4.3}$$

We will see that each of these forms has its own advantages.

Exercise 5.4.1. Prove that $2 < e < 4$. **Hint.** Compute upper and lower estimates for the area under $y = 1/x$.

We now follow Barnes (1984) to establish the number e as the limit of a sequence.

Proposition 5.4.1. *The number e is given by*

$$e = \lim_{n \to \infty} \left(1 + \frac{1}{n}\right)^n. \tag{5.4.4}$$

Proof. Integration by parts produces

$$\int_{1/(n+1)}^{1/n} \ln x \, dx = \frac{1}{n(n+1)} \left(\ln \left(\frac{(n+1)^n}{n^{n+1}} \right) - 1 \right). \tag{5.4.5}$$

The mean value theorem for integrals Thomas and Finney (1996), page 329, yields the existence of a number c_n such that

$$\int_{1/(n+1)}^{1/n} \ln x \, dx = \left(\frac{1}{n} - \frac{1}{n+1} \right) \ln c_n \tag{5.4.6}$$

with

$$\frac{1}{n+1} < c_n < \frac{1}{n}, \tag{5.4.7}$$

so that

$$\lim_{n \to \infty} n c_n = 1. \tag{5.4.8}$$

Then (5.4.5) and (5.4.6) produce

$$\ln\left(\frac{(n+1)^n}{n^{n+1}} \times \frac{1}{c_n}\right) = 1,$$

and thus

$$e = \left(1 + \frac{1}{n}\right)^n \times \frac{1}{nc_n}. \qquad (5.4.9)$$

Observe that $nc_n < 1$, so

$$\left(1 + \frac{1}{n}\right)^n < e$$

and (5.4.8) yields (5.4.4). We actually see that the sequence $(1 + 1/n)^n$ increases to e.   □

The next exercise outlines a different proof of Proposition 5.4.1.

Exercise 5.4.2. Prove that for any $\alpha > 0$ there exists a unique number $e(\alpha)$ such that

$$\int_1^{e(\alpha)} t^{\alpha-1}dt = 1.$$

It is possible to prove that the function $e(\alpha)$ is continuous. Evaluate the integral to obtain

$$e(\alpha) = (1 + \alpha)^{1/\alpha},$$

and conclude that

$$\lim_{\alpha \to 0}(1 + \alpha)^{1/\alpha} = e.$$

Extra 5.4.1. The double inequality

$$\frac{e}{2n + 2} < e - \left(1 + \frac{1}{n}\right)^n < \frac{e}{2n + 1} \qquad (5.4.10)$$

appears in Polya and Szego (1972), problem 170. An extension for $n \in \mathbb{R}$ is discussed in Sandor and Debnath (2000).

Project 5.4.1. We establish the alternative form (5.4.3) following Kazarinoff (1961). The goal is to prove the inequality

$$\left(1 + \frac{1}{n}\right)^n \le \sum_{k=0}^{n} \frac{1}{k!} \le \left(1 + \frac{1}{n}\right)^{n+1} \qquad (5.4.11)$$

from which (5.4.3) will follow.

a) Check that $n(n-1)\cdots(n-k+1) \le n^k$ for $0 \le k \le n$. Conclude that

$$\binom{n}{k} \frac{1}{n^k} \le \frac{1}{k!},$$

and so the inequality on the left of (5.4.11) is established.

b) Prove that $a_n = (1 + 1/n)^{n+1}$ is decreasing, so in order to prove the inequality on the right of (5.4.11), it suffices to prove that for each $n \in \mathbb{N}$ there exists $r = r(n)$ such that

$$\sum_{k=0}^{n} \frac{1}{k!} \le \left(1 + \frac{1}{r}\right)^{r+1} \tag{5.4.12}$$

with $r \ge n$

c) Now fix $n \in \mathbb{N}$. Check that for $r > n$

$$\left(1 + \frac{1}{r}\right)^{r+1} - \sum_{k=0}^{n} \frac{1}{k!} = \sum_{k=1}^{n}\left\{\binom{r+1}{k}r^{-k} - \frac{1}{k!}\right\} + \sum_{k=n+1}^{r}\binom{r+1}{k}r^{-k}. \tag{5.4.13}$$

d) Prove that for $k = 1, \ldots, n$,

$$\binom{r+1}{k}r^{-k} - \frac{1}{k!} = \frac{r^{-k}}{k!}\left(a_1 r^{k-1} + a_2 r^{k-2} + \cdots + a_{k-1}r\right) \tag{5.4.14}$$

for some coefficients a_j that are independent of r. Let M be the largest of the (fixed number of) constants $\{a_1, \ldots, a_{n+1}\}$. Prove that

$$\left|\binom{r+1}{k}r^{-k} - \frac{1}{k!}\right| < \frac{M}{r-1} \tag{5.4.15}$$

so the absolute value of the first sum in (5.4.13) is bounded from above by $Mn/(r-1)$. Now choose $r = r(n)$ sufficiently large so that the upper bound is increased to $\frac{1}{2}(n+1)!$.

e) Prove that the first term in the second sum in (5.4.13) is bounded by $\frac{3}{4}(n+1)!$ for $r = r(n)$ sufficiently large. **Hint.** The term $\binom{r+1}{n+1}r^{-(n+1)}$ tends to 1 as $r \to \infty$ and n is fixed.

This proves (5.4.11).

Note 5.4.2. Exercise (4.2.5) and (5.4.3) show that

$$\sum_{j=0}^{\infty} \frac{(-1)^j}{j!} = \frac{1}{e}. \tag{5.4.16}$$

Exercise 5.4.3. This exercise provides a proof of the existence of the limit (5.4.4) based on the arithmetic–geometric mean inequality.

a) Let $\{a_i\}$ and $\{p_i\}$ be sequences of positive numbers with $a_1 \leq a_2 \leq \cdots \leq a_n$ and $p_1 + \cdots + p_n = 1$. Prove that

$$P_n := \prod_{i=1}^{n} a_i^{p_i} \leq \sum_{i=1}^{n} p_i a_i := S_n,$$

with equality if and only if $a_1 = a_2 = \cdots = a_n$. **Hint** (due to Alzer (1996)): let k be the unique index such that $a_k \leq P_n \leq a_{k+1}$. Then

$$\frac{S_n}{P_n} - 1 = \sum_{i=1}^{k} p_i \int_{a_i}^{P_n} \left(\frac{1}{t} - \frac{1}{P_n}\right) dt + \sum_{i=k+1}^{n} p_i \int_{P_n}^{a_i} \left(\frac{1}{P_n} - \frac{1}{t}\right) dt.$$

b) Choose $p_1 = \cdots = p_n$ to obtain the arithmetic–geometric mean inequality

$$G_n \leq A_n,$$

where

$$A_n := \frac{a_1 + \cdots + a_n}{n} \quad \text{and} \quad G_n := (a_1 \cdots a_n)^{1/n}.$$

c) Give an inductive proof of $G_n \leq A_n$. **Hint** (due to Chong, 1976): assume that $a_1 \leq \cdots \leq a_n$ and prove $a_1 + a_n - A_n > a_1 a_n / A_n$. **Hint**. The inequality $(A_n - a_1)(A_n - a_n)$ is easy to check. Now compute the arithmetic mean of $a_2, \ldots, a_{n-1}$ and $a_1 + a_n - A_n$.

d) This part describes a proof of Mendelson (1951) of the existence of the limit (5.4.4). Use the arithmetic–geometric mean inequality with the $n+1$ numbers $1, 1 + \frac{1}{n}, \ldots, 1 + \frac{1}{n}$ to prove that $a_n = (1 + 1/n)^n$ is increasing. The fact that $b_n = (1 + 1/n)^{n+1}$ is decreasing follows by considering the $n+2$ numbers $1, \frac{n}{n+1}, \cdots, \frac{n}{n+1}$. Now $b_n > a_n$, so they both converge and to the same limit.

Extra 5.4.2. Gauss (1799/1981) observed that the arithmetic mean $a_1 = (a + b)/2$ and the geometric mean $a_2 = \sqrt{ab}$ of two numbers leave the elliptic integral

$$G(a, b) = \int_0^{\pi/2} \frac{d\theta}{\sqrt{a^2 \cos^2 \theta + b^2 \sin^2 \theta}} \tag{5.4.17}$$

invariant, that is, $G(a_1, b_1) = G(a, b)$. The substitution $a, b \mapsto a_1, b_1$ can be repeated, the succesive terms having a common limit $M(a, b)$. This is the **arithmetic–geometric mean** of a and b. The transformations in Extra 2.4.3 represent a rational version of this phenomenon. See Borwein and Borwein (1987) for more details.

Exercise 5.4.4. Newman (1985) found a very clever proof of the invariance of $G(a, b)$. The substitution $x = b \tan \theta$ converts $2G(a, b)$ into

$$\int_{-\infty}^{\infty} \frac{dx}{\sqrt{(a^2 + x^2)(b^2 + x^2)}}.$$

Now make the substitution $x \mapsto x + \sqrt{x^2 + ab}$.

5.5. Arithmetical Properties of e

This section contains the proof of irrationality of the constant e. This is a theme that will reappear throughout the book: we are interested in arithmetical properties of constants appearing in analysis.

Exercise 5.5.1. Prove that the value of an alternating series

$$S = \sum_{n=0}^{\infty} (-1)^n a_n, \tag{5.5.1}$$

with a_n monotonically decreasing to 0, satisfies $a_0 \leq S \leq a_0 + a_1$.

We now present a classical result due to Lambert (1761). This proof has been reproduced by Pennisi (1953).

Theorem 5.5.1. *The number e is irrational.*

Proof. Suppose e is rational and use (5.4.16) to write

$$\frac{b}{a} = e^{-1} = \sum_{n=0}^{\infty} \frac{(-1)^n}{n!}. \tag{5.5.2}$$

Now multiply by $(-1)^{a+1} a!$ to obtain

$$(-1)^{a+1} \left(b(a-1)! - \sum_{n=0}^{a} (-1)^n \frac{a!}{n!} \right) = \frac{1}{a+1} - \frac{1}{(a+1)(a+2)} - \cdots$$

The right-hand side is a convergent alternating series, so by Exercise 5.5.1 its value lies between $1/(a + 1)$ and $1/(a + 2)$. This is impossible because the left hand side is an integer. □

Hermite (1873) proved that e is **transcendental**, that is, there is no polynomial P with integer coefficients such that $P(e) = 0$. The proof of this result can be found in Hardy and Wright (1979), page 172. We present a proof of the weaker result that e does not satisfy a quadratic equation with integer coefficients. The proof is due to Liouville (1840) and has been reproduced by Beatty (1955).

Theorem 5.5.2. *The number e is not quadratic algebraic.*

Proof. Suppose that e satisfies

$$ae^2 + be + c = 0 \text{ with } a, b, c \in \mathbb{Z} \text{ not all zero.} \qquad (5.5.3)$$

From the series (5.4.3) for e we obtain

$$n!\, e = \sum_{k=0}^{n} \frac{n!}{k!} + \sum_{k=n+1}^{\infty} \frac{n!}{k!}.$$

Now

$$\sum_{k=n+1}^{\infty} \frac{n!}{k!} > \frac{n!}{(n+1)!} = \frac{1}{n+1}$$

and

$$\sum_{k=n+1}^{\infty} \frac{n!}{k!} < \sum_{j=1}^{\infty} \frac{1}{(n+1)^j} = \frac{1}{n}, \qquad (5.5.4)$$

so that

$$\sum_{k=n+1}^{\infty} \frac{n!}{k!} = \frac{1}{n+\theta},$$

where $0 < \theta < 1$ depends upon n. Similarly,

$$\frac{n!}{e} = \sum_{k=0}^{n} (-1)^k \frac{n!}{k!} + \sum_{k=n+1}^{\infty} (-1)^k \frac{n!}{k!}$$

and

$$\sum_{k=n+1}^{\infty} (-1)^k \frac{n!}{k!} = \frac{(-1)^{n+1}}{n+1+\phi},$$

where $0 < \phi < 1$ depends upon n. Multiplying (5.5.3) by $n!/e$ we obtain

$$a\left(i + \frac{1}{n+\theta}\right) + bn! + c\left(j + \frac{(-1)^{n+1}}{n+1+\phi}\right) = 0$$

for some integers i, j. Thus

$$\frac{a}{n+\theta} + \frac{c}{n+1+\phi} = 0 \qquad (5.5.5)$$

because the left-hand side is an integer and arbitrarily small for n large. We conclude that $a = -c$ and $0 < \theta < 1 < 1+\phi = \theta$. This is a contradiction. $\square$

5.6. The Exponential Function

In this section we consider the **exponential function** defined by the series

$$e^x := \sum_{k=0}^{\infty} \frac{x^k}{k!}. \tag{5.6.1}$$

Observe that (5.4.3) and (5.4.16) state that $e^1 = e$ and $e^{-1} = 1/e$, so the notation (5.6.1) is consistent.

Note 5.6.1. We have

$$e^x = \lim_{n \to \infty} \mathrm{Ex}(x, n), \tag{5.6.2}$$

where $\mathrm{Ex}(x, n)$ has been introduced in (5.3.7).

Exercise 5.6.1. Prove that

$$e^x = \lim_{n \to \infty} \left(1 + \frac{x}{n}\right)^n. \tag{5.6.3}$$

Hint. Expand by the binomial theorem.

Exercise 5.6.2. This exercise outlines some of the most important properties of the exponential function.
a) Check that

$$\frac{d}{dx} e^x = e^x. \tag{5.6.4}$$

b) Prove that $e^{\ln x} = x$ for all $x > 0$ and $\ln(e^x) = x$ for all $x \in \mathbb{R}$. Therefore these two functions are inverses of each other. **Hint.** Compute the derivatives.
c) Prove that $e^{x+y} = e^x \times e^y$. **Hint.** Use Cauchy's formula given in Exercise 4.2.7 to multiply the two series.
d) Conclude that $e^x \neq 0$.
e) Assume that $E(x) = \sum_{k=0}^{\infty} a_k x^k$ is a power series that satisfies $E'(x) = E(x)$ and $E(0) = 1$. Prove that $a_k = 1/k!$, so that $E(x) = e^x$.

Note 5.6.2. The arbitrary powers of $x \in \mathbb{R}^+$ can be defined in terms of the exponential function. Indeed, we let

$$x^a := e^{a \ln x}. \tag{5.6.5}$$

The differentiation rule $\frac{d}{dx} x^a = a x^{a-1}$ is now valid for $a \in \mathbb{R}$.

Exercise 5.6.3. In this exercise we prove that $\ln x$ and e^x are inverses of each other using the function $e(\alpha)$ introduced in Exercise 5.4.2 by

$$\int_1^{e(\alpha)} t^{\alpha-1}dt = 1. \tag{5.6.6}$$

a) Use a change of variable to establish

$$\int_1^{e(\alpha,x)} t^{\alpha/x-1}dt = x, \tag{5.6.7}$$

where $e(\alpha, x) = e(\alpha)^x$.

b) Let $\alpha \to 0$ to conclude

$$\int_1^{e^x} \frac{dt}{t} = x. \tag{5.6.8}$$

Exercise 5.6.4. Prove that e^x is not a rational function. **Hint.** Reason along the lines of the proof of Theorem 5.2.3.

5.7. Stirling's Formula

The goal of this section is to present an approximation for $n!$, valid for n large, due to Stirling (1730). The first theorem establishes the existence of a certain limit, the exact value of which will be described in future chapters. The proof presented here is due to D. Romik (2000). The reader is referred to Tweedle (1988) for more information on James Stirling's scientific work and to Blyth and Pathak (1986) for another simple proof.

Theorem 5.7.1. *Let* $\varphi(n) = n^{n+1/2}e^{-n}$. *Then the limit*

$$C := \lim_{n \to \infty} \frac{n!}{\varphi(n)} \tag{5.7.1}$$

exists.

Proof. Write

$$\ln n! = \sum_{k=2}^{n} \ln k = \sum_{k=1}^{n} \int_1^k \frac{dx}{x}, \tag{5.7.2}$$

use

$$\int_1^k \frac{dx}{x} = \sum_{j=1}^{k-1} \int_j^{j+1} \frac{dx}{x} \tag{5.7.3}$$

and exchange the order of the two sums to produce

$$\ln n! = \int_1^n \frac{n - \lfloor x \rfloor}{x} \, dx.$$

Now $x = \lfloor x \rfloor + \{x\}$, with $\{x\}$ the fractional part of x, yields

$$\ln n! = (n + 1/2)\ln n - n + 1 + \int_1^n \frac{\{x\} - 1/2}{x} \, dx. \qquad (5.7.4)$$

We now prove that the integral has a limit as $n \to \infty$. Using the fact that $\{x + m\} = \{x\}$ for $m \in \mathbb{N}$ we have

$$\int_1^n \frac{\{x\} - \frac{1}{2}}{x} \, dx = \sum_{k=1}^{n-1} \int_0^1 \frac{\{x\} - \frac{1}{2}}{x + k} \, dx.$$

For $x \in [0, 1]$ we have $\{x\} = x$ and after integrating by parts we get

$$\int_1^n \frac{\{x\} - \frac{1}{2}}{x} \, dx = \frac{1}{2} \sum_{k=1}^{n-1} \int_0^1 \frac{x - x^2}{(x + k)^2} \, dx. \qquad (5.7.5)$$

It is now easy to check that the integral is monotone in n and bounded. We conclude that the limit as $n \to \infty$ exits and is finite:

$$\int_1^\infty \frac{\{x\} - 1/2}{x} \, dx < \infty. \qquad (5.7.6)$$

Exponentiating the identity (5.7.4) establishes the result, with the constant C given by

$$C = \exp\left(1 + \int_1^\infty \frac{\{x\} - 1/2}{x} \, dx\right). \qquad (5.7.7)$$

$\square$

Note 5.7.1. The value $C = \sqrt{2\pi}$ will be established by different methods. For instance see Exercises 6.4.4 and 8.2.8. Thus Stirling's formula states

$$n! \sim \sqrt{2\pi n}(n/e)^n, \qquad (5.7.8)$$

and as a corollary we obtain the evaluation

$$\int_1^\infty \frac{\{x\} - 1/2}{x} \, dx = \ln\sqrt{2\pi} - 1. \qquad (5.7.9)$$

Extra 5.7.1. The proof of Theorem 5.7.1 is an example of the **Euler–MacLaurin summation formula**. Given a function f with continuous derivative on the interval $[1, n]$ we would like to compare

$$\sum_{k=1}^{n} f(k) \text{ with } \int_{1}^{n} f(x)\,dx. \tag{5.7.10}$$

Consider the integral of f on the interval $[k, k+1]$ for $k \in \mathbb{N}$. Integrate by parts and choose the primitive of 1 to be $x - k - 1/2$; this function has integral 0 on $[k, k+1]$. We obtain

$$\int_{k}^{k+1} f(x)\,dx = \frac{1}{2}(f(k+1) + f(k)) - \int_{k}^{k+1} (\{x\} - 1/2)f'(x)\,dx$$

where we have used the fact that $k = \lfloor x \rfloor$ on the interval of integration. Summing from $k = 1$ to $n - 1$ produces

$$\sum_{k=1}^{n} f(k) = \int_{1}^{n} f(x)\,dx + \int_{1}^{n} \{x\}\, f'(x)\,dx + f(1). \tag{5.7.11}$$

This is the Euler–MacLaurin summation formula. The choice $f(x) = \ln x$ leads directly to (5.7.4). The reader will find in Apostol (1999) a detailed account of these ideas.

Exercise 5.7.1. Prove that the central binomial coefficients C_n satisfy

$$C_n \sim \frac{2^{2n}}{\sqrt{\pi n}}. \tag{5.7.12}$$

Exercise 5.7.2. Let $a_n := n!^n$. Prove that

$$\lim_{n \to \infty} (a_n - a_{n-1}) = \frac{1}{e}.$$

Exercise 5.7.3. Check that the radius of convergence of the series

$$\sum_{n=0}^{\infty} \binom{2n}{n} x^n = \frac{1}{\sqrt{1 - 4x}} \tag{5.7.13}$$

is $1/4$.

Exercise 5.7.4. Prove that for $p > 0$

$$\int_{0}^{\infty} e^{-px}\,dx = \frac{1}{p}.$$

This is [G & R] 3.310. Differentiate n times with respect to p to obtain

$$\int_0^\infty x^n e^{-px}\, dx = \frac{n!}{p^{n+1}}. \tag{5.7.14}$$

This is [G & R] 3.351.3.

Project 5.7.1. The proof of Stirling's formula sketched in this project is due to G. and J. Marsaglia (1990).
a) Check that

$$n! = n^{n+1} e^{-n} \int_0^\infty \left(y e^{1-y} \right)^n dy. \tag{5.7.15}$$

Hint. See Exercise 5.7.4.
b) Prove that the equation

$$y e^{1-y} = e^{-z^2/2} \tag{5.7.16}$$

has two branches according to $0 \le y \le 1$ or $1 \le y < \infty$.
c) Use the relation $y'(y-1) = zy$ to prove that the coefficients in the expansion

$$y = 1 + a_1 z + a_2 z^2 + a_3 z^3 + \cdots \tag{5.7.17}$$

satisfy $a_1^2 = 1$, $a_2 = \frac{1}{3}$, $a_3 = \frac{1}{12 a_1}$ and

$$a_1 a_n = a_{n-1} - 2 a_2 a_{n-1} - 3 a_3 a_{n-2} - \cdots \tag{5.7.18}$$

for $n \ge 3$. Conclude that

$$y = 1 + a_1 z + a_2 z^2 + a_3 z^3 + \cdots \tag{5.7.19}$$

solves $y e^{1-y} = e^{-z^2/2}$ for $1 \le y < \infty$ and

$$y = 1 - a_1 z + a_2 z^2 - a_3 z^3 + \cdots \tag{5.7.20}$$

solves $y e^{1-y} = e^{-z^2/2}$ for $0 \le y \le 1$.
d) Verify that (5.7.15) becomes

$$n! = 2 n^{n+1} e^{-n} \sum_{j=0}^\infty (2j+1) a_{2j+1} \left(\frac{2}{n} \right)^j \int_0^\infty u^{2j} e^{-u^2}\, du. \tag{5.7.21}$$

e) Let

$$I_j = \int_0^\infty u^{2j} e^{-u^2}\, du. \tag{5.7.22}$$

Integrate by parts to obtain the recurrence

$$I_j = \frac{2j-1}{2} I_{j-1}. \tag{5.7.23}$$

Conclude that

$$I_j = \frac{(2j)!}{j! \, 2^{2j}} I_0. \tag{5.7.24}$$

The first term of the identity (5.7.21) gives Stirling's formula up to the evaluation of the integral

$$I_0 = \int_0^\infty e^{-u^2} \, du. \tag{5.7.25}$$

This is will done in Chapter 8, where we establish the value $I_0 = \sqrt{\pi}/2$.

Project 5.7.2. The goal of this project is to present a proof of Stirling's formula due to Feller (1967):

a) Define

$$I(x) = \int_0^x \ln y \, dy = x \ln x - x$$

and the numbers

$$a_k = \int_{k-1/2}^k \ln\left(\frac{k}{x}\right) dx \quad \text{and} \quad b_k = \int_k^{k+1/2} \ln\left(\frac{x}{k}\right) dx. \tag{5.7.26}$$

Check that

$$a_k = \frac{1}{2}\ln k - I(k) + I(k-1/2) \quad \text{and} \quad b_k = I(k+1/2) - I(k) - \frac{1}{2}\ln k.$$

b) Prove that

$$a_1 - b_1 + a_2 - b_2 + \cdots + a_n = \ln n! - \frac{1}{2}\ln n - I(n) + I(1/2).$$

c) Establish the identities

$$a_k = -\int_0^{1/2} \ln(1 - t/k) \, dt \quad \text{and} \quad b_k = \int_0^{1/2} \ln(1 + t/k) \, dt$$

and conclude that $a_k > b_k > a_{k+1} > 0$.

d) Define

$$c_k = \begin{cases} a_{(k+1)/2} & \text{if } k \text{ is odd} \\ b_{k/2} & \text{if } k \text{ is even} \end{cases}$$

so that

$$\sum_{k=1}^{2n-1}(-1)^{k-1}c_k = \ln n! - \frac{1}{2}\ln n - I(n) + I(1/2).$$

Check that the series on the left converges giving the existence of the limit (5.7.1) with

$$C = \exp\left(\sum_{k=1}^{\infty}(-1)^{k-1}c_k + \tfrac{1}{2}(\ln 2 + 1)\right).$$

5.8. Some Definite Integrals

In this section we derive several integrals from the basic formula

$$I_n := \int_0^1 x^n\,dx = \frac{1}{n+1}, \tag{5.8.1}$$

valid for $n \in \mathbb{R}$, $n > -1$.

Exercise 5.8.1. Let $n,\,k \in \mathbb{N}$. Check that

$$\int_0^1 x^n \ln^k x\,dx = \frac{(-1)^k\,k!}{(n+1)^{k+1}}. \tag{5.8.2}$$

Hint. Differentiate (5.8.1) with respect to n.

Exercise 5.8.2. Change variables in (5.8.2) to establish

$$\int_0^{\infty} x^k e^{-x}\,dx = k! \tag{5.8.3}$$

The left-hand side makes sense for $k \in \mathbb{R}^+$ and it gives an extension of the factorial function to the positive reals. This is the **gamma function** that will be discussed in Chapter 10.

Mathematica 5.8.1. Integrating (5.8.1) from $n = 0$ to $n = 1$ gives

$$\int_0^1 \frac{x-1}{\ln x}\,dx = \ln 2. \tag{5.8.4}$$

The function $f(x) = (x-1)/\ln x$ does not admit an elementary primitive. A symbolic calculation using Mathematica yields

$$\int \frac{x-1}{\ln x}\,dx = \text{ExpIntegralEi}\,(2\ln x) - \text{LogIntegral}(x),$$

where

$$\mathrm{Ei}(x) = \mathrm{ExpIntegralEi}(x) := -\int_{-x}^{\infty} \frac{e^{-t}}{t}\,dt \qquad (5.8.5)$$

is the **exponential integral** function (if $x > 0$ the definition is slightly modified to avoid the singularity at at $x = 0$) and

$$\mathrm{Li}(x) = \mathrm{LogIntegral}(x) := \int_{2}^{x} \frac{dt}{\ln t} \qquad (5.8.6)$$

is the **logarithmic integral** function. These two functions will be discussed in Volume 2. The introduction to Chapter 11 explains the relation of the function $\mathrm{Li}(x)$ and the distribution of prime numbers.

Exercise 5.8.3. Let $a \in \mathbb{R}^{+}$. Then

$$\int_{0}^{\infty} \frac{e^{-ay} - e^{-2ay}}{y}\,dy = \ln 2. \qquad (5.8.7)$$

Hint. Reduce to (5.8.4). Establish the generalization

$$\int_{0}^{\infty} \frac{e^{-t} - e^{-tx}}{t}\,dt = \ln x. \qquad (5.8.8)$$

Extra 5.8.1. The integral presented in (5.8.8) is an example of a **Frullani integral**. Under some mild conditions on the function f, such that the existence of the limiting value $f(\infty)$, the result

$$\int_{0}^{\infty} [f(ax) - f(bx)] \frac{dx}{x} = [f(0) - f(\infty)]\ln(b/a). \qquad (5.8.9)$$

holds.

Project 5.8.1. In this project we will find the Eulerian polynomials $A_n(x)$ that appeared in Exercise 4.1.3.
a) Prove that for $p > 0$

$$\int_{0}^{\infty} \frac{dx}{1 + e^{px}} = \frac{\ln 2}{p}. \qquad (5.8.10)$$

This is [G & R] 3.311.1.
b) Differentiate the integrand on the left-hand side of (5.8.10) n times with respect to p to conclude there exists a polynomial Q_n of degree $n - 1$, with positive integer coefficients, such that

$$\left(\frac{d}{dp}\right)^{(n)} \frac{1}{1 + e^{px}} = -\frac{x^n e^{px}}{(1 + e^{px})^{n+1}} Q_n(-e^{px}). \qquad (5.8.11)$$

c) Use (5.8.11) to derive the recurrence

$$Q_{n+1}(v) = v(1 - v)\frac{d}{dv}Q_n(v) + (1 + nv)Q_n(v). \qquad (5.8.12)$$

Conclude that Q_n is the Eulerian polynomial A_n.
d) Use (5.8.10) to obtain

$$\int_1^\infty \frac{\ln^n t}{(1 + t)^{n+1}}A_n(-t)dt = (-1)^n n! \ln 2. \qquad (5.8.13)$$

e) Define the coefficients $a_j^{(n)}$ by

$$A_n(v) = \sum_{j=0}^{n-1} a_j^{(n)} v^j. \qquad (5.8.14)$$

Use part a) to derive a recurrence for $a_j^{(n)}$ and prove that $a_j^{(n)} \in \mathbb{N}$.
f) Prove that

$$a_j^{(n)} = \sum_{k=1}^{j+1} (-1)^{j+k+1} k^n \binom{n+1}{j+1-k}. \qquad (5.8.15)$$

Hint. Use Exercise 1.4.6 to complete an inductive step.
g) Prove that A_n is a symmetric polynomial in the sense of Exercise 2.1.5, that is, it satisfies $A_n(x) = x^{n-1}A_n(1/x)$.

5.9. Bernoulli Numbers

In this section we consider the Taylor series expansion

$$\frac{x}{e^x - 1} = 1 - \frac{x}{2} + \sum_{n=1}^\infty B_{2n}\frac{x^{2n}}{(2n)!} \qquad (5.9.1)$$

around $x = 0$. The coefficients B_{2n} are the **Bernoulli numbers** and we discuss some of their arithmetical properties. These numbers will reappear in Chapter 6 in several Taylor series expansions. The reader will find more information about these numbers in K. Dilcher's Web site (2003).

The fact that x is the only odd power in (5.9.1) is justified by the following exercise.

Exercise 5.9.1. Prove that the function

$$f(x) = \frac{x}{e^x - 1} - \left(1 - \frac{x}{2}\right)$$

is even, that is, $f(-x) = f(x)$.

Exercise 5.9.2. Prove that the Bernoulli numbers satisfy

$$\frac{1}{2} - n + \sum_{m=1}^{n} \binom{2n+1}{2m} B_{2m} = 0. \tag{5.9.2}$$

Conclude that B_{2m} are rational numbers. Use the recurrence (5.9.2) to obtain B_{2m} for $m = 1, \cdots, 5$. **Hint.** Use the identity

$$(e^x - 1) \times \frac{x}{e^x - 1} = x. \tag{5.9.3}$$

Mathematica 5.9.1. The Bernoulli numbers are included in Mathematica. For instance the command

```
BernoulliB[50]
```

yields

$$B_{50} = \frac{495057205241079648212477525}{66}. \tag{5.9.4}$$

The next theorem describes a second recurrence for the Bernoulli numbers that will determine their sign. The proof presented here appears in Mordell (1973):

Theorem 5.9.1. *The Bernoulli numbers satisfy*

$$B_{2n} = -\sum_{r=1}^{n-1} \frac{2^{2r} - 1}{2^{2n} - 1} \binom{2n}{2r} B_{2r} B_{2n-2r}, \text{ for } n \geq 2 \tag{5.9.5}$$

Proof. Write

$$\frac{x}{e^x - 1} = \sum_{n=0}^{\infty} b_n \frac{x^n}{n!}, \tag{5.9.6}$$

so that $b_0 = 1$, $b_1 = -1/2$, $b_{2n+1} = 0$ for $n > 1$, and $b_{2n} = B_{2n}$. Then

$$\frac{x}{e^x + 1} = \frac{x}{e^x - 1} - \frac{2x}{e^{2x} - 1}$$

$$= -\sum_{n=0}^{\infty} (2^n - 1) b_n \frac{x^n}{n!}.$$

Multiply by $x/(e^x - 1)$ so the left-hand side becomes $x^2/(e^{2x} - 1)$.

Expanding we obtain

$$\frac{x}{2}\sum_{n=0}^{\infty}b_n 2^n \frac{x^n}{n!} = -\left(\sum_{r=0}^{\infty}(2^r - 1)b_r \frac{x^r}{r!}\right) \times \left(\sum_{s=0}^{\infty}b_s \frac{x^s}{s!}\right).$$

Now equate the coefficients of x^{2n+1} to obtain (5.9.5). $\square$

Corollary 5.9.1. *The Bernoulli numbers are rational numbers and satisfy* $(-1)^{n-1}B_{2n} > 0$.

Extra 5.9.1. The arithmetical properties of the Bernoulli numbers are very interesting. The denominators are completely determined by a theorem discovered independently by von Staudt and Clausen in 1840: If $n \geq 1$ then

$$B_{2n} = I_n - \sum_{p-1|2n}\frac{1}{p},\qquad (5.9.7)$$

where I_n is an integer and the sum runs over all primes p such that $p - 1$ divides n. A proof due to Lucas appears as an exercise in Apostol (1976), page 275. On the other hand the numerators seem to be more mysterious. The mathematical interest in these numerators is mainly due to their connection to *Fermat's last theorem*:

> If $n \geq 3$, then the equation $x^n + y^n = z^n$ has no solutions
> for which $x, y, z \neq 0$.

Kummer introduced in 1847 the concept of a **regular prime number** and established that Fermat's last theorem is true if the exponent n is a regular prime. It turns out that n is not a regular prime if and only if n divides the numerator of one of the Bernoulli numbers $B_2, B_4, \ldots, B_{n-3}$. The only primes $n \leq 100$ that are not regular are 37, 59 and 67. More information about this subject can be found in Ribenboim's books (1979, 1999).

Project 5.9.1. In this project we introduce the **Bernoulli polynomials** $B_n(x)$ by their generating function

$$\frac{te^{xt}}{e^t - 1} = \sum_{n=0}^{\infty}B_n(x)\frac{t^n}{n!}.\qquad (5.9.8)$$

a) Prove that for $n \geq 1$

$$\int_0^1 B_n(x)dx = 0.$$

b) Establish the formula

$$B_n(x) = \sum_{k=0}^{n} \binom{n}{k} B_k x^{n-k}.$$

c) Prove

$$B_n(x + y) = \sum_{k=0}^{n} \binom{n}{k} B_k(x) y^{n-k}.$$

d) Establish the values $B_n(0) = B_n$ and $B_n(1) = (-1)^n B_n$.

e) Prove the symmetry rules $B_n(1 - x) = (-1)^n B_n(x)$ and $B_n(-x) = (-1)^n \left(B_n(x) + n x^{n-1} \right)$.

f) Check the value $B_n(1/2) = -(1 - 2^{1-n}) B_n$.

g) Establish the recurrences $B'_n(x) = n B_{n-1}(x)$ and $B_n(x + 1) - B_n(x) = n x^{n-1}$.

h) Check the identity

$$\sum_{k=0}^{n} \binom{n+1}{k} B_k(x) = (n + 1) x^n.$$

Conclude that for $n \geq 1$

$$\sum_{k=0}^{n} \binom{n+1}{k} B_k = 0.$$

i) Prove that

$$\int_x^y B_n(s)\, ds = \frac{1}{n+1} [B_{n+1}(y) - B_{n+1}(x)]$$

and

$$\int_x^{x+1} B_n(s)\, ds = x^n.$$

Finally deduce

$$\sum_{i=1}^{n} i^p = \frac{B_{p+1}(n+1) - (-1)^p B_{p+1}}{p+1} \tag{5.9.9}$$

Extra 5.9.2. The optimal growth of the Bernoulli numbers has been determined by Alzer (2000): the best possible constants α and β for which

$$\frac{2\,(2n)!}{(2\pi)^{2n}(1 - 2^{\alpha-2n})} \leq |B_{2n}| \leq \frac{2\,(2n)!}{(2\pi)^{2n}(1 - 2^{\beta-2n})} \tag{5.9.10}$$

are $\alpha = 0$ and $\beta = 2 + \ln(1 - \frac{6}{\pi^2})/\ln 2$.

5.10. Combinations of Exponentials and Polynomials

In this section we consider integrals of the form

$$I_{a,b}(\mu) = \int_a^b P(x)e^{\mu x}\, dx \qquad (5.10.1)$$

where $a < b$ and $\mu \in \mathbb{R}$ and P is a polynomial function.

We show that the class of functions

$$\mathfrak{E} := \left\{ f(x) = \sum_{i=0}^n \sum_{j=0}^m a_{i,j} x^i\, e^{\mu_j x} : a_{i,j}, \mu_j \in \mathbb{R};\ n, m \in \mathbb{N} \right\} \qquad (5.10.2)$$

is closed under the formation of primitives.

Exercise 5.10.1. Define $I_m = \int x^m e^{ax}\, dx$. Prove the recursion [G & R] 2.321.1

$$I_m = \frac{1}{a} x^m e^{ax} - \frac{m}{a} I_{m-1} \qquad (5.10.3)$$

and use it to establish

$$I_m = m! e^{ax} \sum_{k=0}^m \frac{(-1)^k x^{m-k}}{(m-k)! a^{k+1}},$$

and [G & R] 2.323: a polynomial P of degree m satisfies

$$\int P(x) e^{ax}\, dx = \frac{1}{a} e^{ax} \sum_{k=0}^m \frac{(-1)^k}{a^k} P^{(k)}(x).$$

Conclude that $\mathfrak{E}$ is closed under the formation of primitives.

Exercise 5.10.2. Evaluate the definite integrals related to Exercise 5.10.1.

Check [G & R] 3.351.1. For $u > 0$

$$\int_0^u x^n e^{-\mu x}\, dx = \frac{n!}{\mu^{n+1}} - e^{-\mu u} \sum_{k=0}^n \frac{n!}{k!} \frac{u^k}{\mu^{n-k+1}}$$

$$= \frac{n!}{\mu^{n+1}} \left(1 - e^{-\mu u}\, \text{Ex}(\mu u, n) \right).$$

(Note that the function Ex has been introduced in (5.3.7)). In particular,

$$\int_0^u xe^{-\mu x}\,dx = \frac{1}{\mu^2} - \frac{1}{\mu^2}e^{-\mu u}(1+\mu u),$$

$$\int_0^u x^2 e^{-\mu x}\,dx = \frac{2}{\mu^3} - \frac{1}{\mu^3}e^{-\mu u}(2+2\mu u + \mu^2 u^2),$$

$$\int_0^u x^3 e^{-\mu x}\,dx = \frac{6}{\mu^4} - \frac{1}{\mu^4}e^{-\mu u}(6+6\mu u + 3\mu^2 u^2 + \mu^3 u^3).$$

These formulas are [G & R] 3.351.7, 3.351.8, 3.351.9, respectively.

Extra 5.10.1. The evaluation of definite integrals that combine exponentials with rational functions require the **exponential integral** function

$$\mathrm{Ei}(x) := -\int_{-x}^{\infty} \frac{e^{-t}}{t}\,dt. \tag{5.10.4}$$

introduced in (5.8.5). Many definite integrals can be expressed in terms of Ei, for instance,

$$\int_a^b \frac{e^x}{x}\,dx = \mathrm{Ei}(b) - \mathrm{Ei}(a), \tag{5.10.5}$$

and

$$\int_0^1 \frac{e^x\,dx}{ax^2+bx+c} = \frac{-e^{(b+u)/2a}}{u}$$
$$\times \left[e^{u/a}\mathrm{Ei}(c_1) - e^{u/a}\mathrm{Ei}(1+c_1) - \mathrm{Ei}(c_2) + \mathrm{Ei}(1+c_2) \right] \tag{5.10.6}$$

where $u = \sqrt{b^2 - 4ac}$, $c_1 = (b-u)/2a$ and $c_2 = (b+u)/2a$.

6

The Trigonometric Functions and π

6.1. Introduction

Chapter 3 described the evaluation of

$$I = \int_0^\infty \frac{P(x)\,dx}{(q_1 x + q_0)^{m+1}}, \quad m \in \mathbb{N}, \tag{6.1.1}$$

in terms of the parameters q_0, q_1, m and the coefficients of the polynomial P. In this chapter we continue with our program to find closed-form evaluations of integrals of rational functions and consider the evaluation of

$$I = \int_0^\infty \frac{P(x)\,dx}{(q_2 x^2 + q_1 x + q_0)^{m+1}}, \quad m \in \mathbb{N}, \tag{6.1.2}$$

for $P(x)$ a polynomial of degree $2m$, in terms of the parameters

$$\mathfrak{P}_2 = \{m; q_0, q_1, q_2\} \cup \{\text{ coefficients of } P\}. \tag{6.1.3}$$

The degree of the polynomial is restricted to ensure convergence of the integral. A restriction on the parameters q_0, q_1 and q_2 to ensure convergence appears in Exercise 2.4.5.

The integral I is a linear combination of the integrals

$$I(m, n; a) := \int_0^\infty \frac{x^n\,dx}{(q_2 x^2 + q_1 x + q_0)^{m+1}}, \quad n \le 2m, \tag{6.1.4}$$

so it suffices to give closed forms of these.

Exercise 6.1.1. Prove that, for $n \le 2m$,

$$I(m, n; a) = C \int_0^\infty \frac{x^n\,dx}{(x^2 + 2ax + 1)^{m+1}} \tag{6.1.5}$$

for $a = q_1/2\sqrt{q_0 q_2}$ and an appropriate constant $C = C(m, n; q_0, q_2)$.

105

Exercise 6.1.2. Let

$$J(m, n; a) := \int_0^\infty \frac{x^n \, dx}{(x^2 + 2ax + 1)^{m+1}}, \quad n \le 2m \qquad (6.1.6)$$

be the integral in the previous exercise.
a) Prove the recurrence

$$J(m, n; a) = J(m - 1, n - 2; a) - 2a J(m, n - 1; a) - J(m, n - 2; a). \qquad (6.1.7)$$

Hint. Divide x^n by $x^2 + 2ax + 1$. Conclude that the value of $J(m, n; a)$ can be determined once $J(m, 0; a)$ and $J(m, 1; a)$ are known.
b) Check that

$$J(m, 1; a) = \frac{1}{2m} - a J(m, 0; a) \qquad (6.1.8)$$

so that the family $J(m, n; a)$ is completely determined from

$$J(m, 0; a) = \int_0^\infty \frac{dx}{(x^2 + 2ax + 1)^{m+1}}. \qquad (6.1.9)$$

In this chapter we develop the basic material required for the evaluation of these integrals.

6.2. The Basic Trigonometric Functions and the Existence of π

The standard approach to trigonometry that appears in calculus texts is to define them by geometric means and use these to establish their analytical properties. For instance, Thomas and Finney (1996), after giving a geometric definition of $\sin x$, proceeds to establish its continuity and the crucial limit

$$\lim_{x \to 0} \frac{\sin x}{x} = 1, \qquad (6.2.1)$$

the latter also by geometric arguments. Thomas states on page 3 that π is irrational and on page 35 claims *since the circumference of the circle is* 2π ... without giving a definition of π. This is a perfectly reasonable approach, given that the students have already heard about π in this way.

A more rigorous approach is described in Spivak (1980). The number π is defined by the integral

$$\pi := 2 \int_{-1}^1 \sqrt{1 - x^2} \, dx. \qquad (6.2.2)$$

Then the area of the sector of angle x is *defined* to be

$$A(x) := \frac{1}{2}x\sqrt{1-x^2} + \int_x^1 \sqrt{1-t^2}\,dt, \qquad (6.2.3)$$

and the trigonometric functions are defined in terms of A. For instance, for $0 \le x \le \pi$, the function $\cos x$ is the unique function satisfying $A(\cos x) = x/2$, and $\sin x$ is defined as $\sqrt{1 - \cos^2 x}$.

In this text we define the **arctangent** function by the integral

$$\tan^{-1} x := \int_0^x \frac{dt}{1+t^2}, \qquad x > 0, \qquad (6.2.4)$$

and we consider this to be our most basic trigonometric function. Similarly we define the **arcsine** function by

$$\sin^{-1} x := \int_0^x \frac{dt}{\sqrt{1-t^2}}, \qquad 0 \le x \le 1. \qquad (6.2.5)$$

The number π is defined by the integral

$$\pi := 2\int_0^\infty \frac{dt}{1+t^2}, \qquad (6.2.6)$$

so that $\pi = 2\tan^{-1}\infty$.

Exercise 6.2.1. Prove the inequalities $\tan^{-1} x \le x$ and $x \le \tan x$. **Hint.** Compare their derivatives.

Proposition 6.2.1. *For $x > 0$ we have*

$$\tan^{-1}\left(\frac{x}{\sqrt{1-x^2}}\right) = \sin^{-1} x, \qquad (6.2.7)$$

so that $\sin \pi/2 = 1$.

Proof. The required identity is

$$\int_0^{x/\sqrt{1-x^2}} \frac{dt}{1+t^2} = \int_0^x \frac{du}{\sqrt{1-u^2}}, \qquad (6.2.8)$$

and this is established by the change of variable $t = u/\sqrt{1-u^2}$.   □

Note 6.2.1. The identity (6.2.7) could have been used to define $\sin^{-1} x$ directly in terms of $\tan^{-1} x$.

Exercise 6.2.2. This exercise outlines the derivation of the formulas for the area and length of a circle of radius r.

a) Check that the area of a circle of radius r is given by

$$A(r) := 4 \int_0^r \sqrt{r^2 - x^2}\, dx. \qquad (6.2.9)$$

Prove that

$$A(r) = r^2 A(1),$$

so that the area of a circle is proportional to the square of its radius.

b) Check that the length of a circle of radius r is given by

$$L(r) := \int_0^r \frac{r\, dx}{\sqrt{r^2 - x^2}}, \qquad (6.2.10)$$

and that

$$L(r) = r L(1).$$

c) Integrate by parts to prove that

$$2A(1) = L(1). \qquad (6.2.11)$$

d) Conclude that $A(r) = \pi r^2$ and $L(r) = 2\pi r$.

The authors, having employed the foregoing argument in their classroom as original, were disappointed to find it in Assmus (1985).

Exercise 6.2.3. Prove, along the lines of Proposition 6.2.1, the identity

$$\tan^{-1}\left(\frac{1-x}{\sqrt{1-x^2}}\right) = \frac{\pi}{4} - \frac{1}{2}\sin^{-1} x. \qquad (6.2.12)$$

Check that the value of π is given by

$$\pi = 2 \int_0^1 \frac{dt}{\sqrt{1-t^2}}. \qquad (6.2.13)$$

Exercise 6.2.4. Check that

$$\int_0^1 \frac{dt}{1+t^2} = \frac{\pi}{4}. \qquad (6.2.14)$$

Hint. The change of variable $x \mapsto 1/x$ should do it. Conclude that $\tan(\pi/4) = 1$. Now use (6.2.7) to obtain the value $\sin(\pi/4) = 1/\sqrt{2}$.

Exercise 6.2.5. Prove that $\tan^{-1} x$ is not a rational function. **Hint.** Write it as P/Q and differentiate to prove first that $1 + x^2$ must divide Q. Then write $Q(x) = (1 + x^2)^m Q_1(x)$ and obtain a contradiction.

Theorem 6.2.1. *The Taylor series expansion of* $\tan^{-1} x$ *is*

$$\tan^{-1} x = \sum_{n=0}^{\infty} \frac{(-1)^n x^{2n+1}}{2n + 1}, \tag{6.2.15}$$

for $|x| < 1$. *According to Ranjan Roy (1990), this was discovered independently by G. W. Leibniz, J. Gregory, and an Indian mathematician of the fourteenth or fifteenth century whose identity is not definitely known.*

Proof. Expand the integrand in (6.2.4) in a geometric series.   □

Exercise 6.2.6. Check that

$$\sum_{n=0}^{\infty} \frac{(-1)^n}{2n + 1} = \frac{\pi}{4}. \tag{6.2.16}$$

Exercise 6.2.7. Prove the identity

$$\int_0^1 \frac{\tan^{-1} x}{x} dx = \sum_{n=0}^{\infty} \frac{(-1)^n}{(2n + 1)^2}. \tag{6.2.17}$$

This number is known as **Catalan's constant**, usually denoted by G. The reader will find in Adamchik (1997) and Bradley (1998) a large collection of formulas for this constant.

Exercise 6.2.8. Establish the hypergeometric representation

$$\tan^{-1} x = x \cdot {}_2F_1 \left(\tfrac{1}{2}, 1, \tfrac{3}{2}; -x^2 \right). \tag{6.2.18}$$

Exercise 6.2.9. Define $\sin x$ as the inverse of $\sin^{-1} x$ and prove (6.2.1). The value

$$\sin \left(\frac{\pi}{2} \right) = 1 \tag{6.2.19}$$

is given in Exercise 6.2.3. Then prove that $y = \sin x$ satisfies $y' = \sqrt{1 - y^2}$. Introduce the function $\cos x$ by $\cos x = \sqrt{1 - \sin^2 x}$ and establish the rule

$$\frac{d}{dx} \cos x = -\sin x. \tag{6.2.20}$$

The special values like $\cos 0 = 1$ and $\cos \pi/2 = 0$ follow from those of the sine.

Exercise 6.2.10. Obtain the Taylor series expansions

$$\sin x = \sum_{k=0}^{\infty} \frac{(-1)^k}{(2k+1)!} x^{2k+1} \quad \text{and} \quad \cos x = \sum_{k=0}^{\infty} \frac{(-1)^k}{(2k)!} x^{2k} \quad (6.2.21)$$

and compute their radius of convergence.

Note 6.2.2. Most of the standard results in trigonometry can be proved directly from the integral representation of the inverse functions. The one exception seems to be the **addition theorem**:

$$\sin(x + y) = \sin x \cos y + \cos x \sin y,$$
$$\cos(x + y) = \cos x \cos y - \sin x \sin y. \quad (6.2.22)$$

This difficulty reappears in the description of **elliptic functions** as inverses of **elliptic integrals**:

$$f(x, k) = \int_0^x \frac{dt}{\sqrt{(1 - t^2)(1 - k^2 t^2)}}. \quad (6.2.23)$$

The case $k = 0$ corresponds to $\sin^{-1} x$. Information about these functions can be obtained in Borwein and Borwein (1987), McKean and Moll (1997) and Whittaker and Watson (1961).

Exercise 6.2.11. Use power series to give a proof of (6.2.22). **Hint.** Use Cauchy's formula to multiply power series given in Exercise 4.2.4.

Exercise 6.2.12. This exercise presents an extension of the identity $\cos(2x) = 2\cos^2 x - 1$.
a) Prove that

$$\cos(nx) = T_n(\cos x) \quad (6.2.24)$$

where T_n is a polynomial of degree n with integer coefficients. The polynomial T_n is called the **Chebyshev polynomial of the first kind**.
b) Show that

$$T_0(x) = 1$$
$$T_1(x) = x$$
$$T_2(x) = 2x^2 - 1$$
$$T_3(x) = 4x^3 - 3x. \quad (6.2.25)$$

c) Prove the recursion formula

$$T_{n+1}(x) = 2x T_n(x) - T_{n-1}(x). \tag{6.2.26}$$

d) Prove that, for n odd, $\sin(nx)$ is a polynomial in $\sin x$.
e) Prove that

$$T_n(x) = \frac{n}{2} \sum_{r=0}^{n/2} \frac{(-1)^r}{n-r} \binom{n-r}{r} (2x)^{n-2r}. \tag{6.2.27}$$

More information about these polynomials appears in Weisstein (1999), page 232.

Note 6.2.3. There is a large literature on the number π, we simply quote our favorites: Blatner (1999), and Arndt and Haenel (2001), chapter 4 in Ebbinghaus et al. (1991), Weisstein (1999), page 1355, and Berggren et al. (1997). There are also plenty of Web sites dedicated to π, with varying degrees of interest. Our favorites are

```
http://www.cecm.sfu.ca/pi
http://mathworld.wolfram.com/Pi.html
http://numbers.computation.free.fr/Constants/
constants.html
```

6.3. Solution of Cubics and Quartics by Trigonometry

In this section we express the roots of the cubic polynomial

$$P_3(x) := x^3 + q_2 x^2 + q_1 x + q_0 = 0 \tag{6.3.1}$$

in terms of trigonometric functions. The identity

$$\sin 3x = -4 \sin^3 x + 3 \sin x \tag{6.3.2}$$

shows that $y = \sin x$ solves a cubic equation, provided we think of $\sin 3x$ as one of the coefficients. The next exercise shows how to transform the general cubic (6.3.1) to a form similar to (6.3.2).

Exercise 6.3.1. In Exercise 2.4.6 we have shown that the general cubic can be transformed to one without the quadratic term.
a) Find a transformation of the form $y = \lambda z$ to convert the general cubic to the form

$$4z^3 - 3z + \beta = 0. \tag{6.3.3}$$

To find the solutions, choose an angle α so that

$$\sin \alpha = \beta. \tag{6.3.4}$$

The roots of the cubic (6.3.3) are then given by

$$x_1 = \sin\left(\frac{\alpha}{3}\right), \quad x_2 = \sin\left(\frac{\alpha}{3} + \frac{2\pi}{3}\right), \quad x_3 = \sin\left(\frac{\alpha}{3} + \frac{4\pi}{3}\right). \tag{6.3.5}$$

This method shows that the three roots are real provided

$$(2q_2^3 - 9q_1q_2 + 27q_0)^2 < 4(q_2 - 3q_1)^3. \tag{6.3.6}$$

In the reduced case $q_2 = 0$ this becomes

$$4q_1^3 + 27q_0^2 < 0. \tag{6.3.7}$$

Extra 6.3.1. The reader will find in McKean and Moll (1997), Chapter 4, a discussion on how to use elliptic functions to solve algebraic equations of degree 5.

6.4. Quadratic Denominators and Wallis' Formula

In this section we consider the evaluation of

$$I := \int_0^\infty \frac{P(x)\,dx}{(x^2 + 1)^{m+1}}, \tag{6.4.1}$$

where $P(x)$ is a polynomial of degree $2m$. This is the special case $a_0 = a_2 = 1$ and $a_1 = 0$ of (6.1.2). The recurrence (6.1.2) shows that it suffices to consider the two cases

$$I_1 = \int_0^\infty \frac{x\,dx}{(x^2 + 1)^{m+1}}, \tag{6.4.2}$$

and

$$J_{2,m} := \int_0^\infty \frac{dx}{(x^2 + 1)^{m+1}}. \tag{6.4.3}$$

The first integral is elementary because the integrand admits a rational primitive:

$$\frac{d}{dx} \frac{-1}{2m(x^2 + 1)^m} = \frac{x}{(x^2 + 1)^{m+1}}, \tag{6.4.4}$$

that yields $I_1 = 1/2m$. This leaves only $J_{2,m}$ for discussion.

The integral $J_{2,m}$ was evaluated by Wallis (1656). The first proof of Theorem 6.4.1 is sometimes found in calculus books (see e.g. Larson et al. (1998), page 492). Different proofs of Wallis' formula (6.4.5) will be presented in Chapter 10 and in the Appendix.

Theorem 6.4.1. Wallis' formula. *Let* $m \in \mathbb{N}$. *Then*

$$J_{2,m} := \int_0^\infty \frac{dx}{(x^2 + 1)^{m+1}}$$

$$= \int_0^{\pi/2} \cos^{2m} \theta \, d\theta = \int_0^{\pi/2} \sin^{2m} \theta \, d\theta = \frac{\pi}{2^{2m+1}} \binom{2m}{m}. \quad (6.4.5)$$

Proof. The change of variable $x = \tan \theta$ (or $x = \cot \theta$) transforms one integral into the other, and integration by parts yields the recursion

$$J_{2,m} = \frac{2m - 1}{2m} J_{2,m-1}. \quad (6.4.6)$$

It is then easy to verify that the right side of (6.4.5) satisfies the same recursion and that both sides yield $\pi/2$ for $m = 0$. □

Exercise 6.4.1. Solve the recursion (6.4.6) directly. **Hint.** Define $b_m = J_{2,m} \binom{2m}{m}^{-1}$ and find a recursion for b_m. Compare with (2.3.18).

Note 6.4.1. Observe that the quadratic integral $J_{2,m}$ is a rational multiple of π:

$$\frac{1}{\pi} J_{2,m} = 2^{-2m-1} \binom{2m}{m} \in \mathbb{Q}. \quad (6.4.7)$$

Note 6.4.2. Wallis' formula appears in [G & R] 3.621.3 expressed as

$$\int_0^{\pi/2} \sin^{2m} x \, dx = \int_0^{\pi/2} \cos^{2m} x \, dx = \frac{(2m - 1)!!}{(2m)!!} \frac{\pi}{2}, \quad (6.4.8)$$

where $(2m - 1)!! = (2m - 1)(2m - 3) \cdots 3 \cdot 1$ and $(2m)!! = (2m - 2) \cdots 4 \cdot 2$.

Exercise 6.4.2. Prove [G & R] 3.621.4:

$$\int_0^{\pi/2} \sin^{2n+1} x \, dx = \frac{(2n)!!}{(2n + 1)!!}, \quad n \geq 0. \quad (6.4.9)$$

Hint. Integrate by parts to produce a recurrence.

Exercise 6.4.3. In this exercise we present a series of integrals that can be evaluated using Wallis' formula (6.4.5). The answer is expressed in terms of

semi-factorials

$$n!! = n \cdot (n-2) \cdot (n-4) \cdots \tag{6.4.10}$$

where the product ends in 2 or 1 according to the parity of n. These semifactorials can be transformed to factorials using

$$(2n)!! = 2^n \, n! \text{ and } (2n-1)!! = \frac{(2n)!}{2^n \, n!}. \tag{6.4.11}$$

a) Prove these identities.
b) Prove [G & R] 3.249.1:

$$\int_0^\infty \frac{dx}{(x^2+a^2)^n} = \frac{(2n-3)!!}{2(2n-2)!!} \frac{\pi}{a^{2n-1}}.$$

c) Prove [G & R] 3.249.2:

$$\int_0^a (a^2 - x^2)^{n-1/2} = \frac{(2n-1)!!\pi \, a^{2n}}{2(2n)!!}.$$

d) Prove [G & R] 3.251.4:

$$\int_0^\infty \frac{x^{2m} \, dx}{(ax^2+c)^n} = \frac{(2m-1)!! \, (2n-2m-3)!! \, \pi}{2(2n-2)!! \, a^m c^{n-m-1} \sqrt{ac}}$$

for $a > 0$, $c > 0$, $n > m + 1$. **Hint.** Do first the case $m = 0$ and then differentiate with respect to the parameter a. What happens at $n = m + 1$?
e) Prove [G & R] 3.251.5:

$$\int_0^\infty \frac{x^{2m+1} \, dx}{(ax^2+c)^n} = \frac{m! \, (n-m-2)!}{2(n-1)! a^{m+1} c^{n-m-1}}, \quad a, c > 0, n > m + 1.$$

Project 6.4.1. This project confirms that the class of functions

$$\mathfrak{T} := \left\{ f(x) = \sum_{i=0}^n \sum_{j=0}^m a_{i,j} \sin^i x \cos^j x \, : \, a_{i,j} \in \mathbb{R}; \, n, m \in \mathbb{N} \right\}$$

is closed under the formation of primitives.

a) Prove that the identities

$$\sin^{2n} x = \frac{1}{2^{2n-1}} \left\{ \sum_{k=0}^{n-1} (-1)^{n-k} \binom{2n}{k} \cos 2(n-k)x + \tfrac{1}{2} \binom{2n}{n} \right\}$$

$$\sin^{2n+1} x = \frac{1}{2^{2n}} \sum_{k=0}^{n} (-1)^{n-k} \binom{2n+1}{k} \sin(2n-2k+1)x$$

$$\cos^{2n} x = \frac{1}{2^{2n-1}} \left\{ \sum_{k=0}^{n-1} \binom{2n}{k} \cos 2(n-k)x + \tfrac{1}{2} \binom{2n}{n} \right\}$$

$$\cos^{2n+1} x = \frac{1}{2^{2n}} \sum_{k=0}^{n} \binom{2n+1}{k} \cos(2n-2k+1)x,$$

are valid for $n \geq 0$. Conclude that $\sin^n x$ and $\cos^m x$ admit a primitive in $\mathfrak{T}$.
b) Prove that in the expression for $f \in \mathfrak{T}$, we may assume $m = 0$ or 1. Conclude then that every function in $\mathfrak{T}$ admits a primitive in that class. The case $m = 0$ is Wallis' formula (6.4.5) and $m = 1$ is elementary.

Exercise 6.4.4. This exercise outlines a proof of the value of the constant C in Stirling's formula described in Theorem 5.7.1. Write (6.4.14) as

$$\frac{\pi}{2} = \frac{2^{4n}}{\binom{2n}{n}^2} \times \frac{Q_n}{2n+1} \tag{6.4.12}$$

and then use $\lim_{n\to\infty} Q_n = 1$ and Exercise 5.7.1 to obtain $C = \sqrt{2\pi}$.

Exercise 6.4.5. Prove Wallis' product

$$\pi = 2 \prod_{k=1}^{\infty} \frac{2k}{2k-1} \cdot \frac{2k}{2k+1} \tag{6.4.13}$$

using Wallis' formula. **Hint.** Divide (6.4.9) by (6.4.8) to produce

$$\frac{\pi}{2} = \frac{2}{1} \cdot \frac{2}{3} \cdot \frac{4}{3} \cdot \frac{4}{5} \cdots \frac{2n}{2n-1} \cdot \frac{2n}{2n+1} \times Q_n \tag{6.4.14}$$

with

$$Q_n = \int_0^{\pi/2} \sin^{2n} x \, dx \Big/ \int_0^{\pi/2} \sin^{2n+1} x \, dx.$$

Show that $1 \leq Q_n \leq 1 + 1/2n$ and pass to the limit.

Extra 6.4.1. The expression

$$\frac{2}{\pi} = \sqrt{\frac{1}{2}} \sqrt{\frac{1}{2} + \frac{1}{2}\sqrt{\frac{1}{2}}} \sqrt{\frac{1}{2} + \sqrt{\frac{1}{2} + \frac{1}{2}\sqrt{\frac{1}{2}}}} \cdots, \qquad (6.4.15)$$

due to Vieta (1970) is one of the oldest representations of π. Osler (1999)
showed that the product

$$\frac{2}{\pi} = \prod_{n=1}^{P} \sqrt{\frac{1}{2} + \frac{1}{2}\sqrt{\frac{1}{2} + \frac{1}{2}\sqrt{\frac{1}{2} + \cdots + \frac{1}{2}\sqrt{\frac{1}{2}}}}} \qquad n \text{ radicals}$$

$$\times \prod_{k=1}^{\infty} \frac{2^{p+1}k - 1}{2^{p+1}k} \cdot \frac{2^{p+1}k + 1}{2^{p+1}k}$$

yields Wallis's formula (6.4.13) for $p = 0$ and Vieta's expression as $p \to \infty$.

Mathematica 6.4.1. A direct symbolic evaluation can be obtained by the
command

```
Integrate [ 1/(x^{2} + 1)^{m+1}, {x, 0, Infinity}]
```

to give

$$\text{If}(\text{Re}[m] > -\tfrac{1}{2}, \frac{\sqrt{\pi}\,(m + \tfrac{1}{2})}{2\Gamma[1 + m]}, \int_0^\infty (1 + x^2)^{-m-1} dx). \qquad (6.4.16)$$

Mathematica indicates that the answer

$$\frac{\sqrt{\pi}\,(m + \tfrac{1}{2})}{2\Gamma[1 + m]} \qquad (6.4.17)$$

is valid only for $\text{Re}[m] > -\tfrac{1}{2}$, and does not evaluate the integral outside
this range. However restrictions on parameters can be introduced via the
Assumptions command. The new input

```
Integrate [ 1/(x^{2} + 1)^{m+1}, {x,0, Infinity},
      Assumptions \rightarrow { Re[m] > -1/2}]
```

yields

$$\frac{\sqrt{\pi}}{2} \frac{\text{Gamma}[\tfrac{1}{2} + m]}{\text{Gamma}[1 + m]}. \qquad (6.4.18)$$

The **gamma function** appearing above will be studied in Chapter 10. There
we show that, for $m \in \mathbb{N}$, the expression (6.4.18) reduces to (6.4.5).

6.5. Arithmetical Properties of π

In this section we discuss a proof of the irrationality of π. This complements the results of Section 5.5 which proved the irrationality of e. There are several other proofs in the literature, see for example Breusch (1954) and Desbrow (1990). The fact that π is transcendental was established by F. Lindemann (1882). A modified proof appears in Hardy and Wright (1979), page 173.

Extra 6.5.1. The question of irrationality of specific numbers is full of subtleties. For instance, it is an open question to decide if $\pi + e$ or πe are irrational numbers.

The proof of the irrationality of π begins with some exercises.

Exercise 6.5.1. Let

$$D_n(r) := \int_{-1}^{1} (1 - x^2)^n \cos(rx)\, dx.$$

Check that

$$D_0(r) = \frac{2 \sin r}{r} \quad \text{and} \quad D_1(r) = \frac{4(\sin r - r \cos r)}{r^3}.$$

Prove the recurrence

$$D_n(r) = \frac{1}{r^2} \left(2n(2n-1)D_{n-1}(r) - 4n(n-1)D_{n-2}(r) \right).$$

Conclude that there exist polynomials P_n and Q_n of degree at most n, with integer coefficients, such that

$$D_n(r) = \frac{n!}{r^{2n+1}} \left(P_n(r) \sin r - Q_n(r) \cos r \right). \tag{6.5.1}$$

We now present a proof of the irrationality of π, based on the polynomials $D_n(r)$, that is due to Niven (1947). The final step of the proof employs the result of the next exercise.

Exercise 6.5.2. Prove that, for any $a > 0$, the term $a^n/n! \to 0$ as $n \to \infty$. **Hint.** Use Stirling's formula (5.7.8).

Theorem 6.5.1. *The number π is irrational.*

Proof. Suppose $\pi = a/b$ and define $r := \pi/2 = a/2b$. Then evaluating (6.5.1) at $\pi/2$ we have

$$\left(\frac{a}{2b}\right)^{2n+1} D_n\left(\frac{a}{2b}\right) = n! P_n\left(\frac{a}{2b}\right).$$

Now clear the denominators and write D_n for $D_n(a/2b)$; we then have that $z = a^{2n+1} D_n/n!$ is an integer. But

$$|D_n(r)| \leq \int_{-1}^{1} |\cos(rx)| \, dx \leq 2$$

independently of n, so that $|z| \leq 2a^{2n+1}/n! \to 0$ as $n \to \infty$, and thus $z = 0$. We conclude that $D_n = 0$ for all n, a contradiction. $\qquad\square$

6.6. Some Expansions in Taylor Series

In this section we describe some functions that contain the central binomial coefficients in their Taylor series. We analyze, for example,

$$f_j(x) = \sum_{n=1}^{\infty} n^j \binom{2n}{n} x^n \qquad (6.6.1)$$

and

$$g_j(x) = \sum_{n=1}^{\infty} \frac{\binom{2n}{n}}{n^j} x^n. \qquad (6.6.2)$$

The specialization of these series to a given value of x leads to some interesting numerical series. The proofs have appeared in Lehmer (1985).

The first formula is the generating function for the central binomial coefficients,

$$\frac{1}{\sqrt{1-4x}} = \sum_{n=0}^{\infty} \binom{2n}{n} x^n, \qquad (6.6.3)$$

which is the result of Exercise 4.2.2, part c).

Project 6.6.1. Prove that there is a polynomial $S_j(x)$ with positive integer coefficients such that

$$\sum_{n=1}^{\infty} n^j \binom{2n}{n} x^n = \frac{2x \, S_j(2x)}{(1-4x)^{j+1/2}}. \qquad (6.6.4)$$

Obtain a recurrence for S_n. The first few are

$$S_1(x) = 1 \tag{6.6.5}$$
$$S_2(x) = 1 + x$$
$$S_3(x) = 1 + 5x + x^2$$
$$S_4(x) = 1 + 15x + 18x^2 + x^3$$
$$S_5(x) = 1 + 37x + 129x^2 + 58x^3 + x^4.$$

Can you find a closed-form for the coefficients of S_n?

The second formula is the Taylor series expansion for $\sin^{-1} x$.

Theorem 6.6.1. *The inverse sine function is given by*

$$\sin^{-1} x = \sum_{n=0}^{\infty} \frac{\binom{2n}{n}}{2^{2n}} \frac{x^{2n+1}}{2n+1}. \tag{6.6.6}$$

Proof. Replace $4x$ by t^2 in (6.6.3) and integrate from 0 to x. $\qquad \square$

Corollary 6.6.1. *Let $x \in (0, 1)$. Then*

$$\sum_{n=1}^{\infty} \frac{1}{n} \binom{2n}{n} x^n = 2 \ln \left(\frac{1 - \sqrt{1 - 4x}}{2x} \right). \tag{6.6.7}$$

In particular

$$\sum_{n=1}^{\infty} \frac{1}{n} \binom{2n}{n} 2^{-2n} = 2 \ln 2. \tag{6.6.8}$$

Proof. Divide (6.6.3) by x and integrate from 0 to x. $\qquad \square$

We now continue with our discussion of special Taylor series.

Theorem 6.6.2.

$$\frac{2x \sin^{-1} x}{\sqrt{1 - x^2}} = \sum_{n=1}^{\infty} \frac{(2x)^{2n}}{n \binom{2n}{n}}. \tag{6.6.9}$$

Proof. We start with Gregory's series (6.2.15) and set $t = x/\sqrt{1-x^2}$ so that, according to (6.2.7), we have $\tan^{-1} t = \sin^{-1} x$. Then

$$
\frac{x \sin^{-1} x}{\sqrt{1-x^2}} = \sum_{n=1}^{\infty} \frac{(-1)^{n-1} x^{2n}}{(2n-1)(1-x^2)^n}
$$

$$
= \sum_{n=1}^{\infty} \frac{(-1)^{n-1}}{2n-1} \sum_{j=0}^{\infty} (-1)^j x^{2(j+n)} \binom{-n}{j}
$$

$$
= \sum_{n=1}^{\infty} \frac{(-1)^{n-1}}{2n-1} \sum_{j=0}^{\infty} \binom{n+j-1}{j} x^{2(j+n)}
$$

$$
= \sum_{m=1}^{\infty} x^{2m} \sum_{k=1}^{m} \frac{(-1)^{k-1}(m-1)!}{(k-1)!\,(m-k)!\,(2k-1)}.
$$

The identity (6.6.9) is thus reduced to proving

$$
m \binom{2m}{m} \sum_{j=0}^{m-1} \frac{(-1)^j (m-1)!}{j!\,(m-j-1)!(2j+1)} = 2^{2m-1}. \qquad (6.6.10)
$$

To check this identity write the sum as

$$
\sum_{j=0}^{m-1} \frac{(-1)^j}{2j+1} \binom{m-1}{j} = \int_0^1 \sum_{j=0}^{m-1} (-1)^j \binom{m-1}{j} y^{2j}\, dy
$$

$$
= \int_0^1 (1-y^2)^{m-1}\, dy
$$

$$
= \int_0^{\pi/2} \sin^{2m-1}\theta\, d\theta.
$$

The value of (6.6.10) now follows from Wallis' formula (6.4.5). □

Exercise 6.6.1. Use Theorem 6.6.2 to establish the sums

$$
\sum_{n=1}^{\infty} \frac{1}{n \binom{2n}{n}} = \frac{\pi}{3\sqrt{3}} \qquad (6.6.11)
$$

$$
\sum_{n=1}^{\infty} \frac{2^n}{n \binom{2n}{n}} = \frac{\pi}{2} \qquad (6.6.12)
$$

$$\sum_{n=1}^{\infty} \frac{3^n}{n\binom{2n}{n}} = \frac{2\pi}{\sqrt{3}} \tag{6.6.13}$$

$$\sum_{n=1}^{\infty} \frac{(5-\sqrt{5})^n}{n\,2^n\binom{2n}{n}} = \frac{2\pi}{5}\sqrt{5-2\sqrt{5}}. \tag{6.6.14}$$

Hint. The last sum requires the value

$$\sin\left(\frac{\pi}{5}\right) = \frac{1}{2}\sqrt{\frac{5-\sqrt{5}}{2}}. \tag{6.6.15}$$

To establish this value, let $\theta = \pi/10$, so that $\sin 2\theta = \cos 3\theta$ and check that $u = \sin\theta$ satisfies $4u^2 + 2u - 1 = 0$.

Extra 6.6.1. The problem of determining the rational numbers m/n such that the sine of the angle $\theta = \frac{m\pi}{n}$ is expressed by radicals was considered by Gauss. The following information appears in

http://mathworld.wolfram.com/
TrigonometryAngles.html

The angles $\frac{m\pi}{n}$ for which the trigonometric functions may be expressed in terms of radicals of real numbers are those n for which the regular polygon of n sides is constructible (with compass and ruler). Gauss proved that n must be of the form

$$n = 2^k p_1 p_2 \cdots p_s \tag{6.6.16}$$

where $k \in \mathbb{N}$ and p_i are distinct Fermat primes (a prime of the form $2^{2^m} + 1$). The only known Fermat primes are 3, 5, 17, 257 and 65537. Therefore, the value $\sin(\pi/7)$ cannot be expressed by radicals.

Exercise 6.6.2. Prove that

$$\sum_{n=1}^{\infty} \frac{2^{2n}}{\binom{2n}{n}} x^{2n} = \frac{x^2}{1-x^2} + \frac{x\sin^{-1}x}{(1-x^2)^{3/2}} \tag{6.6.17}$$

and obtain the value

$$\sum_{n=1}^{\infty} \frac{1}{\binom{2n}{n}} = \frac{2\pi\sqrt{3}+9}{27}. \tag{6.6.18}$$

Hint. Apply $x^2 \frac{d}{dx}\frac{1}{x}$ to both sides of (6.6.9).

Exercise 6.6.3. Find a closed from for the sum

$$\sum_{n=1}^{\infty} \frac{nx^n}{\binom{2n}{n}}. \tag{6.6.19}$$

In particular, check that

$$\sum_{n=1}^{\infty} \frac{n2^n}{\binom{2n}{n}} = \pi + 3. \tag{6.6.20}$$

This expression for π was employed by Plouffe (2003) to develop an efficient method to compute it.

Exercise 6.6.4. Establish the expansion

$$\left(\sin^{-1} x\right)^2 = \frac{1}{2} \sum_{n=1}^{\infty} \frac{(2x)^{2n}}{n^2 \binom{2n}{n}}. \tag{6.6.21}$$

Hint. Divide (6.6.9) by x and integrate.

Exercise 6.6.5. Prove

$$\sum_{n=1}^{\infty} \frac{2^n}{n^2 \binom{2n}{n}} = \frac{\pi^2}{8}. \tag{6.6.22}$$

Exercise 6.6.6. Establish the value

$$\sum_{n=1}^{\infty} \frac{1}{n^2 \binom{2n}{n}} = \frac{\pi^2}{18}. \tag{6.6.23}$$

Mathematica 6.6.1. Continuation of this process yields

$$\sum_{n=1}^{\infty} \frac{(2x)^{2n}}{n^3 \binom{2n}{n}} = 4 \int_0^x \frac{(\sin^{-1} t)^2}{t} \, dt. \tag{6.6.24}$$

This function is not elementary. A symbolic evaluation using Mathematica yields

$$\sum_{n=1}^{\infty} \frac{(2x)^{2n}}{n^3 \binom{2n}{n}} = 2x^2 \text{Hypergeometric}_4 F_3 \left[\{1, 1, 1, 1\}, \{\tfrac{3}{2}, 2, 2\}, x^2\right].$$

The hypergeometric functions appearing here are defined in (3.5.5). Mathematica also gives special values of the identity (6.6.24). For instance

$$\int_0^1 \frac{(\sin^{-1} t)^2}{t} \, dt = \frac{\pi^2 \ln 2}{4} - \frac{7}{8}\zeta(3). \tag{6.6.25}$$

The number $\zeta(3)$ is called **Apery's constant** and is discussed in Chapter 11.

Project 6.6.2. Study the power series expansion of $(\sin^{-1} x)^n$. The first non-trivial case is

$$\left(\sin^{-1} x\right)^3 = \sum_{k=1}^{\infty} \frac{3!}{(2k+1)!} \prod_{m=1}^{k}(2m+1)^2 \sum_{j=0}^{k-1} \frac{1}{(2j+1)^2} x^{2k+1},$$

that appears in [G & R] 1.645.3. Hauss (1994) provides some information about these expansions.

Exercise 6.6.7. This exercise presents some formulas for π^2 given by Ewell (1992).

a) Integrate by parts to prove that

$$\int_0^x \sqrt{1-t^2}\, dt = \frac{1}{2} \sin^{-1} x + \frac{x}{2}\sqrt{1-x^2}$$

b) Integrate the expansion (6.6.6) to produce

$$x - \sum_{k=1}^{\infty} \frac{\binom{2k-2}{k-1}}{k\, 2^{2k-1}} \cdot \frac{x^{2k+1}}{2k+1} = \frac{1}{2}\sin^{-1} x + \frac{x}{2}\sqrt{1-x^2}.$$

Let $x = \sin t$ and integrate from 0 to $\pi/2$ to yield

$$\pi^2 = 12 - 16 \sum_{k=1}^{\infty} \frac{1}{(2k-1)(2k+1)^2}. \qquad (6.6.26)$$

c) Using a similar technique to parts a) and b), prove

$$\pi^2 = \frac{128}{9} - 128 \sum_{k=1}^{\infty} \frac{k+1}{(4k^2-1)(2k+3)^2}$$

and

$$\pi^2 = 4 + 32 \sum_{k=1}^{\infty} \frac{k}{(2k-1)(2k+1)^2}$$

by using

$$t^2\sqrt{1-t^2} = t^2 - \sum_{j=2}^{\infty} \frac{\binom{2j-4}{j-2}}{(j-1)\, 2^{2j-3}} t^{2j}$$

and

$$\frac{t^2}{\sqrt{1-t^2}} = \sum_{j=1}^{\infty} \frac{\binom{2j-2}{j-1}}{2^{2j-2}} t^{2j}.$$

6.7. A Sequence of Polynomials Approximating $\tan^{-1} x$

The goal of this section is to describe a sequence of polynomials that provide a very good approximation for $\tan^{-1} x$ on the interval $[0, 1]$. The details appear in Medina (2003).

The Taylor series for $\tan^{-1} x$ yields

$$\tan^{-1} x = \int_0^x \frac{dt}{1 + t^2} = \int_0^x \sum_{k=0}^{\infty} (-1)^k t^{2k} dt$$

$$= f_n(x) + (-1)^{n+1} \int_0^x \frac{t^{2n+2}}{1 + t^2} dt,$$

where

$$f_n(x) = \sum_{k=0}^{n} \frac{(-1)^k}{2k + 1} x^{2k+1}. \tag{6.7.1}$$

Exercise 6.7.1. Prove that

$$|\tan^{-1} x - f_n(x)| \geq \frac{x^{2n+3}}{2(2n + 3)}. \tag{6.7.2}$$

In particular $|\pi/4 - f_n(1)| \geq 1/(2(2n + 3))$. **Hint.** Write the difference as an integral and obtain an inequality for the denominator.

Project 6.7.1. Let

$$p_1(x) = x^6 - 4x^5 + 5x^4 - 4x^2 + 4 \tag{6.7.3}$$

and

$$p_m(x) = x^4 (1 - x)^4 p_{m-1}(x) + (-4)^{m-1} p_1(x) \tag{6.7.4}$$

for $m \geq 2$. Prove that

$$\frac{x^{4m}(1 - x)^{4m}}{1 + x^2} = p_m(x) + \frac{(-4)^m}{1 + x^2}. \tag{6.7.5}$$

Hint. Use induction.
b) Use the definition of $p_m(x)$ to establish

$$p_m(x) = p_1(x) \sum_{k=0}^{m-1} (-4)^{m-1-k} x^{4k} (1 - x)^{4k}. \tag{6.7.6}$$

c) Prove that

$$p_m(t) = (-1)^{m+1} 4^m \sum_{k=0}^{2m-1} (-1)^k t^{2k} + t^{4m}(5 - 4t + t^2)$$

$$\times \sum_{k=1}^{m} (1 - t)^{4(m-k)} (-4)^{k-1}.$$

Hint. Use (6.7.5) and

$$\frac{1}{1+t^2} = \sum_{k=0}^{2m-1} (-1)^k t^{2k} + \frac{t^{4m}}{1+t^2}.$$

d) Define

$$h_m(x) = f_{2m-1}(x) + \frac{(-1)^{m+1}}{4^m} \int_0^x t^{4m}(5 - 4t + t^2)$$

$$\times \sum_{k=1}^{m} (1 - t)^{4(m-k)} (-4)^{k-1} \, dt.$$

and prove that

$$\int_0^1 \frac{x^4(1-x)^4}{1+x^2} dx = \frac{22}{7} - \pi. \tag{6.7.7}$$

In particular this evaluation proves that $\pi \neq \frac{22}{7}$. Write (6.7.7) as

$$h_1(1) - \tan^{-1} 1 = \frac{1}{4}\left(\frac{22}{7} - \pi\right). \tag{6.7.8}$$

e) Evaluate $h_7(1)$ and compute the number of digits of π obtained from $4h_7(1)$.

Exercise 6.7.2. Prove that

$$\left| \int_0^x p_m(t) - \frac{(-1)^{m+1} 4^m}{1+t^2} dt \right| \leq 2^{-8m}. \tag{6.7.9}$$

Hint. Use $x(1 - x) \leq 1/4$ for $0 < x < 1$.

Theorem 6.7.1. *Define*

$$h_m(x) = \frac{(-1)^{m+1}}{2^{2m}} \int_0^x p_m(t) \, dt. \tag{6.7.10}$$

Then the polynomials $h_m(x)$ satisfy

$$\left| h_m(x) - \tan^{-1} x \right| \leq \left(\frac{1}{2^{5/4}}\right)^{\deg(h_m)+1} \tag{6.7.11}$$

for all $x \in [0, 1]$.

Proof. This is the result of Exercise 6.7.2. □

f) Establish the closed-form expression

$$h_m(x) = \sum_{j=1}^{2m} \frac{(-1)^{j+1}}{2j-1} x^{2j-1} + \sum_{j=0}^{4m-2} \frac{a_j}{(-1)^{m+1}4^m(4m+j+1)} x^{4m+j+1},$$

where

$$a_{2i} = (-1)^{i+1} \sum_{k=i+1}^{2m} (-1)^k \binom{4m}{2k} \quad \text{and} \quad a_{2i-1} = (-1)^{i+1} \sum_{k=i}^{2m-1} (-1)^k \binom{4m}{2k+1}.$$

Hint. Prove that

$$\frac{(1-t)^{4m}}{1+t^2} = \sum_{j=0}^{4m-2} a_j t^j + \frac{r_m(t)}{1+t^2}, \tag{6.7.12}$$

where r_m is a polynomial of degree at most 1.

Project 6.7.2. Find a rational function R such that

$$\int_0^1 R(x)\,dx = \pi - \frac{333}{106}.$$

The number $333/106$ is the second convergent of the continued fraction of π, $22/7$ being the first. General information about continued fractions can be found in Hardy and Wright (1979). This may be a difficult project, F. Beukers (2000) states that

It is not clear whether there exists a natural choice of F which produces the approximation 333/106.

The author is seeking rational approximations to π in the form

$$J(F) = \int_0^1 \frac{F(t)\,dt}{1+t^2}. \tag{6.7.13}$$

The reader should analyze Beukers' approximations

$$J_n = \int_0^1 \frac{t^{2n}(1-t^2)^{2n}}{(1+t^2)^{3n+1}} \cdot \left[(1+it)^{3n+1} + (1-it)^{3n+1}\right] dt, \tag{6.7.14}$$

where $i^2 = -1$ is the imaginary unit.

6.8. The Infinite Product for sin x

In this section we discuss the product representations for trigonometric functions that appeared in Euler's treatise in 1748 (1988). The reader will find in Nahim (1998) historical information about these topics. The products discussed here generalize the factorization (2.4.4) to the case in which the polynomial Q is replaced by $\sin x$.

Theorem 6.8.1. *The product representations for* sin *x and* cos *x are given by*

$$\sin x = x \prod_{k=1}^{\infty} \left(1 - \frac{x^2}{(\pi k)^2} \right) \tag{6.8.1}$$

and

$$\cos x = \prod_{k=1}^{\infty} \left(1 - \frac{x^2}{(\pi (k - \frac{1}{2}))^2} \right). \tag{6.8.2}$$

Proof. The argument given here appears in Venkatachaliengar (1962). Start with

$$I_n(x) := \int_0^{\pi/2} \cos xt \, \cos^n t \, dt$$

and integrate by parts to obtain

$$n(n - 1)I_{n-2}(x) = (n^2 - x^2)I_n(x).$$

Since $I_n(0) > 0$ we get for $n \geq 2$

$$\frac{I_{n-2}(x)}{I_{n-2}(0)} = \left(1 - \frac{x^2}{n^2} \right) \frac{I_n(x)}{I_n(0)}. \tag{6.8.3}$$

Using the values $I_0(0) = \pi/2$ and $I_1(0) = 1$ we have

$$\sin \left(\frac{\pi x}{2} \right) = \frac{\pi x}{2} \frac{I_0(x)}{I_0(0)} \quad \text{and} \quad \cos \left(\frac{\pi x}{2} \right) = (1 - x^2) \frac{I_1(x)}{I_1(0)}.$$

Now

$$|I_n(0) - I_n(x)| = \left| \int_0^{\pi/2} (1 - \cos xt) \cos^n t \, dt \right|$$

$$\leq \frac{1}{2} x^2 \int_0^{\pi/2} t^2 \cos^n t \, dt$$

$$\leq \frac{1}{2} x^2 \int_0^{\pi/2} t \cos^{n-1} t \, \sin t \, dt$$

$$= \frac{1}{n} I_n(0),$$

where we have used $t \leq \tan t$. Thus

$$\lim_{n \to \infty} \frac{I_n(x)}{I_n(0)} = 1. \qquad (6.8.4)$$

Now replace $\pi x/2$ by x, so (6.8.1) follows from (6.8.4). $\qquad \square$

Exercise 6.8.1. Use the recurrence (6.8.3) to obtain a closed form for the integral $I_n(x)$.

Exercise 6.8.2. Use the product (6.8.1) to derive the Wallis product

$$\pi = 2 \prod_{k=1}^{\infty} \frac{2k}{2k-1} \cdot \frac{2k}{2k+1}. \qquad (6.8.5)$$

Exercise 6.8.3. This exercise outlines a proof of the product representation (6.8.1).
a) Prove the identity

$$\sin(nx) = K(n) \sin x \times \prod_{r=1}^{(n-1)/2} \left(1 - \frac{\sin^2 x}{\sin^2(\pi r/n)} \right),$$

for n odd. This is a representation for the polynomial requested in Exercise 6.2.12. **Hint.** Locate the zeros of $\sin(nx)$.
b) Let $x \to 0$ to obtain $K(n) = n$.
c) Conclude that

$$\sin x = n \sin(x/n) \prod_{r=1}^{\infty} (1 + f_r(n, x)), \qquad (6.8.6)$$

where

$$f_r(n, x) = \begin{cases} 0 & r > (n-1)/2 \\ -\dfrac{\sin^2(x/n)}{\sin^2(r\pi/n)} & r \leq (n-1)/2 \end{cases} \qquad (6.8.7)$$

d) Let $n \to \infty$ to obtain (6.8.1). The representation (6.8.2) follows from the identity $\cos x = \sin(2x)/2 \sin x$.

Extra 6.8.1. The convergence of an infinite product can be treated in parallel to that of infinite series. Given a sequence of positive numbers $\{a_n\}$, we form the partial products

$$p_n = (1 + a_1)(1 + a_2) \cdots (1 + a_n) \qquad (6.8.8)$$

and if p_n converges to a limit p then we write

$$p = \prod_{n=1}^{\infty}(1 + a_n). \tag{6.8.9}$$

It turns out that p_n converges if and only if the series $\sum a_n$ converges. See Hijab (1997) for details and examples.

The naive extension of (2.4.4), that gives the factorization of a polynomial in terms of its roots, fails. The construction of a function f with roots at $\{a_1, a_2, \cdots\}$ via

$$f(x) = \prod_{k=1}^{\infty}\left(1 - \frac{x}{a_k}\right) \tag{6.8.10}$$

might not be convergent. Weierstrass introduced *elementary factors*

$$E_0(z) = 1 - z,$$
$$E_p(z) = (1 - z)\exp\left(z + z^2/2 + \cdots + z^p/p\right)$$

and showed that it is possible to choose indices p_k so that the modified product

$$P(x) = \prod_{k=1}^{\infty}E_{p_k}\left(\frac{z}{a_k}\right)$$

gives an honest function with the desired zeros. Greene and Krantz (2002) give complete details.

6.9. The Cotangent and the Riemann Zeta Function

In this section we discuss some elementary properties of the **cotangent function**

$$\cot x = \frac{\cos x}{\sin x}. \tag{6.9.1}$$

Exercise 6.9.1. Check that $\cot x$ is the **logarithmic derivative** of $\sin x$, that is,

$$\cot x = \frac{d}{dx}\ln\sin x. \tag{6.9.2}$$

The product representation given in Theorem 6.8.1 yields a similar one for $\cot x$. Replacing x by πx simplifies the form of the factors.

Exercise 6.9.2. Check the product representation

$$\cot \pi x = \frac{1}{\pi x} \prod_{k=1}^{\infty} \frac{1 - \left(\frac{2x}{2k-1}\right)^2}{1 - \left(\frac{x}{k}\right)^2}. \tag{6.9.3}$$

We next describe the Taylor series expansion of $\cot x - 1/x$.

Proposition 6.9.1. *The expansion of cotangent is given by*

$$\cot x = \frac{1}{x} - \sum_{n=1}^{\infty} (-1)^{n-1} B_{2n}\, 2^{2n} \frac{x^{2n-1}}{(2n)!}. \tag{6.9.4}$$

Proof. The expansion (5.9.1) and the identity

$$\frac{e^y + 1}{e^y - 1} = 1 + \frac{2}{e^y - 1}$$

yield

$$\frac{x}{\tan x} = 1 - \sum_{n=1}^{\infty} (-1)^{n-1} B_{2n}\, 2^{2n} \frac{x^{2n}}{(2n)!}, \tag{6.9.5}$$

after letting $y = 2ix$. This is equivalent to (6.9.4). $\qquad\qquad\square$

We now compare two expansions of $\cot x$ to obtain a relation between the Bernoulli numbers B_{2n} and the values of the Riemann zeta function

$$\zeta(s) = \sum_{n=1}^{\infty} \frac{1}{n^s} \tag{6.9.6}$$

at the even integers. This is due to Euler. The function $\zeta(s)$ will be discussed in Chapter 11.

Proposition 6.9.2. *The expansion of cotangent is*

$$\cot x = \frac{1}{x} - 2 \sum_{n=1}^{\infty} \frac{x^{2n-1}\, \zeta(2n)}{\pi^{2n}}. \tag{6.9.7}$$

Proof. Logarithmic differentiation of (6.8.1) yields

$$\cot x = \frac{1}{x} + \sum_{k=1}^{\infty} \frac{2x}{x^2 - \pi^2 k^2}$$

$$= \frac{1}{x} + \sum_{k=1}^{\infty} \frac{1}{x - \pi k} + \frac{1}{x + \pi k},$$

where to keep the sum convergent we sum the terms k and $-k$ together. Expanding $1/(x - \pi k)$ in power series we get

$$\cot x = \frac{1}{x} + \sum_{k=1}^{\infty} \frac{1}{\pi k} \sum_{j=0}^{\infty} \frac{(-1)^j x^j}{\pi^j k^j} - \sum_{k=-\infty}^{-1} \frac{1}{\pi k} \sum_{j=0}^{\infty} \frac{x^j}{\pi^j k^j}.$$

Now observe that the terms with j even cancel out and we are lead to

$$\cot x = \frac{1}{x} - \sum_{k=1}^{\infty} \sum_{r=1}^{\infty} \frac{2x^{2r-1}}{\pi^{2r} k^{2r}}$$

$$= \frac{1}{x} - 2 \sum_{r=1}^{\infty} \frac{x^{2r-1}}{\pi^{2r}} \sum_{k=1}^{\infty} \frac{1}{k^{2r}}$$

and this is (6.9.7). □

Corollary 6.9.1. *The Bernoulli numbers are given by*

$$B_{2n} = (-1)^{n-1} \frac{\zeta(2n)}{\pi^{2n}} \times \frac{(2n)!}{2^{2n-1}}. \tag{6.9.8}$$

In particular, $\zeta(2n)$ is a rational multiple of π^{2n}.

Exercise 6.9.3. Integrate the expansion of $\cot x$ to derive

$$\ln \sin x = \ln x - \sum_{n=1}^{\infty} \frac{\zeta(2n)}{n \pi^{2n}} x^{2n}. \tag{6.9.9}$$

Confirm that the special case $x = \pi/2$ yields

$$\sum_{n=1}^{\infty} \frac{\zeta(2n)}{n \, 2^{2n}} = \ln\left(\frac{\pi}{2}\right).$$

Exercise 6.9.4. Prove the identity

$$\frac{1}{\sin^2 x} = \sum_{n=-\infty}^{\infty} \frac{1}{(x + \pi n)^2} \qquad (6.9.10)$$

Hint. Compute $(\ln \sin x)''$. See 10.7.2 for a related calculation.

Exercise 6.9.5. Check the expansion

$$\tan x = \sum_{n=1}^{\infty} \frac{2^{2n}(2^{2n} - 1)}{(2n)!} |B_{2n}| x^{2n-1}. \qquad (6.9.11)$$

Hint. Use $\tan x = \cot x - 2\cot 2x$. This appears in [G & R] 1.411.5.

Exercise 6.9.6. Check the expansion

$$\operatorname{cosec} x = \frac{1}{x} + \sum_{n=1}^{\infty} \frac{2}{(2n)!} |B_{2n}| x^{2n-1}. \qquad (6.9.12)$$

Hint. Use $\operatorname{cosec} x = \cot x + \tan(x/2)$. This appears in [G & R] 1.411.11.

Project 6.9.1. The expansion of $\sec x$ requires the introduction of a new class of numbers.
a) Prove that the Taylor series for secant has the form

$$\sec x = \sum_{n=0}^{\infty} \frac{(-1)^n E_{2n}}{(2n)!} x^{2n}, \qquad (6.9.13)$$

where the **Euler numbers** E_{2n} are defined recursively by

$$E_0 = 1 \quad \text{and} \quad E_{2j} := -\sum_{k=0}^{j-1} \binom{2j}{2k} E_{2k}. \qquad (6.9.14)$$

Hint. Use the identity $\cos x \times \sec x = 1$.
b) Check that the Euler numbers E_{2n} are integers and show that $E_0 = 1$, $E_2 = -1$, $E_4 = 5$, $E_6 = -61$, $E_8 = 1385$.
c) Use the identity $\frac{d}{dx} \tan x = \sec^2 x$ to obtain

$$B_{2n} = \frac{(-1)^{n-1} n}{2^{2n-1}(2^{2n} - 1)} \sum_{r=0}^{n-1} \binom{2n - 2}{2r} E_{2r} E_{2n-2r-2}. \qquad (6.9.15)$$

d) What does one get from $\frac{d}{dx} \sec x = \tan x \sec x$?

Note 6.9.1. The expansion of secant appears in [G & R] 1.411.9. The reader
will find in Atkinson (1986) a description of the expansion of sec x and tan x.

6.10. The Case of a General Quadratic Denominator

The reader is familiar with the decomposition

$$R(x) = R_e(x) + R_o(x) \tag{6.10.1}$$

where

$$R_e(x) = \frac{R(x) + R(-x)}{2} \quad \text{and} \quad R_o(x) = \frac{R(x) - R(-x)}{2} \tag{6.10.2}$$

are the even and odd parts of R respectively. In this section we describe the
relation of this decomposition and the integration of rational functions.

Let R be a rational function. Then (6.10.1) yields

$$\int_0^\infty R(x)\,dx = \int_0^\infty R_e(x)\,dx + \int_0^\infty R_o(x)\,dx, \tag{6.10.3}$$

where we assume that all the integrals are finite. In the integral of the odd
part, let $x = \sqrt{t}$ to obtain

$$\int_0^\infty R_o(x)\,dx = \frac{1}{2}\int_0^\infty \frac{R(\sqrt{t}) - R(-\sqrt{t})}{2\sqrt{t}}\,dt. \tag{6.10.4}$$

Exercise 6.10.1. Let R be a rational function. Prove that

$$\mathfrak{F}(R(x)) = \frac{R(\sqrt{x}) - R(-\sqrt{x})}{2\sqrt{x}} \tag{6.10.5}$$

is also rational.

Therefore (6.10.3) yields

$$\int_0^\infty R(x)\,dx = \int_0^\infty R_e(x)\,dx + \frac{1}{2}\int_0^\infty \mathfrak{F}(R(x))\,dx, \tag{6.10.6}$$

and the question of integration of rational functions is reduced to the integration of *even* functions and the study of the map $\mathfrak{F}$.

Extra 6.10.1. The map $\mathfrak{F}$ has many interesting properties. We have studied
in Boros et al. (2003, 2004) the action of $\mathfrak{F}$ on rational functions of the form

$R(x) = P(x)/Q_n(x)$, where P is a polynomial and

$$Q_n(x) = \prod_{k=1}^{n}(x^{m_k} - 1) \qquad (6.10.7)$$

where $n \in \mathbb{N}$ and m_k are odd positive integers. This choice of denominator was motivated by the fact that if R has a pole at x_0 then $\mathfrak{F}(R)$ has a pole at x_0^2. Therefore the existence of a pole of modulus different than 1 leads to growth of the coefficients of $\mathfrak{F}^{(n)}(R)$. The reader can verify that the rational function $\mathfrak{F}(R)$ has the same denominator as R.

In the case $n = 1$ we have

$$\mathfrak{F}\left(\frac{x^j}{x^m - 1}\right) = \frac{x^{\gamma_m(j)}}{x^m - 1} \qquad (6.10.8)$$

with

$$\gamma_m(j) = m \left\lfloor \frac{j}{2} \right\rfloor - \frac{(m-1)(j-1)}{2}. \qquad (6.10.9)$$

The study of the iterates of $\mathfrak{F}$ is therefore reduced to that of $\gamma_m : \mathbb{Z} \to \mathbb{Z}$. The iterates of γ_m reach the set $\mathfrak{A}_m = \{0, 1, 2, \ldots, m-2\}$ in a finite a number of steps and leave this set invariant. The behavior of γ_m inside $\mathfrak{A}_m$ is determined by arithmetical properies of m. For instance, for m prime, there is a single orbit precisely when 2 is a **primitive root** modulo m, that is, the numbers $\{2^j : 0 \le j \le m - 1\}$ reduced modulo m are all distinct. Artin (1964) conjectured that 2 is a primitive root for infinitely many primes. See Murty (1988) for an update on this conjecture. The primes $m \le 100$ for which 2 is a primitive root are

$$\{3, 5, 11, 13, 19, 29, 37, 53, 59, 61, 67, 83\}.$$

In the case of $n > 1$ the iterates of R, appropriately normalized, converge to a limit: we state the result only in the case when the integers $m_1, \ldots, m_n$ are relatively prime, the general case appears in Boros et al. (2003). Let $L = 1/(m_1 \cdots m_n)$, then

$$\lim_{j \to \infty} \frac{\mathfrak{F}^{(j)}(R(x))}{2^{(n-1)j}} = \frac{L A_{n-1}(x)}{(1-x)^n (n-1)!} \qquad (6.10.10)$$

where $A_{n-1}(x)$ is the Eulerian polynomial defined in (4.1.13).

In Chapter 2, Project 2.3.1 we have analyzed the form of the integral

$$L_m(a) = \int_0^\infty \frac{dx}{(x^2 + 2ax + 1)^{m+1}}. \qquad (6.10.11)$$

The even part of the integrand can be written as a linear combination of integrals of the form

$$\int_0^\infty \frac{x^{2j}\,dx}{(x^4 + 2(1 - 2a^2)x^2 + 1)^{m+1}}$$

that will be studied in the next chapter. The map $\mathfrak{F}$ yields linear combinations of integrals of the form

$$\int_0^\infty \frac{t^n\,dt}{(t^2 + 2(1 - 2a^2)t + 1)^{m+1}}.$$

Exercise 6.10.2. Check that it suffices to consider the case $n = 0$. **Hint.** Divide t^n by the quadratic denominator.

Project 6.10.1. The process described above and the results of Chapter 7 give a relation between the integrals $L_m(a)$ and $L_m(1 - 2a^2)$. Describe it.

6.11. Combinations of Trigonometric Functions and Polynomials

This section contains the analog of Section 5.10 for the classes of functions

$$\mathfrak{S} = \left\{ f(x) = \sum_{i=0}^n \sum_{j=0}^m s_{i,j} x^i \sin^j x : s_{i,j} \in \mathbb{R},\ n, m \in \mathbb{N} \right\} \quad (6.11.1)$$

and

$$\mathfrak{C} = \left\{ f(x) = \sum_{i=0}^n \sum_{j=0}^m c_{i,j} x^i \cos^j x : c_{i,j} \in \mathbb{R},\ n, m \in \mathbb{N} \right\}. \quad (6.11.2)$$

The conclusion of the exercise is that $\mathfrak{S}$ and $\mathfrak{C}$ are closed under the formation of primitives.

Exercise 6.11.1. Let $m,\ n \in \mathbb{N}$. Prove the recurrences

$$\int x^m \sin^n x\,dx = \frac{x^{m-1} \sin^{n-1} x}{n^2} (m \sin x - nx \cos x)$$

$$+ \frac{n-1}{n} \int x^m \sin^{n-2} x\,dx - \frac{m(m-1)}{n^2} \int x^{m-2} \sin^n x\,dx$$

and

$$\int x^m \cos^n x \, dx = \frac{x^{m-1}\cos^{n-1}x}{n^2}(m\cos x + nx\sin x)$$
$$+ \frac{n-1}{n}\int x^m \cos^{n-2} x \, dx - \frac{m(m-1)}{n^2}\int x^{m-2}\cos^n x \, dx.$$

These expressions appear in [G & R] 2.631.2 and 2.631.3 respectively.

The integration of the product of a trigonometric function and a rational one requires the **sine integral**

$$\text{si}(x) = -\int_x^\infty \frac{\sin t}{t} dt \qquad (6.11.3)$$

and the **cosine integral**

$$\text{ci}(x) = -\int_x^\infty \frac{\cos t}{t} dt. \qquad (6.11.4)$$

For instance

$$\int_0^\infty \frac{\sin mx}{x+b} dx = \left(\frac{\pi}{2} - \text{si}(bm)\right)\cos(bm) + \text{ci}(bm)\sin(bm)$$

and

$$\int_0^\infty \frac{\sin(mx)\,dx}{ax^2+bx+c} = \frac{1}{2y}(2\text{ci}(r_1)\sin(r_1) - 2\text{ci}(r_2)\sin(r_2)$$
$$+ \cos(r_1)(\pi - 2\text{si}(r_1)) - \cos(r_2)(\pi - 2\text{si}(r_2)),$$

where

$$y = \sqrt{b^2 - 4ac}, \ r_1 = \frac{(b-y)m}{2a} \text{ and } r_2 = \frac{(b+y)m}{2a}.$$

7

A Quartic Integral

7.1. Introduction

Chapter 3 described the evaluation of

$$\int_0^\infty \frac{P(x)\,dx}{(q_1 x + q_0)^{m+1}} \qquad (7.1.1)$$

and Chapter 6 continued the program of evaluating integrals of rational functions with a discussion of

$$\int_0^\infty \frac{P(x)\,dx}{(q_2 x^2 + q_1 x + q_0)^{m+1}}. \qquad (7.1.2)$$

The natural next step of

$$\int_0^\infty \frac{P(x)\,dx}{(q_3 x^3 + q_2 x^2 + q_1 x + q_0)^{m+1}} \qquad (7.1.3)$$

is left for to the reader to explore. In this chapter we consider the evaluation of the quartic integral

$$\int_0^\infty \frac{P(x)\,dx}{(q_4 x^4 + q_2 x^2 + q_0)^{m+1}} \qquad (7.1.4)$$

where P is a polynomial and $m \in \mathbb{N}$. The convergence of the integral requires the degree of P to be at most $4m + 2$. Dividing P by the denominator $q_4 x^4 + q_2 x^2 + q_0$ expresses (7.1.4) as a sum of integrals of the same type in which the numerator is only of degree 3. The identity

$$\int_0^\infty \frac{p_3 x^3 + p_1 x}{(q_4 x^4 + q_2 x^2 + q_0)^{m+1}}\,dx = \frac{1}{2}\int_0^\infty \frac{p_3 u + p_1}{(q_4 u^2 + q_2 u + q_0)^{m+1}}\,du \quad (7.1.5)$$

shows that the odd part of the polynomial P yields the elementary integral considered in Chapter 6. For example, the evaluation of

$$I = \int_0^\infty \frac{x^8 + 3x^3 + 1}{(x^4 + 4x^2 + 1)^5}\,dx \qquad (7.1.6)$$

137

uses

$$x^8 + 3x^3 + 1 = (x^4 - 4x^2 + 15) \times (x^4 + 4x^2 + 1) + (3x^3 - 56x^2 - 14)$$

and

$$x^4 - 4x^2 + 15 = 1 \times (x^4 + 4x^2 + 1) + (14 - 8x^2)$$

to obtain

$$I = \int_0^\infty \frac{3x^3 - 56x^2 - 14}{(x^4 + 4x^2 + 1)^5} dx + \int_0^\infty \frac{-8x^2 + 14}{(x^4 + 4x^2 + 1)^4} dx$$
$$+ \int_0^\infty \frac{dx}{(x^4 + 4x^2 + 1)^3}.$$

Exercise 7.1.1. Use the method of partial fractions to confirm the value

$$\int_0^\infty \frac{3x^3 \, dx}{(x^4 + 4x^2 + 1)^5} = \frac{5\left(60 + 7\sqrt{3} \ln(7 - 4\sqrt{3})\right)}{20736}.$$

Mathematica 7.1.1. The Mathematica commands

$$\texttt{PolynomialQuotient[p,q,x]}$$

and

$$\texttt{PolynomialRemainder[p,q,x]}$$

give the quotient and remainder of the divison of the polynomials p and q, respectively.

Therefore the problem is reduced to the study of the integrals

$$N_{0,4}(q_4, q_2, q_0; m) := \int_0^\infty \frac{dx}{(q_4 x^4 + q_2 x^2 + q_0)^{m+1}} \qquad (7.1.7)$$

and

$$N_{1,4}(q_4, q_2, q_0; m) := \int_0^\infty \frac{x^2 \, dx}{(q_4 x^4 + q_2 x^2 + q_0)^{m+1}} \qquad (7.1.8)$$

The next exercise shows the normalization of both families (7.1.7) and (7.1.8) in terms of the integrals

$$N_{0,4}(a; m) = \int_0^\infty \frac{dx}{(x^4 + 2ax^2 + 1)^{m+1}} \qquad (7.1.9)$$

and

$$N_{1,4}(a;m) = \int_0^\infty \frac{x^2\,dx}{(x^4 + 2ax^2 + 1)^{m+1}}. \qquad (7.1.10)$$

Exercise 7.1.2. Check the identities

$$N_{0,4}(q_4, q_2, q_0; m) := \frac{1}{q_0^{m+3/4} q_4^{1/4}} N_{0,4}\left(\frac{q_2}{2\sqrt{q_0 q_4}}; m\right)$$

and

$$N_{1,4}(q_4, q_2, q_0; m) := \frac{1}{q_0^{m+1/4} q_4^{3/4}} N_{1,4}\left(\frac{q_2}{2\sqrt{q_0 q_4}}; m\right).$$

Exercise 7.1.3. Prove that

$$N_{0,4}(1;m) = \frac{\pi}{2^{4m+3}}\binom{4m+2}{2m+1}. \qquad (7.1.11)$$

In the next section we present an explicit formula for the quartic integral $N_{0,4}(a;m)$; the corresponding expressions for $N_{1,4}(a;m)$ are obtained in similar form.

7.2. Reduction to a Polynomial

In this section we establish a preliminary closed form evaluation for a quartic integral. More refined versions appear in Section 7.9.

Theorem 7.2.1. *Let*

$$N_{0,4}(a;m) = \int_0^\infty \frac{dx}{(x^4 + 2ax^2 + 1)^{m+1}} \qquad (7.2.1)$$

and define

$$P_m(a) = \frac{1}{\pi} 2^{m+3/2}(a + 1)^{m+1/2} N_{0,4}(a;m). \qquad (7.2.2)$$

Then $P_m(a)$ is a polynomial in a of degree m with rational coefficients given *by*

$$P_m(a) = \sum_{j=0}^{m} \binom{2m+1}{2j}(a+1)^j \sum_{k=0}^{m-j} \binom{m-j}{k}$$

$$\times \binom{2(m-k)}{m-k} 2^{-3(m-k)}(a-1)^{m-k-j}. \tag{7.2.3}$$

Proof. In order to evaluate $N_{0,4}(a;m)$ for a nonnegative integer m, we start with the change of variable $x = \tan\theta$, yielding

$$N_{0,4}(a;m) = \int_0^{\pi/2} \left(\frac{\cos^4\theta}{\sin^4\theta + 2a\sin^2\theta\cos^2\theta + \cos^4\theta}\right)^{m+1} \times \frac{d\theta}{\cos^2\theta}.$$

After the substitution $u = 2\theta$, the integral becomes

$$N_{0,4}(a;m) = 2^{-(m+1)} \int_0^{\pi} \left(\frac{(1+\cos u)^2}{(1+a)+(1-a)\cos^2 u}\right)^{m+1} \times \frac{du}{1+\cos u}.$$

The substitution $y = 1/x$ in (7.2.1) produces a different expression for $N_{0,4}(a;m)$:

$$N_{0,4}(a;m) = \int_0^{\infty} \left(\frac{y^4}{y^4 + 2ay^2 + 1}\right)^{m+1} \frac{dy}{y^2},$$

and, as before, the substitutions $y = \tan\theta$, $u = 2\theta$ produce

$$N_{0,4}(a;m) = 2^{-(m+1)} \int_0^{\pi} \left(\frac{(1-\cos u)^2}{(1+a)+(1-a)\cos^2 u}\right)^{m+1} \times \frac{du}{1-\cos u}.$$

These two expressions are averaged in order to obtain an integral representation for $N_{0,4}(a;m)$ that contains only even powers of $\cos u$. Thus

$$N_{0,4}(a;m) = 2^{-m-2} \int_0^{\pi} \frac{(1-\cos u)^{2m+1} + (1+\cos u)^{2m+1}}{((1+a)+(1-a)\cos^2 u)^{m+1}} du,$$

where the integrand is a function of $\cos^2 u$. Indeed, expanding the powers by the binomial theorem yields

$$N_{0,4}(a;m) = 2^{-(m+1)} \sum_{j=0}^{m} \binom{2m+1}{2j}$$

$$\times \int_0^{\pi} ((1+a)+(1-a)\cos^2 u)^{-(m+1)} \cos^{2j} u \, du. \tag{7.2.4}$$

We now compute the integral appearing in (7.2.4). Let

$$I_m^j(a) = \int_0^\pi \left((1+a) + (1-a)\cos^2 u \right)^{-(m+1)} \cos^{2j} u \; du.$$

Then

$$I_m^j(a) = 2^{m-j} \int_0^{2\pi} \left((3+a) + (1-a)\cos v \right)^{-(m+1)} (1+\cos v)^j \; dv$$

$$= 2^{m-j+1} \int_0^\pi \left((3+a) + (1-a)\cos v \right)^{-(m+1)} (1+\cos v)^j \; dv,$$

where we have made the substitution $v = 2u$ and used the symmetry of the cosine in the last step.

For each fixed value of the index j, the integrand is a rational function of $\cos v$, so the substitution $z = \tan(v/2)$ is a natural one. It yields

$$I_m^j(a) = 2 \int_0^\infty \left[2 + (1+a)z^2 \right]^{-(m+1)} \times (1+z^2)^{m-j} dz$$

$$= 2(1+a)^{-(m+1)} \int_0^\infty \left(z^2 + \frac{2}{1+a} \right)^{-(m+1)}$$

$$\times \left(z^2 + \frac{2}{1+a} + \frac{a-1}{a+1} \right)^{m-j} dz$$

$$= 2(1+a)^{-(m+1)} \int_0^\infty \sum_{k=0}^{m-j} \binom{m-j}{k} \left(z^2 + \frac{2}{1+a} \right)^{-m-1+k}$$

$$\times \left(\frac{a-1}{a+1} \right)^{m-j-k} dz.$$

Finally let $z = \sqrt{2/(1+a)} \tan \varphi$ in order to scale the last integral. Using Wallis' formula (6.4.5), we obtain

$$I_m^j(a) = \pi \times 2^{-1/2-3m}(1+a)^{-m-1/2}$$

$$\times \sum_{k=0}^{m-j} \binom{2(m-k)}{m-k} \binom{m-j}{k} 2^{3k}(a+1)^j (a-1)^{m-j-k}. \quad (7.2.5)$$

Thus

$$N_{0,4}(a;m) = \pi \times \sum_{j=0}^{m} \binom{2m+1}{2j} (a+1)^{-m-1/2+j}$$

$$\times \sum_{k=0}^{m-j} \binom{m-j}{k} \binom{2(m-k)}{m-k} 2^{3k-4m-3/2}(a-1)^{m-j-k}. \quad (7.2.6)$$

□

Note 7.2.1. Gradshteyn and Ryzhik's table of integrals (1994) [G & R] contains Wallis' formula as 3.249.1. We were surprised not to find (7.2.6).

Corollary 7.2.1. *Let $a \in \mathbb{Q}$. Then*

$$\frac{1}{\pi\sqrt{2(1+a)}} \int_0^\infty \frac{dx}{(x^4 + 2ax^2 + 1)^{m+1}} \in \mathbb{Q}. \qquad (7.2.7)$$

Proof. The expression in (7.2.7) is $P_m(a)$ divided by $2^{m+2}(1+a)^{m+1}$. □

The expression given for $N_{0,4}(a; m)$ allows the explicit evaluation of this integral for a given value of the parameter a. This is efficient only for small values of m.

Exercise 7.2.1. Check that

$$P_0(a) = 1$$
$$P_1(a) = \tfrac{1}{2}(2a + 3)$$
$$P_2(a) = \tfrac{3}{8}(4a^2 + 10a + 7)$$
$$P_3(a) = \tfrac{1}{16}(40a^3 + 140a^2 + 172a + 77).$$

Exercise 7.2.2. Show that

$$N_{0,4}(a, 0) = \int_0^\infty \frac{dx}{(x^4 + 2ax^2 + 1)} = \frac{\pi}{2^{3/2}(a+1)^{1/2}} \times 1, \qquad (7.2.8)$$

$$N_{0,4}(a; 1) = \int_0^\infty \frac{dx}{(x^4 + 2ax^2 + 1)^2} = \frac{\pi}{2^{7/2}(a+1)^{3/2}} \times (2a+3). \qquad (7.2.9)$$

Exercise 7.2.3. Check the values

$$\int_0^\infty \frac{x^2\, dx}{x^4 + (2a-1)x^2 + 1} = \frac{\pi}{2\sqrt{2a+1}}$$
$$\int_0^\infty \frac{dx}{bx^4 + 2ax^2 + 1} = \frac{\pi}{2\sqrt{2}}\frac{1}{\sqrt{a+\sqrt{b}}} \qquad (7.2.10)$$

and

$$\int_0^\infty \frac{x^2\, dx}{bx^4 + 2ax^2 + 1} = \frac{\pi}{2\sqrt{2b}}\frac{1}{\sqrt{a+\sqrt{b}}}.$$

Hint. Use $x \to 1/x$ in the first integral to reduce it to an $N_{0,4}$ case.

7.3. A Triple Sum for the Coefficients

In this section we discuss the first formula for the coefficients $d_l(m)$ in

$$P_m(a) = \sum_{l=0}^{m} d_l(m)a^l. \tag{7.3.1}$$

The formula developed here is a consequence of the elementary evaluation of $N_{0,4}(a;m)$. The expression in Theorem 7.3.1 can be used to evaluate $d_l(m)$ efficiently if l is small compared to m. We illustrate this with the calculation of $d_0(m)$ and $d_1(m)$.

Theorem 7.3.1. *The coefficients $d_l(m)$ are given by*

$$d_l(m) = \sum_{j=0}^{l} \sum_{s=0}^{m-j} \sum_{k=s+l}^{m} \frac{(-1)^{k-l-s}}{2^{3k}} \binom{2k}{k} \binom{2m+1}{2(s+j)} \binom{m-s-j}{m-k}$$

$$\times \binom{s+j}{j} \binom{k-s-j}{l-j}.$$

Proof. We start by reversing the order of summation in the expression for $P_m(a)$ in Theorem 7.2.1 and replacing k by $m-k$ to write

$$P_m(a) = \sum_{k=0}^{m} \binom{2k}{k} 2^{-3k} \sum_{v=0}^{k} \binom{2m+1}{2v} \binom{m-v}{m-k} (a+1)^v (a-1)^{k-v}.$$

We now expand the terms $(a+1)^v$ and $(a-1)^{k-v}$, giving

$$P_m(a) = \sum_{k=0}^{m} \sum_{v=0}^{k} \sum_{j=0}^{v} \sum_{r=0}^{k-v} 2^{-3k} \binom{2k}{k} \binom{2m+1}{2v} \binom{m-v}{m-k}$$

$$\times \binom{v}{j} \binom{k-v}{r} (-1)^{k-v-r} a^{j+r}$$

$$= \sum_{k=0}^{m} \sum_{v=0}^{m} \sum_{j=0}^{m} \sum_{r=0}^{m} 2^{-3k} \binom{2k}{k} \binom{2m+1}{2v} \binom{m-v}{m-k}$$

$$\times \binom{v}{j} \binom{k-v}{r} (-1)^{k-v-r} a^{j+r},$$

where we have extended all the sums to m, the added terms vanishing. Now replace r by $l - j$ to obtain

$$P_m(a) = \sum_{l=0}^{m}\sum_{k=0}^{m}\sum_{v=0}^{m}\sum_{j=0}^{m} 2^{-3k} \binom{2k}{k}\binom{2m+1}{2v}\binom{m-v}{m-k}$$

$$\times \binom{v}{j}\binom{k-v}{l-j}(-1)^{k-v-l+j}a^l.$$

We will now rewrite the sums over the ranges where the coefficients are non-zero. First consider the coefficient $\binom{v}{j}$. Its presence restricts j to the range $0 \le j \le v$. The appearance of $\binom{k-v}{l-j}$ then yields $0 \le l - j \le k - v$, so that $0 \le k - v$. From $\binom{2m+1}{2v}$ we obtain the restriction $0 \le v \le m$, and from $\binom{m-v}{m-k}$ we get $v \le m$, $k \le m$ and $v \le k$, so that $v \le k \le m$. Finally, $\binom{k-v}{l-j}$ leads to $k \ge v$, $l \ge j$ and $k - v \ge l - j$.

Now that we have derived the new ranges for the indices we proceed to perform the inversion. First choose l in the range $0 \le l \le m$. Then choose j so that $0 \le j \le l$. Next we pick v in the range $j \le v \le m$, and finally k is chosen so that $l - j + v \le k \le m$. This completes the change of variable. $\square$

Exercise 7.3.1. Check the evaluations of the constant and linear term:

$$d_0(m) = \sum_{s=0}^{m}\sum_{k=s}^{m} \frac{(-1)^{k-s}}{2^{3k}}\binom{2k}{k}\binom{2m+1}{2s}\binom{m-s}{m-k} \qquad (7.3.2)$$

and

$$d_1(m) = \sum_{s=0}^{m-1}\sum_{k=s+1}^{m} \frac{(-1)^{k-s-1}}{2^{3k}}\binom{2k}{k} \times (m-s)\binom{2m+2}{2s+1}\binom{m-s-1}{m-k}. \qquad (7.3.3)$$

Use WZ theory described in the Appendix to obtain recurrences for these sums.

Exercise 7.3.2. Use Mathematica to create a list of the coefficients of $P_m(a)$ and observe that, in spite of the alternating sign appearing in the expression for $d_l(m)$ given in (7.3.1), these coefficients seem to be positive. A proof of this result appears in Section 7.9, Corollary 7.9.1.

7.4. The Quartic Denominators: A Crude Bound

The expression in Theorem 7.3.1 shows that $d_l(m)$ is a rational number whose denominator is a power of 2. In this section we establish a bound for the 2-adic valuation of the coefficients $d_l(m)$. See Section 1.2 for the notation.

Exercise 7.4.1. Prove that the 2-adic valuation of $d_l(m)$ satisfies the lower bound $\mu_2(d_l(m)) \geq -3m$.

The worst possible case for the value of $\mu_2(d_l(m))$ appears from the term $k = m$ in the sum in Theorem 7.3.1. But this coefficient is multiplied by the central binomial coefficient C_m. Using

$$\mu_2(C_m) = m - \mu_2(m!)$$

we obtain an improvement of the bound given in Exercise 7.4.1.

Proposition 7.4.1. *The 2-adic valuation of $d_l(m)$ satisfies*

$$\mu_2(d_l(m)) \geq D := -2m - \sum_{i=1}^{\infty} \left\lfloor \frac{m}{2^i} \right\rfloor. \qquad (7.4.1)$$

Extra 7.4.1. Optimal bounds for

$$\mu_p(n!) = \sum_{k=1}^{\infty} \left\lfloor \frac{n}{p^k} \right\rfloor \qquad (7.4.2)$$

have been given by Berndt and Bhargava (1993), page 593, in an expository paper about Ramanujan's work. They have established that

$$\frac{n}{p-1} - \frac{\ln(n+1)}{\ln p} \leq \mu_p(n!) \leq \frac{n-1}{p-1}. \qquad (7.4.3)$$

Exercise 7.4.2. Check that the bound D in (7.4.1) satisfies

$$3m - \frac{\ln(m+1)}{\ln 2} \leq -D \leq 3m - 1.$$

In particular, $\lim\limits_{n \to \infty} -D/m = 3$.

We conclude that the contribution of the central binomial coefficient to the value of $\mu_2(d_l(m))$ is asymptotically negligible as $m \to \infty$. In Corollary 7.9.1 we establish that $\mu_2(d_l(m)) = 2m - 1$. This requires a new method.

7.5. Closed-Form Expressions for $d_l(m)$.

It is now possible to evaluate explicitly an expression for the first few leading coefficients of $P_m(a)$. These evaluations require the value of the binomial sums discussed in Section 1.4.

A Quartic Integral

Proposition 7.5.1. *The leading coefficient $d_m(m)$ is given by*

$$d_m = 2^{-m}\binom{2m}{m}. \tag{7.5.1}$$

The next term is

$$d_{m-1}(m) = (2m+1)2^{-(m+1)}\binom{2m}{m}. \tag{7.5.2}$$

Proof. The sum in Theorem 7.3.1 yields

$$d_m(m) = 2^{-3m}\binom{2m}{m}\sum_{l=0}^{m}\binom{2m+1}{2l}$$

$$= 2^{-m}\binom{2m}{m}.$$

This gives (7.5.1). To establish (7.5.2), the corresponding sum is

$$\sum_{l=0}^{m-1}\sum_{v=l}^{l+1}\sum_{k=m-1-l+v}^{m}2^{-3k}\binom{2k}{k}\binom{2m+1}{2v}\binom{m-v}{m-k}$$

$$\times\binom{v}{l}\binom{k-v}{m-1-l}(-1)^{k-v-m+l+1}.$$

The inner sum in v contains only two terms and a simple calculation produces

$$d_{m-1}(m) = \left\{2^{-3(m-1)}\binom{2m-2}{m-1} - 2^{-3m}\binom{2m}{m}\right\}\sum_{l=0}^{m-1}(m-l)\binom{2m+1}{2l}$$

$$+ 2^{-3m}\binom{2m}{m}\sum_{l=1}^{m}l\binom{2m+1}{2l}$$

$$= (2m+1)2^{-(m+1)}\binom{2m}{m}. \qquad \square$$

Exercise 7.5.1. Compute the next two terms. The results are

$$d_{m-2}(m) = \frac{m-1}{2m-1}2^{-(m+2)}(4m^2+2m+1)\binom{2m}{m}$$

and

$$d_{m-3}(m) = \frac{(m-2)}{3(2m-1)}2^{-(m+3)}(8m^3+4m+3)\binom{2m}{m}.$$

Hint. Use the results of Exercise 1.4.7 and

$$\sum_{k=0}^{n} k^3 \binom{2n+1}{2k} = (n+2)(2n+1)^2 2^{2n-5}. \qquad (7.5.3)$$

7.5.1. Scaling of Coefficients

Based on the structure of the first few coefficients described above we introduce the scaling

$$d_l(m) = \left\{ \frac{(m+l)!}{(m-l)!\, l!\, m!} 2^{-m} \right\} e_l(m). \qquad (7.5.4)$$

Project 7.5.1. Define

$$Q_l(m) = \frac{m!}{(m-l)!}\, {}_2F_1\left[\tfrac{1}{2}, -l, -m; 2\right]. \qquad (7.5.5)$$

Check that for l fixed and independent of m, the expression $Q_l(m)$ is a *polynomial of degree l* with *integer* coefficients. Confirm the values

$$
\begin{aligned}
Q_0(m) &= 1\\
Q_1(m) &= m + 1\\
Q_2(m) &= m^2 + m + 1\\
Q_3(m) &= m^3 + 2m + 3\\
Q_4(m) &= m^4 - 2m^3 + 5m^2 + 8m + 9\\
Q_5(m) &= m^5 - 5m^4 + 15m^3 + 5m^2 + 29m + 45 \qquad (7.5.6)
\end{aligned}
$$

Verify that $e_l(m)$ defined in (7.5.4) satisfies

$$e_l(m) = Q_{m-l}(2m) \qquad (7.5.7)$$

and prove that, for l odd, $Q_l(m)$ is divisible by $m + 1$.

7.6. A Recursion

In this section we prove a recursion for the integrals $N_{0,4}(a; m)$. The argument is based on Hermite's reduction procedure for the indefinite integration of rational functions. Bronstein (1997) contains a detailed description of these ideas.

Let $V(x) = x^4 + 2ax^2 + 1$. Then V and V' have no common factor so the Euclidean algorithm produces polynomials B and C such that

$$-\frac{1}{m} = CV + BV'. \qquad (7.6.1)$$

Indeed, a simple calculation yields

$$B(x) = -\frac{1}{4m}\frac{1}{a^2 - 1}\left((1 - 2a^2)x - ax^3\right) \quad \text{and}$$

$$C(x) = -\frac{1}{m}\left(1 + \frac{a}{a^2 - 1}x^2\right).$$

This is the answer to Exercise 4.3.1.

Divide (7.6.1) by V^{m+1} and integrate from 0 to ∞ to produce

$$N_{0,4}(a;m) = \left(1 + \frac{1 - 2a^2}{4m(a^2 - 1)}\right)N_{0,4}(a;m - 1)$$

$$+ \frac{(4m - 3)a}{4m(a^2 - 1)}N_{1,4}(a;m - 1).$$

This recursion can be also be written as

$$N_{0,4}(a;m) = \left(1 + \frac{1 - 2a^2}{4m(a^2 - 1)}\right)N_{0,4}(a;m - 1)$$

$$- \frac{(4m - 3)a}{8m(m - 1)(a^2 - 1)}\frac{d}{da}N_{0,4}(a;m - 2). \quad (7.6.2)$$

Proposition 7.6.1. *The polynomials* $P_m(a)$ *satisfy*

$$P_m(a) = \frac{(2m - 3)(4m - 3)a}{4m(m - 1)(a - 1)}P_{m-2}(a) - \frac{(4m - 3)a(a + 1)}{2m(m - 1)(a - 1)}\frac{d}{da}P_{m-2}(a)$$

$$+ \frac{4m(a^2 - 1) + 1 - 2a^2}{2m(a - 1)}P_{m-1}(a). \quad (7.6.3)$$

Proof. Use (7.2.2) in (7.6.2). □

Exercise 7.6.1. The goal of this exercise is to provide an evaluation of the integral

$$N_{0,4}(0;m) = \int_0^\infty \frac{dx}{(x^4 + 1)^{m+1}} \quad (7.6.4)$$

as

$$N_{0,4}(0;m) = \frac{\pi}{m!}2^{-2m-3/2} \times \prod_{l=1}^m (4l - 1). \quad (7.6.5)$$

a) Verify that the recursion (7.6.2) reduces to

$$N_{0,4}(0;m) = \frac{4m - 1}{4m}N_{0,4}(0;m - 1). \quad (7.6.6)$$

Therefore the proof of (7.6.5) reduces to the evaluation of

$$N_{0,4}(0;0) = \int_0^\infty \frac{dx}{x^4 + 1}. \tag{7.6.7}$$

b) Check the factorization

$$x^4 + 1 = (x^2 + \sqrt{2}x + 1)(x^2 - \sqrt{2}x + 1) \tag{7.6.8}$$

so (7.6.7) can be evaluated by partial fractions. The more general case can be done by the decomposition

$$\frac{1}{(x^2 + cx + 1)(x^2 - cx + 1)} = \frac{2x + c}{4c(x^2 + cx + 1)} + \frac{1}{4(x^2 + cx + 1)}$$
$$- \frac{2x - c}{4c(x^2 - cx + 1)} + \frac{1}{4(x^2 - cx + 1)}.$$

Check it and obtain from it the value of the integral

$$\int_0^\infty \frac{dx}{(x^2 + cx + 1)(x^2 - cx + 1)}.$$

c) The particular value $c = \sqrt{2}$ yields

$$\int_0^\infty \frac{dx}{x^4 + 1} = \frac{\pi}{2\sqrt{2}}. \tag{7.6.9}$$

Mathematica 7.6.1. The command

```
Factor[x^4 + 1]
```

returns the polynomial $x^4 + 1$ unfactored. To instruct Mathematica to factor using different types of algebraic numbers use

```
Factor[x^4 + 1, Extension -> Sqrt[2]]
```

The result is (7.6.8).

Exercise 7.6.2. Confirm the special values

$$P_m(1) = 2^{-2m} \binom{4m + 1}{2m} \quad \text{and} \quad P'_m(1) = \frac{m(m + 1)}{2m + 3} P_m(1) \tag{7.6.10}$$

and use them to show that the right-hand side of (7.6.3) is, in spite of its appearance, a polynomial in a. These special values will reappear in Exercise 10.5.3.

Project 7.6.1. a) Use the recurrence (7.6.3) to obtain a recurrence for the coefficients $d_l(m)$.

b) Obtain formulas for $d_0(m)$ and $d_1(m)$.

c) Produce a recurrence for the polynomial $Q_l(m)$ defined in Project 7.5.1.

7.7. The Taylor Expansion of the Double Square Root

We now present the results given by Boros and Moll (2001) to evaluate the coefficients of the Taylor expansion of $h(c) := \sqrt{a + \sqrt{1 + c}}$. The particular case $a = 1$ is a standard example often used to illustrate Lagrange's inversion formula. Berndt (1985), pages 71, 72 and 304–307, gives a complete history of this problem.

Lemma 7.7.1. *Let*

$$g(c) = \int_0^\infty \frac{dx}{x^4 + 2ax^2 + 1 + c}$$

and $h(c) = \sqrt{a + \sqrt{1 + c}}$*. Then* $g(c) = \pi\sqrt{2}h'(c)$*. In particular,*

$$h'(0) = \frac{1}{\pi\sqrt{2}} N_{0,4}(a; 0).$$

Proof. Write $g(c)$ as

$$g(c) = \frac{1}{1 + c} \int_0^\infty \frac{dx}{x^4/(1 + c) + (2a/(1 + c))x^2 + 1},$$

and now use Exercise 7.2.3 to evaluate $g(c)$. □

Note 7.7.1. Compare with part j) of Project 2.3.1.

Theorem 7.7.1. *The Taylor expansion of* $h(c) = \sqrt{a + \sqrt{1 + c}}$ *is given by*

$$\sqrt{a + \sqrt{1 + c}} = \sqrt{a + 1} + \frac{1}{\pi\sqrt{2}} \sum_{k=1}^\infty \frac{(-1)^{k-1}}{k} N_{0,4}(a; k - 1)c^k. \quad (7.7.1)$$

Proof. Evaluate $h^{(k)}(0)$ using Lemma 7.7.1. □

Corollary 7.7.1. *We have*

$$\sqrt{1 + \sqrt{1 + c}} = \sqrt{2}\left[1 + \sum_{k=1}^\infty \frac{(-1)^{k-1}}{k} \frac{1}{2^{4k}} \binom{4k - 2}{2k - 1} c^k\right].$$

Proof. Use the value $2^{4k-1}N_{0,4}(1;k-1) = \pi \binom{4k-2}{2k-1}$. $\qquad\qquad\qquad\square$

This appears in Bromwich (1926), page 192, exercise 21, and is a special case of [G & R] 1.114.1.

7.8. Ramanujan's Master Theorem and a New Class of Integrals

We establish a connection between $N_{0,4}(a;m)$ and a new family of integrals. This is used to establish a single sum formula for $P_m(a)$.

Theorem 7.8.1. *Define*

$$J_m(a) := \int_0^\infty \frac{x^{m-1}dx}{(a+\sqrt{1+x})^{2m+1/2}}. \qquad (7.8.1)$$

Then

$$J_m(a) = \frac{1}{\pi}2^{6m+3/2}\left[m\binom{4m}{2m}\binom{2m}{m}\right]^{-1} \times N_{0,4}(a;m). \qquad (7.8.2)$$

The proof of Theorem 7.8.1 is based on Ramanujan's Master Theorem stated below.

Theorem 7.8.2. *Suppose F has a Taylor expansion around $c = 0$ of the form*

$$F(c) = \sum_{m=0}^\infty \frac{(-1)^m}{m!}\varphi(m)c^m.$$

Then the moments of F, defined by

$$M_m = \int_0^\infty c^{m-1}F(c)\,dc, \qquad (7.8.3)$$

can be computed via

$$M_m = (m-1)!\varphi(-m). \qquad (7.8.4)$$

Berndt (1985) provides a proof and exact hypotheses for the validity of the Master Theorem. Observe that the expression (7.8.4) requires the ability to compute the function φ outside its original range, namely at negative indices.

Proof of Theorem 7.8.1. We apply the Master Theorem to the expansion in Theorem 7.7.1. Differentiate the integral $N_{0,4}(a;k-1)$ j times and replace

x by $1/x$ to produce

$$\left(\frac{d}{da}\right)^j N_{0,4}(a; k-1) = \frac{(-1)^j 2^j (k+j-1)!}{(k-1)!} \times \int_0^\infty \frac{x^{4k+2j-2} dx}{(x^4 + 2ax^2 + 1)^{k+j}}.$$

From (7.7.1) we obtain

$$\left(\frac{d}{da}\right)^j \sqrt{a + \sqrt{1+c}} = \left(\frac{d}{da}\right)^j \sqrt{a+1} + \sum_{k=1}^\infty \frac{(-1)^k}{k!} \varphi(k) c^k$$

with

$$\varphi(k) = (-1)^{j+1} \frac{1}{\pi\sqrt{2}} (k+j-1)! \, 2^j \times \int_0^\infty \frac{x^{4k+2j-2} dx}{(x^4 + 2ax^2 + 1)^{k+j}}.$$

Now replace k by $-m$ to produce

$$\varphi(-m) = (-1)^{j+1} \frac{1}{\pi\sqrt{2}} (-m+j-1)! \, 2^j \times \int_0^\infty \frac{x^{-4m+2j-2} dx}{(x^4 + 2ax^2 + 1)^{-m+j}}.$$

The choice $j = 2m + 1$ yields

$$\varphi(-m) = \frac{m! \, 2^{2m+1}}{\pi\sqrt{2}} N_{0,4}(a; m). \tag{7.8.5}$$

The moments of the function $H(c) := \left(\frac{d}{da}\right)^j \sqrt{a + \sqrt{1+c}}$ are computed directly as

$$M_k = \frac{(-1)^{j+1}(2j-3)!}{2^{2(j-1)} (j-2)!} \times \int_0^\infty \frac{c^{k-1} dc}{\left(a + \sqrt{1+c}\right)^{j-1/2}},$$

and the choices $j = 2m + 1$ and $k = m$ produce

$$M_m = \frac{(4m-1)!}{2^{4m} (2m-1)!} \times J_m(a).$$

Ramanujan's Master Theorem now yields (7.8.2).

Exercise 7.8.1. Check the details.

Note 7.8.1. The Ramanujan Master Theorem will be used in Chapter 10, Exercise 10.4.6 to give a new proof of a classical identity of Legendre.

7.9. A Simplified Expression for $P_m(a)$

In this section we evaluate the integrals $J_m(a)$ defined in the previous section. This will prove that the function

$$P_m(a) := \frac{1}{\pi} 2^{m+3/2}(a+1)^{m+1/2} N_{0,4}(a;m) \qquad (7.9.1)$$

is a polynomial in a. The expression for $P_m(a)$ will show that $P_m(a)$ is a Jacobi polynomial with parameters $m + \frac{1}{2}$, $-(m + \frac{1}{2})$. This classical family of polynomials is defined by

$$(1-x)^\alpha (1+x)^\beta P_n^{(\alpha,\beta)}(x) = \frac{(-1)^n}{2^n n!} \left(\frac{d}{dx}\right)^n \left[(1-x)^{n+\alpha}(1+x)^{n+\beta}\right].$$

Lemma 7.9.1. Let $f_m(u) = u(u^2 - 1)^{m-1}$. Then the integral $J_m(a)$ in (7.8.1) is given by

$$J_m(a) = \frac{2^{2m+1}(2m)!}{(4m)!} \sum_{j=0}^{m} 2^{2j} \frac{(2m-2j)!}{(m-j)!} f_m^{(j+m-1)}(1) \times (1+a)^{-(2m-2j+1)/2}.$$

Proof. The substitution $u = \sqrt{1+x}$ yields

$$J_m(a) = 2 \int_1^\infty f_m(u)(a+u)^{-(2m+1/2)} du. \qquad (7.9.2)$$

The result now follows by repeated integration by parts. The derivatives $f_m^{(j)}(u)$ vanish identically for $j \geq 2m$, and they also vanish at $u = 1$ for $0 \leq j \leq m - 2$. ☐

Lemma 7.9.2. *The polynomial $P_m(a)$ is given by*

$$P_m(a) = \frac{m}{2^{3m-1}(m!)^2} \sum_{k=0}^{m} 2^{2k} \frac{(2m-2k)!}{(m-k)!} f_m^{(k+m-1)}(1) \times (1+a)^k.$$

Proof. Substitute the formula in Proposition 7.9.1 into (7.8.2) and use (7.9.1). ☐

We now find a closed form for the derivatives of f_m at $u = 1$.

Proposition 7.9.1. *Let $0 \leq k \leq m$. Then*

$$f_m^{(k+m-1)}(1) = 2^{m-k-1} \frac{(m-1)!(m+k)!}{(m-k)! k!}. \qquad (7.9.3)$$

Proof. Expanding $f_m(u)$ and differentiating we have

$$f_m^{(k+m-1)}(1) = \sum_{j\geq 0}(-1)^j \binom{m-1}{j} \times \frac{(2m-2j-1)!}{(m-2j-k)!}.$$

It suffices to prove

$$b_{m,k} := \sum_{j\geq 0}(-1)^j \binom{m-1}{j}\binom{2m-2j-1}{k+m-1} = 2^{m-k-1}\binom{m}{k}\left(1+\frac{k}{m}\right),$$

(7.9.4)

which is equivalent to (7.9.3). Indeed:

$$\sum_k b_{m,k}x^{k+m-1} = \sum_k \left(\sum_{j\geq 0}(-1)^j\binom{m-1}{j}\binom{2m-2j-1}{k+m-1}\right)x^{k+m-1}$$

$$= \sum_{j\geq 0}(-1)^j\binom{m-1}{j}\sum_k\binom{2m-2j-1}{k+m-1}x^{k+m-1}$$

$$= \sum_{j\geq 0}(-1)^j\binom{m-1}{j}\sum_k\binom{2m-2j-1}{k}x^k$$

$$= \sum_{j\geq 0}(-1)^j\binom{m-1}{j}(x+1)^{2m-2j-1}$$

$$= (x+1)^{2m-1}\sum_{j\geq 0}(-1)^j\binom{m-1}{j}(x+1)^{-2j}$$

$$= (x+1)^{2m-1} \times \left[1-(x+1)^{-2}\right]^{m-1}$$

$$= x^{m-1}(x+2)^m - x^{m-1}(x+2)^{m-1}$$

$$= \sum_{k=0}^m\binom{m}{k}2^{m-k-1}\left(1+\frac{k}{m}\right)x^{k+m-1}.$$

Thus (7.9.4) holds and the proof is complete. □

Theorem 7.9.1. *The polynomial $P_m(a)$ is given by*

$$P_m(a) = 2^{-2m}\sum_{k=0}^m 2^k\binom{2m-2k}{m-k}\binom{m+k}{m}(a+1)^k.$$ (7.9.5)

Proof. This follows directly from Propositions 7.9.2 and 7.9.1. □

Exercise 7.9.1. Check the hypergeometric representation

$$P_m(a) = 2^{-2m} \binom{2m}{m} {}_2F_1\left(-m, m+1; \tfrac{1}{2} - m; \tfrac{1+a}{2}\right). \qquad (7.9.6)$$

Extra 7.9.1. The expression (7.9.5) fits the pattern of the explicit form of the Jacobi polynomials

$$P_m^{(\alpha,\beta)}(a) := \sum_{k=0}^{m} (-1)^{m-k} \binom{m+\beta}{m-k} \binom{m+k+\alpha+\beta}{k} \left(\frac{a+1}{2}\right)^k$$

with parameters $\alpha = m + \tfrac{1}{2}$ and $\beta = -(m + \tfrac{1}{2})$. Therefore

$$P_m(a) = P_m^{(m+1/2, -m-1/2)}(a). \qquad (7.9.7)$$

The reader should be very careful on the interpretation of this identity. There are several properties of the Jacobi polynomials $P_m^{(\alpha,\beta)}(a)$ that are established under the assumption that the parameters α, β are independent of m. For instance, the recursion

$$2(m+1)(m+\alpha+\beta+1)(2m+\alpha+\beta)P_{m+1}^{(\alpha,\beta)}(a) = \qquad (7.9.8)$$
$$(2m+\alpha+\beta+1)\left\{(2m+\alpha+\beta)a + \alpha^2 - \beta^2\right\}P_m^{(\alpha,\beta)}(a)$$
$$-2(m+\alpha)(m+\beta)(2m+\alpha+\beta+2)P_{m-1}^{(\alpha,\beta)}(a)$$

does not reduce to (7.6.3) after replacing $\alpha = m + 1/2$ and $\beta = -(m + 1/2)$.

Corollary 7.9.1. *The coefficients $d_l(m)$ are given by*

$$d_l(m) = 2^{-2m} \sum_{k=l}^{m} 2^k \binom{2m-2k}{m-k} \binom{m+k}{m} \binom{k}{l}. \qquad (7.9.9)$$

Extra 7.9.2. Let $a_l(m) = m - l + 2$, $b_l(m) = 8m^2 - 4l^2 + 24m + 19$ and $c_l(m) = (4m+5)(4m+3)(m+l+1)$. Define

$$d_l^*(m) := m! 2^m d_l(m). \qquad (7.9.10)$$

The WZ method described in the Appendix yields the recurrence

$$a_l(m)d_l^*(m+2) + b_l(m)d_l^*(m+1) + c_l(m)d_l^*(m) = 0. \qquad (7.9.11)$$

It would be interesting to explore properties of the coefficients $d_l(m)$ that can be obtained from this recursion.

Project 7.9.1. Reconsider the polynomials $Q_l(m)$ defined in Project 7.5.1. In particular, find an expression for its coefficients.

Exercise 7.9.2. Use the value of $N_{0,4}(0; m)$ to obtain the identity

$$\sum_{k=0}^{m} 2^k \binom{2m-2k}{m-k} \binom{m+k}{m} = \frac{2^m}{m!} \prod_{k=1}^{m} (4k-1). \qquad (7.9.12)$$

Conclude that the odd part of $m!$ divides the product $\prod_{k=1}^{m} (4k-1)$.

Exercise 7.9.3. Use the values of $P_m(1)$ obtained from (7.2.2) and (7.9.5) to prove that

$$\sum_{k=0}^{m} 2^{-2k} \binom{2k}{k} \binom{2m-k}{m} = \sum_{k=0}^{m} 2^{-2k} \binom{2k}{k} \binom{2m+1}{2k}. \qquad (7.9.13)$$

Use the WZ method described in the Appendix to check that both sides of (7.9.13) satisfy

$$(2m+3)(2m+2)f(m+1) = (4m+5)(4m+3)f(m) \qquad (7.9.14)$$

and that they agree at $m = 1$. This gives an automatic proof of (7.9.13).

Corollary 7.9.2. *The integral $N_{0,4}(a; m)$ is given by*

$$N_{0,4}(a; m) = \frac{\pi}{2^{3m+3/2} (1+a)^{m+1/2}} \sum_{k=0}^{m} 2^k \binom{2m-2k}{m-k} \binom{m+k}{m} (a+1)^k.$$

Corollary 7.9.3. *The integral $J_m(a)$ in (7.8.1) is given by*

$$J_m(a) = 2^{3m} \frac{m! \, (m-1)! \, (2m)!}{(4m)! \, (1+a)^{m+1/2}} \sum_{k=0}^{m} 2^k \binom{2m-2k}{m-k} \binom{m+k}{m} (a+1)^k.$$

Project 7.9.2. Give a direct proof of (7.9.13).

Extra 7.9.3. The coefficients $d_l(m)$ have many interesting properties that need to be explored. We present information about a couple of them.

7.9.1. Unimodality

A finite sequence of real numbers $\{c_j : 0 \le j \le m\}$ is said to be **unimodal** if there exists an index $0 \le j \le m$ such that c_i increases up to $i = j$ and decreases from then on, that is, $c_0 \leq c_1 \leq \cdots \leq c_j$ and $c_j \geq c_{j+1} \geq \cdots \geq c_m$. A polynomial is said to be unimodal if its sequence of coefficients is unimodal. Unimodal polynomials arise often in combinatorics, geometry and algebra, and have been the subject of considerable research in recent years. Techniques employed in the discussion of unimodality are surveyed by Brenti (1994) and Stanley (1989).

A stronger concept is that of **logconcavity**: a sequence of positive real numbers $\{c_0, c_1, \ldots, c_m\}$ is said to be *logarithmically concave* (or *logconcave* for short) if $c_{j+1}c_{j-1} \leq c_j^2$ for $1 \leq j \leq m - 1$.

The original motivation for this definition came from studying roots of polynomials. A sufficient condition for logconcavity of a polynomial is given by the location of its zeros: a polynomial, all of whose zeros are real and negative, is logconcave and therefore unimodal (Wilf, 1990). A second criteria for the logconcavity of a polynomial was determined by Brenti (1994). A sequence of real numbers is said to have *no internal zeros* if whenever a_i, $a_k \neq 0$ and $i < j < k$ then $a_j \neq 0$. Brenti's criteria state that if $P(x)$ is a logconcave polynomial with nonnegative coefficients and no internal zeros, then $P(x + 1)$ is logconcave. In this spirit we have produced an elementary proof of (Boros and Moll, 1999).

Theorem 7.9.2. *The polynomial* $P(x + 1)$ *is unimodal if the coefficients of* $P(x)$ *are positive and nondecreasing.*

It follows from here that the coefficients $d_l(m)$ are unimodal. It is an open problem to establish if they form a logconcave sequence. Much more is conjectured: introduce an operator on sequences by the rule

$$\mathfrak{L}\{a_i\} = \{a_i^2 - a_{i-1}a_{i+1}\} \tag{7.9.15}$$

with the understanding that if the sequence $\{a_i\}$ is finite, say from $i = 1$ to $i = n$, then we declare $a_i = 0$ for $i < 0$ and $i > n$. Thus $\mathfrak{L}$ maps logconcave sequences to positive ones. We say that $\{a_i\}$ is ∞-logconcave if $\mathfrak{L}^{(k)}\{a_i\}$ is always positive. We have conjectured that the sequence $\{d_l(m)\}$, which motivated all these ideas, has this property.

The binomial coefficients are the usual sequence on which properties related to unimodality are tested. The next project is a first step towards the ∞-logconcavity of $d_l(m)$.

Project 7.9.3. Prove that the binomial coefficients are ∞-logconcave.

The next exercise presents a connection between logconcavity and series expansions. The details were given by L. Carlitz as a response to a question proposed by D. Newman; see Newman and Carlitz (1959).

Exercise 7.9.4. Let $f(x) = c_0 + c_1 x + c_2 x^2 + \cdots$ be a power series with $c_n > 0$ and $c_{n+1} c_{n-1} > c_n^2$. Prove that the expansion of $1/f(x)$ has all negative coefficients d_n (except for the constant term). **Hint.** Check that

$$0 = c_n + \sum_{j=1}^{n} d_j c_{n-j},$$

$$d_{n+1} = -c_{n+1} - \sum_{j=1}^{n} d_j c_{n-j+1}.$$

Multiply the first equation by $-c_{n+1}$ and the second one by c_n and add.

Divisibility properties. The p-adic valuation of the coefficients $d_l(m)$ was studied by Boros et al. (2000) in the case $p = 2$. The value

$$v_2(d_0(m)) = -(m + v_2(m!)) \qquad (7.9.16)$$

is elementary. The result

$$v_2(d_1(m)) = 1 - 2m + v_2\left(\binom{m+1}{2}\right) + s_2(m) \qquad (7.9.17)$$

where $s_2(m)$ is the sum of the binary digits of m was established by Boros et al. (2000).

The problem for odd primes seems more difficult. Extensive symbolic calculations suggest the existence of a sequence of positive integers m_j such that $v_3(m_j) = 0$. These integers satisfy

$$m_{j+1} - m_j \in \{2, 7, 20, 61, 182, \dots\} \qquad (7.9.18)$$

where the sequence $\{q_j\}$ in (7.9.18) is defined by $q_1 = 2$ and $q_{j+1} = 3q_j + (-1)^{j+1}$.

Project 7.9.4. Discover formulas for $v_p(d_l(m))$.

7.10. The Elementary Evaluation of $N_{j,4}(a;m)$

The goal of this section is to provide elementary expressions for

$$N_{j,4}(a;m) = \int_0^\infty \frac{x^{2j}\,dx}{(x^4 + 2ax^2 + 1)^{m+1}}$$

based on the formula for the coefficients of $P_m(a)$. The values of $N_{j,4}(a;m)$ for $m+1 \le j \le 2m+1$ can be obtained by using the symmetry relation

$$N_{j,4}(a;m) = N_{2m+1-j,4}(a;m),$$

so we may restrict to the case $0 \le j \le m$.

Exercise 7.10.1. Establish the formula

$$\left(\frac{d}{da}\right)^r N_{0,4}(a;m) = (-1)^r 2^r \frac{(m+r)!}{m!} \int_0^\infty \frac{x^{2r i}\,dx}{(x^4 + 2ax^2 + 1)^{m+r+1}}$$

and use it to obtain an expression for $N_{j,4}(a;m)$ in terms of $N_{0,4}$.

Exercise 7.10.2. Use the finite sum expression for $N_{0,4}(a;m)$ to obtain a similar formula for the family of integrals $N_{j,4}(a;m)$. **Answer**:

$$N_{j,4}(a;m) = \frac{\pi}{2^{3m+3/2}(a+1)^{m+1/2}}$$

$$\times \sum_{k=0}^{m-j} 2^k \binom{2m-2k}{m-k}\binom{m-j+k}{2k}\binom{2k}{k}\binom{m}{k}^{-1}(a+1)^k.$$

Exercise 7.10.3. Let $b > 0$, $c > 0$, $a > -\sqrt{bc}$, $m \in \mathbb{N}$, and $0 \le j \le m$. Then

$$N_{j,4}(a,b,c;m) = \int_0^\infty \frac{x^{2j}\,dx}{\left(bx^4 + 2ax^2 + c\right)^{m+1}}$$

$$= \pi \left[c(c/b)^{m-j} \left\{ 8(a + \sqrt{bc}) \right\}^{2m+1} \right]^{-1/2}$$

$$\times \sum_{k=0}^{m-j} 2^k \binom{2m-2k}{m-k}\binom{m-j+k}{2k}\binom{2k}{k}$$

$$\times \binom{m}{k}^{-1}\left(\frac{a}{\sqrt{bc}}+1\right)^k.$$

Hint. Use the scaling $u = \lambda x$ with a carefully chosen value of λ.

Exercise 7.10.4. Let $b > 0$, $c > 0$, $a > -\sqrt{bc}$, $m \in \mathbb{N}$, and $m + 1 \le j \le 2m + 1$. Prove that

$$
N_{j,4}(a, b, c; m) = \int_0^\infty \frac{x^{2j}\, dx}{\left(bx^4 + 2ax^2 + c\right)^{m+1}}
$$

$$
= \pi \left[b(b/c)^{j-m-1} \left\{ 8(a + \sqrt{bc}) \right\}^{2m+1} \right]^{-1/2}
$$

$$
\times \sum_{k=0}^{j-m-1} 2^k \binom{2m - 2k}{m - k} \binom{m - j + k}{2k} \binom{2k}{k}
$$

$$
\times \binom{m}{k}^{-1} \left(\frac{a}{\sqrt{bc}} + 1 \right)^k.
$$

7.11. The Expansion of the Triple Square Root

This section consists of a single project that produces the Taylor series expansion of the **triple square root**

$$
h_{a,b}(c) := \sqrt{a + \sqrt{b + \sqrt{1 + c}}}. \tag{7.11.1}
$$

The formula proposed in (7.11.6) was discovered by symbolic manipulations. It involves the idea of the homogenization of a polynomial: given

$$
P(x) = a_0 x^n + a_1 x^{n-1} + \cdots + a_n \tag{7.11.2}
$$

it is often useful to construct the polynomial in two variables

$$
P^*(x, y) = a_0 x^n + a_1 x^{n-1} y + \cdots + a_n y^n, \tag{7.11.3}
$$

in which every monomial has the form $x^{n-k}y^k$. These polynomials satisfy $P^*(tx, ty) = t^n P^*(x, y)$, so that if (x, y) is a zero of P^* and $t \in \mathbb{R}$, then (tx, ty) is also a zero. The reader will find this expressed as *the zeros of a homogeneous polynomial make projective sense*. These ideas are presented by Cox et al. (1998). The interesting part of the project below is to explain why these formulas occur.

Project 7.11.1. Prove that the coefficients of the Taylor series expansion

$$
h_{a,b}(c) = \sum_{n=0}^\infty \beta_n(a, b)c^n \tag{7.11.4}
$$

are given by

$$\beta_0(a, b) = \sqrt{a + \sqrt{1 + b}} \qquad (7.11.5)$$

and

$$\beta_n(a, b) = \frac{(-1)^{n-1}}{n \, 2^{2n+1}} \sum_{k=0}^{n-1} \binom{2n - 2 - k}{n - 1} q^{-k-1/2} P_k^*(a, \sqrt{1 + b}), \qquad (7.11.6)$$

where $q := (1 + b)(a + \sqrt{1 + b})$ and

$$P_k^*(a, z) = z^k P_k(a/z)$$

is the homogenization of P_k.

8

The Normal Integral

8.1. Introduction

The evaluation

$$\int_0^\infty e^{-x} \, dx = 1 \qquad (8.1.1)$$

is elementary because the integrand $f(x) = e^{-x}$ admits a primitive. In this chapter we discuss several evaluations of the *normal integral*

$$I := \int_0^\infty e^{-x^2} \, dx. \qquad (8.1.2)$$

Most of the calculus texts discuss this problem in a chapter on improper integrals and postpone its evaluation to the section on several variables. For instance Thomas and Finney (1996) state in Exercise 28, page 364, that *the error function, important in probability and in the theory of heat flow and signal transmission, must be evaluated numerically because there is no elementary expression for the antiderivative of e^{-x^2}*. The exercise continues with a numerical evaluation of the *error function*

$$\mathrm{erf}(x) = \frac{2}{\sqrt{\pi}} \int_0^x e^{-t^2} \, dt. \qquad (8.1.3)$$

The fact that e^{-x^2} does not have an elementary primitive is a consequence of Liouville's work (1835) on integration in finite terms. The reader will find in Marchisotto and Zakeri (1994) an elementary introduction to these ideas.

The case of the normal integral is settled by the following result.

8.1.1. Strong Liouville Theorem (Special Case, 1835)

If $f(x)$ and $g(x)$ are rational functions with $g(x)$ nonconstant, then $f(x)e^{g(x)}$ has an elementary primitive if and only if there exists a rational function $R(x)$ such that $f(x) = R'(x) + R(x)g'(x)$.

162

The next result appears in Marchisotto and Zakeri (1994). The next exercise is used in the proof.

Exercise 8.1.1. Suppose the polynomials A, B, q satisfy

$$A(x)q(x) = B(x)\frac{d}{dx}q(x), \tag{8.1.4}$$

and q and B have no common zeros. Prove that q has no zeros.

Theorem 8.1.1. *The integral*

$$I_n = \int x^{2n} e^{-x^2} dx, \; for \; n \in \mathbb{N} \tag{8.1.5}$$

is nonelementary.

Proof. We use Liouville's theorem with $f(x) = x^{2n}$ and $g(x) = -x^2$. Then we must have $x^{2n} = R'(x) - 2xR(x)$, where $R(x) = p(x)/q(x)$, so that,

$$\left[x^{2n}q(x) - p'(x) + 2xp(x)\right]q(x) = -p(x)q'(x). \tag{8.1.6}$$

It follows from Exercise 8.1.1 that $q(x)$ has no zeros, so we may assume $q(x) \equiv 1$. Then (8.1.6) becomes

$$x^{2n} = p'(x) - 2xp(x). \tag{8.1.7}$$

The degree of p must be $2n - 1$, and (8.1.7) produces

$$x^{2n} = c_1 + \sum_{j=1}^{2n-2} \left[(j+1)c_{j+1} - 2c_{j-1}\right] x^j - 2c_{2n-2}x^{2n-1} - 2c_{2n-1}x^{2n}$$

Therefore $c_{2n-1} = -1/2$ and $c_1 = 0$, $(j+1)c_{j+1} - 2c_{j-1} = 0$ for $j = 1, 2, \ldots, 2n - 2$, from where we conclude that $c_3 = 0$, $c_5 = 0, \ldots, c_{2n-1} = 0$. This is a contradiction. $\square$

Exercise 8.1.2. Let $n \in \mathbb{N} \cup \{0\}$ and $p > 0$. Prove that $x^{2n+1}e^{-px^2}$ has an elementary primitive and confirm the evaluation

$$\int_0^\infty x^{2n+1} e^{-px^2} dx = \frac{n!}{2p^{n+1}}. \tag{8.1.8}$$

This is [G & R] 3.461.3. The companion formula

$$\int_0^\infty x^{2n} e^{-px^2} dx = \frac{(2n)!}{2^{2n+1}n!p^{n+1/2}} \sqrt{\pi} \tag{8.1.9}$$

appears in [G & R] (1994), 3.461.2 and in Project 5.7.1. Check it by reducing to the case $n = 0$.

8.2. Some Evaluations of the Normal Integral

In this section we present several proofs of the identity

$$\int_0^\infty e^{-x^2}\, dx = \frac{\sqrt{\pi}}{2}. \tag{8.2.1}$$

8.2.1. The Squaring Trick

The usual trick is to square the integral and evaluate the resulting double integral by polar coordinates:

$$
\begin{aligned}
I^2 &= \int_0^\infty \int_0^\infty e^{-(x^2+y^2)}\, dx\, dy \\
&= \int_0^{\pi/2} \int_0^\infty e^{-r^2} r\, dr\, d\theta \\
&= \frac{\pi}{4}.
\end{aligned}
$$

8.2.2. Small Variation

Start as above and square (8.1.2) to produce

$$
\begin{aligned}
I^2 &= \int_0^\infty e^{-x^2} \int_0^\infty e^{-u^2}\, du\, dx = \int_0^\infty e^{-x^2} \int_0^\infty e^{-x^2 y^2} x\, dy\, dx \\
&= \int_0^\infty \int_0^\infty x e^{-x^2(1+y^2)}\, dx\, dy \\
&= \frac{1}{2} \int_0^\infty \frac{dy}{1+y^2} \\
&= \frac{\pi}{4}.
\end{aligned}
$$

This proof yields the basic identity

$$\left(\int_0^\infty e^{-x^2}\, dx \right)^2 = \frac{1}{2} \int_0^\infty \frac{dy}{1+y^2}. \tag{8.2.2}$$

Exercise 8.2.1. This appears in Borwein and Borwein (1987), p. 27, Ex. 3: Let

$$f(x) := \left(\int_0^x e^{-t^2}\, dt \right)^2, \qquad g(x) := \int_0^1 e^{-x^2(1+t^2)} \frac{dt}{1+t^2}.$$

Check that $f'(x) + g'(x) = 0$ so $f(x) + g(x) \equiv g(0) = \pi/4$. Conclude that $I = \sqrt{f(\infty)} = \sqrt{\pi}/2$.

8.2.3. A Proof Using Wallis' Formula

This proof uses Wallis' formula (6.4.5). We start with an exercise.

Exercise 8.2.2. Prove that

$$1 - x^2 \le e^{-x^2} \text{ for } 0 \le x \le 1$$

and

$$e^{-x^2} \le \frac{1}{1 + x^2} \text{ for } x \ge 0.$$

Proof. The inequalities in Exercise 8.2.2 yield

$$\int_0^1 (1 - x^2)^n \, dx \le \int_0^1 e^{-nx^2} \, dx = \frac{1}{\sqrt{n}} \int_0^{\sqrt{n}} e^{-y^2} \, dy < \frac{1}{\sqrt{n}} \int_0^\infty e^{-y^2} \, dy.$$

Similarly

$$\int_0^\infty \frac{dx}{(1 + x^2)^n} \ge \int_0^\infty e^{-nx^2} \, dx = \frac{1}{\sqrt{n}} \int_0^\infty e^{-y^2} \, dy.$$

Wallis's formula yields

$$\int_0^\infty \frac{dx}{(1 + x^2)^n} = \frac{\pi n}{2^{2n} (2n - 1)} \binom{2n}{n}$$

and

$$I_n := \int_0^1 (1 - x^2)^n \, dx = \frac{2^{2n}}{(2n + 1) \binom{2n}{n}}. \tag{8.2.3}$$

This evaluation is outlined in the next exercise. Therefore

$$\frac{\sqrt{n} \, 2^{2n}}{(2n + 1) \binom{2n}{n}} \le \int_0^\infty e^{-y^2} \, dy \le \frac{\pi n^{3/2} \binom{2n}{n}}{2^{2n} (2n - 1)}$$

and now use (5.7.12) to pass to the limit.

Exercise 8.2.3. This exercise provides a proof of (8.2.3).
a) Check that I_n satisfies

$$I_{n+1} = \frac{2n + 2}{2n + 3} I_n, \tag{8.2.4}$$

and that the right-hand side of (8.2.3) satisfies the same recursion and both sides agree at $n = 0$.

b) Solve (8.2.4) and produce a closed form expression for I_n.

Exercise 8.2.4. Prove the identity

$$\sum_{j=0}^{n} \frac{(-1)^j}{2j+1} \binom{n}{j} = \frac{2^{2n}}{(2n+1)\binom{2n}{n}}. \tag{8.2.5}$$

Hint. Expand $(1 - x^2)^n$ and use (8.2.3).

Give a proof of (8.2.5) using the WZ method described in the Appendix.

8.2.4. Reduction to a Special Value of Γ

The **gamma function**

$$\Gamma(x) = \int_0^\infty t^{x-1} e^{-t} \, dt \tag{8.2.6}$$

will be studied in Chapter 10. In particular the special value

$$\Gamma(\tfrac{1}{2}) = \sqrt{\pi} \tag{8.2.7}$$

will appear in (10.1.22). The integral definition of Γ gives

$$\int_0^\infty e^{-t} t^{-1/2} dt = \sqrt{\pi} \tag{8.2.8}$$

and the change of variable $x \to x^2$ yields the value of the normal integral.

8.2.5. A Proof of Kortram and Sums of Two Squares

The next proof is due to Kortram (1993). This proof connects the normal integral to sums of two squares. This is a classical problem in number theory. The number of solutions of the equation

$$n_1^2 + n_2^2 = n \tag{8.2.9}$$

in integers n_1, n_2 is denoted by $r_2(n)$. For example, $r_2(2) = 4$, corresponding to $(\pm 1)^2 + (\pm 1)^2 = 2$ and $r_2(3) = 0$ as the reader can easily check. There are many interesting expressions for $r_2(n)$. Jacobi proved in 1829 that $r_2(n)$ is four times the excess (if any) of its positive divisors $d \equiv 1 \mod 4$ over its positive divisors $d \equiv 3 \mod 4$. Fermat's theorem (1640), that odd primes congruent to 1 modulo 4 are sums of two squares but primes congruent to 3 modulo 4 are not, is a special case.

Exercise 8.2.5. Establish the identity

$$\left(\sum_{n=-\infty}^{\infty} x^{n^2} \right)^2 = \sum_{m=0}^{\infty} r_2(m) x^m. \tag{8.2.10}$$

Let $0 < x < 1$ and consider the decreasing function

$$t \mapsto x^{t^2} = e^{t^2 \ln x}.$$

Then

$$\int_0^{\infty} e^{t^2 \ln x} dt \le \sum_{n=0}^{\infty} x^{n^2} \le 1 + \int_0^{\infty} e^{t^2 \ln x} dt.$$

The change of variables $u \mapsto t \times \sqrt{-\ln x}$ shows that

$$\lim_{x \uparrow 1} \sqrt{-\ln x} \sum_{n=0}^{\infty} x^{n^2} = \int_0^{\infty} e^{-u^2} du.$$

Now use $\lim_{x \to 1} \ln x / (x - 1) = 1$ to conclude

$$\lim_{x \uparrow 1} \sqrt{1 - x} \sum_{n=0}^{\infty} x^{n^2} = \int_0^{\infty} e^{-u^2} du.$$

Squaring we get

$$\left(\int_0^{\infty} e^{-u^2} du \right)^2 = \lim_{x \uparrow 1} \tfrac{1}{4}(1 - x) \left(\sum_{n=0}^{\infty} x^{n^2} \right)^2 \tag{8.2.11}$$

$$= \lim_{x \uparrow 1} (1 - x) \sum_{n=0}^{\infty} r_2(n) x^n.$$

In order to evaluate the limit we require a lemma:

Lemma 8.2.1. *Let $A_n > 0$ such that the power series $\sum_{n=0}^{\infty} A_n x^n$ converges for $0 \le x < 1$ and diverges at $x = 1$. Let B_n be such that $\lim_{n \to \infty} B_n / A_n = \lambda$. Then*

$$\lim_{x \uparrow 1} \frac{\sum B_n x^n}{\sum A_n x^n} = \lambda.$$

Proof. Choose $\epsilon > 0$ such that $|B_n/A_n - \lambda| < \epsilon/2$. Then

$$\left| \frac{\sum B_n x^n}{\sum A_n x^n} - \lambda \right| = \sum_{n=0}^{\infty} (B_n - \lambda A_n) x^n \times \left(\sum_{n=0}^{\infty} A_n x^n \right)^{-1}$$

$$\leq \left(\sum_{n=0}^{\infty} A_n x^n \right)^{-1} \left(\sum_{n=0}^{N} |B_n - \lambda A_n| + \frac{\epsilon}{2} \sum_{n=N+1}^{\infty} A_n x^n \right).$$

$$< \left(\sum_{n=0}^{\infty} A_n x^n \right)^{-1} \left(\sum_{n=0}^{N} |B_n - \lambda A_n| + \frac{\epsilon}{2} \right)$$

$$< \epsilon$$

for x sufficiently close to 1. $\qquad\qquad\qquad\qquad\qquad\qquad\qquad\qquad\square$

Now we apply the lemma to $A_n = n + 1$ and $B_n = \frac{1}{4} \sum_{k=0}^{n} r_2(k)$. The number B_n counts the number of lattice points inside a quarter circle of radius $\sqrt{n}$. Then associate to each lattice point the unit northeast square to obtain the bound

$$\frac{\pi}{4}(\sqrt{n})^2 \leq B_n \leq \frac{\pi}{4}(\sqrt{n} + \sqrt{2})^2.$$

It follows that

$$\lim_{n \to \infty} \frac{B_n}{A_n} = \frac{\pi}{4}.$$

The lemma now gives

$$\frac{\pi}{4} = \lim_{x \uparrow 1} \frac{\sum B_n x^n}{\sum A_n x^n}$$

$$= \lim_{x \uparrow 1} \frac{(1 - x) \sum B_n x^n}{(1 - x) \sum (n + 1) x^n}.$$

Using part b) of Exercise 5.2.7 this reduces to

$$\frac{\pi}{4} = \frac{1}{4} \lim_{x \uparrow 1} (1 - x) \sum r_2(n) x^n.$$

The value of the normal integral now follows from (8.2.11).

Exercise 8.2.6. This exercise outlines a proof by Yzeren (1979).
a) Define for $t \geq 0$

$$f(t) := \int_{-\infty}^{\infty} \frac{e^{-t(x^2+1)}}{x^2 + 1} \, dx.$$

b) Use $ph < e^{ph} - 1 < phe^{ph}$ for $p = x^2 + 1$ and integrate over x to obtain

$$\frac{Ie^{-(t+h)}}{(t+h)^{1/2}} < \frac{f(t) - f(t+h)}{h} < \frac{Ie^{-t}}{t^{1/2}},$$

with

$$I = \int_{-\infty}^{\infty} e^{-x^2}\, dx.$$

c) Let $h \to 0$ to produce

$$f'(t) = -It^{-1/2}e^{-t}.$$

d) Prove $f(t) \le I\, e^{-t}$ so $f(t) \to 0$ as $t \to \infty$. Thus

$$f(t) = I \int_t^{\infty} u^{-1/2}e^{-u}\, du = 2I \int_{\sqrt{t}}^{\infty} e^{-x^2}\, dx.$$

Evaluating at $t = 0$ to obtain the value of I.

Exercise 8.2.7. This exercise outlines a proof by Coleman (1954).
a) Prove that

$$I_{n,m} := \int_0^{\infty} x^m e^{-nx^2/2}\, dx \tag{8.2.12}$$

satifies the recursion

$$I_{n,m} = \frac{m-1}{n} I_{n,m-2}. \tag{8.2.13}$$

b) The inequality

$$\int_0^{\infty} x^m (x - t)^2 e^{-nx^2/2}\, dx = t^2 I_{n,m} - 2t I_{n,m+1} + I_{n,m+2} > 0$$

yields

$$I_{n,m+1} < \sqrt{I_{n,m}\, I_{n,m+2}}. \tag{8.2.14}$$

c) Choose $m = n$ in (8.2.14) and use (8.2.13) to get

$$I_{n,n+1} < \sqrt{I_{n,n}\, I_{n,n+2}} = \sqrt{\frac{n}{n+1}} I_{n,n+2} < I_{n,n+2}.$$

d) Choose $m = n - 1$ to obtain

$$I_{n,n} < \sqrt{I_{n,n-1}\, I_{n,n+1}} = I_{n,n+1}.$$

Conclude that $I_{n,n} < I_{n,n+1} < I_{n,n+2}$.

e) Let $n = 2k + 1$ and use the reduction formula (8.2.13) to obtain the value of $I_{2k+1,2k+1}$, $I_{2k+1,2k+2}$ and $I_{2k+1,2k+3}$ in terms of $I_{n,0}$.

f) Check that

$$I_{n,0} = \int_0^\infty e^{-nx^2/2} \, dx = \frac{1}{\sqrt{2n+1}} \int_0^\infty e^{-u^2/2} \, du.$$

Let

$$f(n) := \frac{2 \cdot 4 \cdot 6 \cdots (2n)}{1 \cdot 3 \cdot 5 \cdots (2n+1)} \sqrt{2n+1}$$

and check that

$$f(n) < \int_0^\infty e^{-u^2/2} \, du < \left(1 + \frac{1}{2n+1}\right) f(n) \qquad (8.2.15)$$

Conclude that

$$\int_0^\infty e^{-u^2/2} \, du = \lim_{n \to \infty} \frac{2 \cdot 4 \cdot 6 \cdots (2n)}{1 \cdot 3 \cdot 5 \cdots (2n+1)} \sqrt{2n+1}.$$

Now recognize the limit from (6.8.5).

Exercise 8.2.8. This exercise describes the evaluation of Romik (2000) of the constant C in Stirling's formula (5.7.1). Recall that the constant C is given by

$$C = \lim_{n \to \infty} \frac{2^{2n+1/2}}{\binom{2n}{n} \sqrt{n}}. \qquad (8.2.16)$$

a) Prove that for a polynomial f and $n \in \mathbb{N}$

$$f(x) = f(0) + f'(0)x + \frac{1}{2} f''(0)x^2 + \cdots + \frac{1}{n!} f^{(n)}(0)x^n + R_n(x) \qquad (8.2.17)$$

with

$$R_n(x) = \frac{1}{n!} \int_0^x f^{(n+1)}(t)(x - t)^n \, dt. \qquad (8.2.18)$$

Hint. Integrate by parts.

b) Use (8.2.18) with $f(x) = (1 + x)^{2n+1}$ to prove that

$$\lim_{n \to \infty} \frac{R_n(1)}{2^{2n+1}} = \frac{\sqrt{\pi}}{\sqrt{2}C}. \qquad (8.2.19)$$

Hint. Use the change of variable $t \to u/\sqrt{n}$ to produce the normal integral.

c) Use (8.2.17) to evaluate $R_n(1) = 2^{2n}$. Conclude that $C = \sqrt{2\pi}$.

8.3. Formulae from Gradshteyn and Rhyzik (G & R)

This section contains a list of integrals appearing in [G & R] that can be evaluated in terms of the normal integral.

Exercise 8.3.1. Prove [G & R] 3.321.3. For $q > 0$

$$\int_0^\infty e^{-q^2 x^2} \, dx = \frac{\sqrt{\pi}}{2q}.$$

Exercise 8.3.2. Prove [G & R] 3.323.2:

$$\int_{-\infty}^\infty e^{-p^2 x^2 \pm qx} \, dx = \exp\left(\frac{q^2}{4p^2}\right) \frac{\sqrt{\pi}}{|p|}.$$

Exercise 8.3.3. Prove [G & R] 3.361.2. For $q > 0$ and $a \in \mathbb{R}$:

$$\int_{-a}^\infty \frac{e^{-qx}}{\sqrt{x+a}} \, dx = \sqrt{\frac{\pi}{q}} e^{aq}.$$

The special cases $a = 0$, 1 and -1 appear in [G & R]: 3.361.2, 3.361.3 and 3.362.1, respectively.

The next exercise presents an *indefinite* integral that can be reduced to the error function erf(x) introduced in (8.1.3).

Exercise 8.3.4. Check [G & R] 2.33;

$$\int e^{-(ax^2 + 2bx + c)} \, dx = \frac{1}{2}\sqrt{\frac{\pi}{a}} \operatorname{erf}\left(\sqrt{a}x + b/\sqrt{a}\right).$$

Exercise 8.3.5. Check [G & R] 3.462.7:

$$\int_0^\infty x^2 e^{-\mu x^2 - 2\nu x} \, dx = -\frac{\nu}{2\mu^2} + \sqrt{\frac{\pi}{\mu^5}} \frac{2\nu^2 + \mu}{4} e^{\nu^2/\mu} \left[1 - \operatorname{erf}\left(\frac{\nu}{\sqrt{\mu}}\right)\right].$$

Exercise 8.3.6. Check [G & R] 3.468.2:

$$\int_0^\infty \frac{x e^{-\mu x^2}}{\sqrt{a^2 + x^2}} \, dx = \frac{1}{2}\sqrt{\frac{\pi}{\mu}} e^{a^2 \mu} \left[1 - \operatorname{erf}(a\sqrt{\mu})\right].$$

8.4. An Integral of Laplace

In this section we present the evaluation of an integral due to Laplace. It appears in [G & R] 3.325. The technique of the proof is due to Schlomilch and more examples of it will be discussed in Chapter 13.

172   *The Normal Integral*

Example 8.4.1. Let $a, b \in \mathbb{R}^+$. Then

$$L(a,b) := \int_0^\infty e^{-ax^2 - b/x^2}\, dx = \frac{1}{2}\sqrt{\frac{\pi}{a}}\, e^{-2\sqrt{ab}}. \tag{8.4.1}$$

Proof. The change of variable $t = \sqrt{a}x$ shows that $L(a,b) = f(ab)/\sqrt{a}$ where

$$f(c) := \int_0^\infty e^{-t^2 - c/t^2}\, dt. \tag{8.4.2}$$

To evaluate $f(c)$ we make the change of variable $y = \sqrt{c}/t$ in (8.4.2) and add the resulting integral to the original one to produce

$$f(c) = \frac{1}{2}\int_0^\infty e^{-(t^2 + c/t^2)}\left(1 + \frac{\sqrt{c}}{t^2}\right) dt$$

$$= \frac{1}{2}\int_{-\infty}^\infty e^{-s^2 - 2\sqrt{c}}\, ds$$

by introducing $s = t - \sqrt{c}/t$. The result now follows from the value of the normal integral. $\qquad\square$

Exercise 8.4.1. Let $a, b \in \mathbb{R}^+$. Prove that

$$\int_0^\infty e^{-(ax + b/x)} \frac{dx}{\sqrt{x}} = \frac{\sqrt{\pi}}{\sqrt{a}}\, e^{-2\sqrt{ab}}. \tag{8.4.3}$$

In particular

$$\int_0^\infty e^{-(x + b/x)} \frac{dx}{\sqrt{x}} = \sqrt{\pi}\, e^{-2\sqrt{b}}. \tag{8.4.4}$$

Exercise 8.4.2. The four integrals in this exercise are in Section 3.472 of [G & R]. Check them by reducing them to Laplace's example.

$$\int_0^\infty \left(e^{-a/x^2} - 1\right) e^{-\mu x^2}\, dx = \frac{1}{2}\sqrt{\frac{\pi}{\mu}}\left[e^{-2\sqrt{a\mu}} - 1\right],$$

$$\int_0^\infty x^2 \exp\left(-a/x^2 - \mu x^2\right) dx = \frac{1}{4}\sqrt{\frac{\pi}{\mu^3}}(1 + 2\sqrt{a\mu})e^{-2\sqrt{a\mu}},$$

$$\int_0^\infty \exp\left(-a/x^2 - \mu x^2\right) \frac{dx}{x^2} = \frac{1}{2}\sqrt{\frac{\pi}{a}}\, e^{-2\sqrt{a\mu}},$$

$$\int_0^\infty \exp\left[-\frac{1}{2a}(x^2 + 1/x^2)\right] \frac{dx}{x^4} = \sqrt{\frac{a\pi}{2}}(1 + a)e^{-1/a}.$$

9

Euler's Constant

The Euler–Mascheroni constant γ defined by

$$\gamma := \lim_{n \to \infty} \sum_{k=1}^{n} \frac{1}{k} - \ln n$$

is regarded as an important constant of analysis, shadowed only by e and π in significance. Havil (2003) provides an excellent history of this constant.

9.1. Existence of Euler's Constant

In this section we present an elementary proof of the existence of Euler's constant. The notation

$$H_n = \sum_{k=1}^{n} \frac{1}{k} \tag{9.1.1}$$

for the **harmonic numbers** is employed throughout.

Lemma 9.1.1. *The limit*

$$\gamma := \lim_{n \to \infty} H_n - \ln n$$

exists.

Proof. Let $a_n = H_n - \ln n$. To prove the existence of γ first observe that

$$a_{n+1} - a_n = \frac{1}{n+1} - \ln(n+1) + \ln n = \frac{1}{n+1} + \ln\left(\frac{n}{n+1}\right) < 0$$

because $\ln(1 - x) + x < 0$. On the other hand (5.2.17) shows that $\frac{1}{n} < a_n < 1$. The sequence a_n is, therefore, decreasing and bounded. The bounds on a_n show $0 \leq \gamma \leq 1$. □

9.2. A Second Proof of the Existence of Euler's Constant

In the proof of the existence of e given in Section 5.2 we have established in (5.4.8) the existence of a sequence c_n satisfying

$$\lim_{n \to \infty} nc_n = 1.$$

Now write

$$e = a_n \left(1 + \frac{1}{n}\right)^n \tag{9.2.1}$$

where $a_n := \frac{1}{nc_n}$ satisfies $1 < a_n < 1 + \frac{1}{n}$ so that

$$\lim_{n \to \infty} a_n = 1.$$

Now from (9.2.1) we have

$$1 = \ln a_n + n \left(\ln(n+1) - \ln n\right)$$

and now replace $n = 1, 2, \cdots$ to obtain

$$
\begin{aligned}
1 &= \ln a_1 + \ln 2 - \ln 1 \\
\frac{1}{2} &= \ln a_2^{1/2} + \ln 3 - \ln 2 \\
\frac{1}{3} &= \ln a_3^{1/3} + \ln 4 - \ln 3 \\
&\cdots\cdots\cdots \\
\frac{1}{n} &= \ln a_n^{1/n} + \ln(n+1) - \ln n.
\end{aligned}
$$

Adding we get

$$H_n = \ln \left(a_1 a_2^{1/2} a_3^{1/3} \cdots a_n^{1/n}\right) + \ln(n+1).$$

Let

$$b_n = \ln \left(a_1 a_2^{1/2} a_3^{1/3} \cdots a_n^{1/n}\right) \tag{9.2.2}$$

then b_n is increasing and bounded from above because

$$
\begin{aligned}
b_n &= \ln a_1 + \frac{1}{2} \ln a_2 + \cdots + \frac{1}{n} \ln a_n \\
&< \ln(1 + 1/1) + \frac{1}{2} \ln(1 + 1/2) + \cdots + \frac{1}{n} \ln(1 + 1/n) \\
&< 1 + \frac{1}{2^2} + \frac{1}{3^2} + \cdots + \frac{1}{n^2} \\
&< \sum_{n=1}^{\infty} \frac{1}{n^2}.
\end{aligned}
$$

It follows from the integral test that the last series converges and we conclude that $H_n - \ln(n+1)$ is increasing and bounded from above. Therefore

$$\gamma := \lim_{n \to \infty} H_n - \ln(n+1)$$

exists.

Note 9.2.1. In Chapter 11 we give proofs of Euler's remarkable formula

$$\zeta(2) := \sum_{n=1}^{\infty} \frac{1}{n^2} = \frac{\pi^2}{6}, \tag{9.2.3}$$

therefore the sequence b_n is bounded by $\pi^2/6$.

Extra 9.2.1. Alzer and Brenner (1992) have studied the sequences

$$x_n = \frac{H_n}{\ln(n+1)} \quad \text{and} \quad y_n = \frac{H_{n+1} - 1}{\ln(n+1)} \tag{9.2.4}$$

and have shown that $y_n < 1 < x_n$ and that x_n is strictly increasing and converges to 1. Similarly, y_n decreases to 1.

Exercise 9.2.1. This exercise outlines a proof by Johnsonbaugh (1981) of the existence of Euler's constant.
a) Apply the trapezoidal rule (Thomas and Finney, 1996), page 346, to $f(t) = 1/t$ to obtain

$$\int_{k-1}^{k} \frac{dt}{t} = \frac{1}{2} \left(\frac{1}{k} + \frac{1}{k-1} \right) - \frac{1}{6\xi_k^3} \tag{9.2.5}$$

with $k - 1 < \xi_k < k$.
b) Sum (9.2.5) and pass to the limit as $n \to \infty$ to prove that

$$\lim_{n \to \infty} \ln n - H_n = -\frac{1}{2} - \frac{1}{6} \sum_{n=2}^{\infty} \frac{1}{\xi_k^3}. \tag{9.2.6}$$

Check that $r_n := \frac{1}{6} \sum_{k=n+1}^{\infty} \xi_k^{-3}$ satisfies

$$\frac{1}{12(n+1)^2} < r_n < \frac{1}{12(n-1)^2}.$$

Hint. Estimate the series r_n by an integral.
c) Estimate γ in terms of the Apery constant

$$\zeta(3) = \sum_{k=1}^{\infty} \frac{1}{k^3}. \tag{9.2.7}$$

In other words, obtain bounds on γ in terms of $\zeta(3)$ using the inequality from part b). The Apery constant will be considered in Chapter 11.

9.3. Integral Forms for Euler's Constant

We now establish the first of many integral representation for γ.

Exercise 9.3.1. Prove that

$$\int_0^1 \frac{1 - (1 - x)^n}{x}\, dx = H_n \tag{9.3.1}$$

Proposition 9.3.1. *The Euler constant γ is given by*

$$\gamma = \int_0^1 \frac{1 - e^{-x}}{x} dx - \int_1^\infty \frac{e^{-x}}{x} dx. \tag{9.3.2}$$

Proof. Use (9.3.1) to obtain

$$H_n = \int_0^n \left(1 - (1 - x/n)^n\right) \frac{dx}{x}$$

$$= \int_0^1 \left(1 - (1 - x/n)^n\right) \frac{dx}{x} + \int_1^n \left(1 - (1 - x/n)^n\right) \frac{dx}{x}$$

$$= \int_0^1 \left(1 - (1 - x/n)^n\right) \frac{dx}{x} - \int_1^n (1 - x/n)^n \frac{dx}{x} + \ln n.$$

Therefore

$$\gamma := \lim_{n \to \infty} H_n - \ln n$$

$$= \lim_{n \to \infty} \int_0^1 \frac{(1 - (1 - x/n)^n)}{x} dx - \int_1^n \frac{(1 - x/n)^n}{x} dx$$

$$= \int_0^1 \frac{1 - e^{-x}}{x} dx - \int_1^\infty \frac{e^{-x}}{x} dx.$$

where we have used the basic limit for e^{-x} given in (5.6.3). $\square$

The next expression for γ appears in [G & R]: 4.229.

Proposition 9.3.2. *Euler's constant is given by*

$$\int_0^\infty e^{-x} \ln x\, dx = -\gamma. \tag{9.3.3}$$

Proof. Start with

$$\int_0^\infty e^{-x} \ln x \, dx = \int_0^1 e^{-x} \ln x \, dx + \int_1^\infty e^{-x} \ln x \, dx.$$

Now integrate by parts in the first integral to obtain

$$\int_0^1 e^{-x} \ln x \, dx = -\int_0^1 \frac{d}{dx}(e^{-x} - 1) \times \ln x \, dx$$

$$= \int_0^1 \frac{e^{-x} - 1}{x} \, dx,$$

similar for second integral. Now use (9.3.2).   □

Exercise 9.3.2. Check that

$$\int_0^1 \ln(-\ln x) \, dx = -\gamma. \tag{9.3.4}$$

Hint. Reduce to (9.3.3).

Proposition 9.3.3. *The Euler's constant is given by*

$$\gamma = \int_0^\infty e^{-x} \left(\frac{1}{1 - e^{-x}} - \frac{1}{x} \right) dx. \tag{9.3.5}$$

Proof. We follow Rao (1956). Start with

$$\int_0^\infty e^{-rx} dx = \frac{1}{r} \tag{9.3.6}$$

and integrate (9.3.6) from $r = 1$ to n to produce

$$\int_0^\infty \frac{e^{-x} - e^{-nx}}{x} \, dx = \ln n. \tag{9.3.7}$$

Now use (9.3.6) for $r = 1, 2, \ldots, n$ and (9.3.7) to obtain

$$a_n := H_n - \ln n = \int_0^\infty \left(e^{-x} + e^{-2x} + \cdots + e^{-nx} \right) dx - \int_0^\infty \frac{e^{-x} - e^{-nx}}{x} \, dx.$$

Now rewrite this as

$$a_n = \int_0^\infty e^{-x} \left(\frac{1}{1 - e^{-x}} - \frac{1}{x} \right) dx + \int_0^\infty e^{-nx} \left(\frac{1}{x} - \frac{1}{e^x - 1} \right) dx$$

and now both integrals are convergent. Now use the inequalities

$$\frac{1}{2} - \frac{x}{8} < \frac{1}{x} - \frac{1}{e^x - 1} < \frac{1}{2} \tag{9.3.8}$$

to obtain

$$\int_0^\infty e^{-nx} \left(\frac{1}{x} - \frac{1}{e^x - 1} \right) dx < \frac{1}{2} \int_0^\infty e^{-nx} \, dx = \frac{1}{2n}$$

so as $n \to \infty$, we get the representation (9.3.5). $\qquad\square$

Note 9.3.1. The evaluation (9.3.7) appeared in Exercise 5.8.3 as an example of a Frullani integral.

Exercise 9.3.3. Check that the Euler's constant is given by

$$\gamma = \int_0^1 \left(\frac{1}{1-t} + \frac{1}{\ln t} \right) dt \qquad (9.3.9)$$

Proposition 9.3.4. *Let* $n \in \mathbb{N}$. *Then*

$$\gamma = H_n - \ln n - \int_n^\infty \frac{\{x\}}{x^2} \, dx. \qquad (9.3.10)$$

In particular

$$\gamma = 1 - \int_1^\infty \frac{\{x\}}{x^2} \, dx. \qquad (9.3.11)$$

Here $\{x\}$ is the fractional part of x. Compare with (5.7.9).

Proof. For $m > n$ we have

$$\int_n^m \frac{\{x\}}{x^2} dx = \sum_{j=n}^{m-1} \int_j^{j+1} \frac{\{x\}}{x^2} dx$$

$$= \sum_{j=n}^{m-1} \int_j^{j+1} \left(\frac{1}{x} - \frac{j}{x^2} \right) dx$$

$$= \ln m - \ln n - \sum_{j=n}^{m-1} \frac{1}{j+1}$$

$$= \ln m - \ln n - H_m + H_n.$$

Now let $m \to \infty$ to obtain the result. $\qquad\square$

We now present an integral representation that appears in Volume V of Berndt's book on Ramanujan's Notebooks Berndt (1994). This is part of the proof of Entry 21 on Chapter 36.

Proposition 9.3.5. Let $f(x) = e^{-\alpha e^x} + e^{-\alpha e^{-x}} - 1$. Then

$$\int_0^\infty f(x)\,dx = -\gamma - \ln \alpha.$$

In particular

$$\gamma = -\int_0^\infty \left(\exp(-e^x) + \exp(-e^{-x}) - 1 \right)\,dx.$$

Proof. Observe that f is even so that

$$2\int_0^\infty f(x)\,dx = \int_{-\infty}^\infty \left(\exp(-\alpha e^x) + \exp(-\alpha e^{-x}) \right)\,dx$$

$$= \int_0^\infty \frac{e^{-\alpha u} + e^{-\alpha/u} - 1}{u}\,du$$

$$= -\int_0^{1/\alpha} \frac{1 - e^{-\alpha u}}{u}\,du + \int_{1/\alpha}^\infty \frac{e^{-\alpha u}}{u}\,du + \int_0^{1/\alpha} \frac{e^{-\alpha/u}}{u}\,du$$

$$\quad - \int_{1/\alpha}^\infty \frac{1 - e^{-\alpha/u}}{u}\,du.$$

$$= -\int_0^1 \frac{1 - e^{-x}}{x}\,dx + \int_1^\infty \frac{e^{-x}}{x}\,dx + \int_{\alpha^2}^\infty \frac{e^{-x}}{x}\,dx$$

$$\quad - \int_0^{\alpha^2} \frac{1 - e^{-x}}{x}\,dx$$

$$= -\gamma + \int_1^\infty \frac{e^{-x}}{x}\,dx - \int_0^1 \frac{1 - e^{-x}}{x}\,dx - \int_1^{\alpha^2} \frac{dx}{x}$$

$$= -2(\gamma + \ln \alpha).$$

$\square$

Exercise 9.3.4. Prove the representation

$$\gamma = \frac{ab}{a - b} \int_0^\infty \frac{e^{-t^a} - e^{-t^b}}{t}\,dt \qquad (9.3.12)$$

valid for $a, b > 0$ with $a \neq b$.

Exercise 9.3.5. Establish Catalan's formula

$$\gamma = 1 - \int_0^1 \left(\sum_{k=1}^\infty x^{2^k} \right) \frac{dx}{1 + x} \qquad (9.3.13)$$

and read about the generalization

$$\gamma = \int_0^1 \left(\frac{n}{1-x^n} - \frac{1}{1-x} \right) \sum_{k=1}^{\infty} x^{n^k-1} \, dx \qquad (9.3.14)$$

due to Berndt and Bowman (2000).

9.4. The Rate of Convergence to Euler's Constant

In the next theorem we establish that the rate of convergence of a_n to γ is comparable to $1/n$.

Theorem 9.4.1. *The sequence $a_n := H_n - \ln n$ converges to γ and*

$$\lim_{n \to \infty} n \, (a_n - \gamma) = \frac{1}{2}. \qquad (9.4.1)$$

Proof. Start with

$$a_n - \gamma = \int_0^\infty e^{-nx} \left(\frac{1}{x} - \frac{1}{e^x - 1} \right) \, dx$$

so that from (9.3.8) we have

$$\frac{1}{2n} - \frac{1}{8n^2} < a_n - \gamma < \frac{1}{2n}.$$

This shows that $a_n \to \gamma$ and (9.4.1). $\qquad \square$

A proof by Young. The following proof of Theorem 9.4.1 is due to Young (1991). Let R_n be the area of between $y = 1/x$ and the horizontal line $y = 1/(n+1)$ from $x = n$ to $x = n+1$. Then

$$R_n = \int_n^{n+1} \frac{dx}{x} - \frac{1}{n+1} = \ln(n+1) - \ln n - \frac{1}{n+1}.$$

Therefore

$$\sum_{k=n}^{n+j} R_k = a_n - a_{n+j+1}$$

where, as usual,

$$a_n = H_n - \ln n.$$

Observe that R_n is bounded from above by the area of the triangle connecting the endpoints and pushing all the regions R_{n+j+1} to the region $n \le x \le n+1$

we obtain $a_n - \gamma \leq 1/2n$. To obtain a lower bound for R_m bound it from below by the area of the triangle T_2 connecting the points $(m + 1, \frac{1}{m+1})$ and $(m + 2, \frac{1}{m+2})$ and with base on $m + 1 \leq x \leq m + 2$. To see that R_m has area bigger that T_2 translate it along its diagonal to have base over $m \leq x \leq m + 1$. We conclude

$$R_m > \frac{1}{2}\left(\frac{1}{m+1} - \frac{1}{m+2}\right)$$

so that

$$a_n - \gamma = \sum_{m=n}^{\infty} R_m \geq \frac{1}{2}\sum_{m=n}^{\infty}\left(\frac{1}{m+1} - \frac{1}{m+2}\right) = \frac{1}{2(n+1)}.$$

We have shown

$$\frac{1}{2n+2} \leq a_n - \gamma \leq \frac{1}{2n} \qquad (9.4.2)$$

and the proof is complete.

Extra 9.4.1. The inequalities (9.4.2) show that $a_n - \gamma$ is comparable to $1/2n$. The optimal values of a and b for the inequalities

$$\frac{1}{2n+a} < a_n - \gamma < \frac{1}{2n+b} \qquad (9.4.3)$$

are $a = (2\gamma - 1)/(1 - \gamma)$ and $b = 1/3$. See Chen and Q_i: (2003) for details.

9.4.1. A Quicker Convergence to Euler's Constant

In this section we follow De Temple (1993) to prove that a small modification of the sequence defining the Euler's constant, produces a sequence that converges to γ with error of the order $1/n^2$.

Proposition 9.4.1. *Let*

$$b_n := H_n - \ln(n + \tfrac{1}{2}).$$

Then

$$\frac{1}{24(n+1)^2} < b_n - \gamma < \frac{1}{24n^2}$$

so that

$$\lim_{n\to\infty} n^2(b_n - \gamma) = \frac{1}{24}.$$

Proof. Observe that

$$b_n - b_{n+1} = f(n)$$

with

$$f(x) = -\frac{1}{x+1} + \ln\left(x + \tfrac{3}{2}\right) - \ln\left(x + \tfrac{1}{2}\right),$$

then

$$f'(x) = \frac{-1}{4(x+1)^2(x+\tfrac{1}{2})(x+\tfrac{3}{2})}.$$

Then

$$\frac{1}{4(x+1)^4} < -f'(x) < \frac{1}{4(x+\tfrac{1}{2})^4}.$$

The upper bound is clear, for the lower bound use

$$\left(x + \tfrac{1}{2}\right)\left(x + \tfrac{3}{2}\right) = x^2 + 2x + \tfrac{3}{4} < (x+1)^2.$$

Now

$$f(k) = -\int_k^\infty f'(x)\,dx < \frac{1}{4}\int_k^\infty \frac{dx}{(x+\tfrac{1}{2})^4} = \frac{1}{12(k+\tfrac{1}{2})^3}.$$

The proof follows now from the inequality

$$\left(k + \tfrac{1}{2}\right)^{-3} < \int_k^{k+1} x^{-3}\,dx$$

that can be verified directly. Now

$$b_n - \gamma = \sum_{k=n}^\infty b_n - b_{n+1} = \sum_{k=n}^\infty f(k)$$

$$< \frac{1}{12}\sum_{k=n}^\infty \frac{1}{(k+1/2)^3} < \frac{1}{12}\int_n^\infty \frac{dx}{x^3} = \frac{1}{24n^2}.$$

For the lower bound use

$$f(k) > \frac{1}{4}\int_k^\infty \frac{dx}{(x+1)^4} = \frac{1}{12}(k+1)^{-3}$$

to conclude that

$$b_n - \gamma > \frac{1}{12}\sum_{k=n}^\infty (k+1)^{-3} > \frac{1}{12}\int_{n+1}^\infty \frac{dx}{x^3} = \frac{1}{24(n+1)^2}. \qquad \square$$

9.5. Series Representations for Euler's Constant

In this section we discuss several representations of γ in terms of infinite series.

The first class of series are modifications of the fundamental definition

$$\gamma = \lim_{n\to\infty} H_n - \ln n. \tag{9.5.1}$$

Introduce the notation

$$s_n = \sum_{k=1}^{2^n} \frac{1}{k} = H_{2^n} \quad \text{and} \quad \sigma_n = \sum_{k=1}^{2^n} \frac{(-1)^{k+1}}{k}$$

Exercise 9.5.1. a) The Euler constant is given by

$$\gamma = \lim_{n\to\infty} s_n - n \ln 2. \tag{9.5.2}$$

Hint. Replace n by 2^n in (9.5.1).

b) Write $\ln 2 = \sigma_n + r_n$ with

$$r_n = \sum_{k=2^n+1}^{\infty} \frac{(-1)^{k+1}}{k}.$$

Show that $\lim_{n\to\infty} n r_n = 0$ so

$$\gamma = \lim_{n\to\infty} s_n - n\,\sigma_n. \tag{9.5.3}$$

c) Let $t_n = s_n - n\sigma_n$. Prove that the identity

$$\sum_{n=1}^{N} t_{n+1} - t_n = t_{N+1} - t_1$$

yields

$$\gamma = 1 - \sum_{n=1}^{\infty} n(\sigma_{n+1} - \sigma_n) \tag{9.5.4}$$

and

$$\gamma = 1 - \sum_{n=1}^{\infty} \sum_{m=2^{n-1}+1}^{2^n} \frac{n}{2m(2m-1)}.$$

d) Addison (1967) improved the result in part c) with

$$\gamma = \frac{1}{2} + \sum_{n=1}^{\infty} \sum_{m=2^{n-1}}^{2^n-1} \frac{n}{2m(2m+1)(2m+2)}. \tag{9.5.5}$$

Prove this by expanding the right-hand side in partial fractions.

Exercise 9.5.2. The series

$$\gamma = \sum_{i=1}^{\infty} \frac{(-1)^i}{i} \left\lfloor \frac{\ln i}{\ln 2} \right\rfloor \tag{9.5.6}$$

is due to Vacca (1910). Prove it by reducing it to (9.5.5). This identity is reminiscent of the classical

$$\ln \tfrac{1}{2} = \sum_{i=1}^{\infty} \frac{(-1)^i}{i}. \tag{9.5.7}$$

9.6. The Irrationality of γ

The question of whether γ is a rational number is still open. Sondow (2003) has developed a criterion based on the integral

$$F_n(t) := \int_t^{\infty} \left(\frac{n!}{(x)_{n+1}} \right)^2 dx \tag{9.6.1}$$

and the sequence

$$L_n := \sum_{j=n+1}^{\infty} F_n(j) - \binom{2n}{n} \gamma + \sum_{i=0}^{n} \binom{n}{i}^2 H_{n+i}. \tag{9.6.2}$$

The criterion states that $\gamma \notin \mathbb{Q}$ if

$$\overline{\lim_{n \to \infty}} \frac{\{L_n d_{2n}\}}{t^n} > 0$$

for some $t \in (0, e^2/16)$. Here d_n is the least common multiple of the first n integers. The appearance of this number theoretical function is due to its relation with the denominators of the harmonic numbers $H_n = 1 + \frac{1}{2} + \cdots + \frac{1}{n}$.

Project 9.6.1. Write the harmonic number in reduced form $H_n = a_n/b_n$. Explore the relation between b_n and the least common multiple of $\{1, 2, \cdots, n\}$.

The question of irrationality of γ has motivated its numerical evaluation to high precision. Knuth (1962) obtained 1271 places of Euler's constant in 1962, using the Euler–MacLaurin summation. Sweeney (1963) obtained 3683 places the following year, which was subsequently extended by Brent (1977) to 20700 places in 1977. Brent and McMillan (1980) computed 30100 places in 1980 using certain identities involving modified Bessel functions. They also calculated the first 29200 partial quotients in the regular continued fraction expansion of γ and deduced that if $\gamma \in \mathbb{Q}$, then its integer denominator must exceed 10^{15000}. J. Borwein used a variant of Brent's algorithm to compute 172,000 digits of γ in December 1993. Then they conclude that if γ is rational, then its denominator must have at least 60,000 digits. T. Papanikolaou in 1997 improved this to 242080 digits. *It seems unlikely that* γ *is rational.*

10

Eulerian Integrals: The Gamma and Beta Functions

10.1. Introduction

The origin of the gamma function is found in the works of Leonard Euler in his search for an interpolating function for $n!$. In this book, the gamma function has appeared in Chapter 3 in the evaluation of a finite sum

$$\sum_{j=0}^{n} \frac{(-1)^{n-j}}{m-j} \binom{n}{j} = \frac{(-1)^{n+1} n\Gamma(-m)\Gamma(n)}{\Gamma(1-m+n)} \tag{10.1.1}$$

in Corollary 3.4.1. The established value of this sum yields the identity

$$\frac{(-1)^{n+1} n\Gamma(-m)\Gamma(n)}{\Gamma(1-m+n)} = \left[\binom{m}{n} (m-n) \right]^{-1}, \tag{10.1.2}$$

for $m > n \in \mathbb{N}$. This function has also appeared in the symbolic evaluation of the definite integral

$$\int_0^\infty \frac{x^n \, dx}{(a_1 x + a_2)^{m+1}} = \frac{1}{a_1^{n+1} a_2^{m-n}} \frac{\Gamma(m-n)\Gamma(n+1)}{\Gamma(m+1)} \tag{10.1.3}$$

and Wallis' formula (6.4.5):

$$\int_0^\infty \frac{dx}{(x^2+1)^{m+1}} = \frac{\sqrt{\pi}\,\Gamma(\frac{1}{2}+m)}{2\Gamma(m+1)}. \tag{10.1.4}$$

In this chapter we will study the gamma function that appears in these symbolic answers.

The modern definition of the **gamma function**

$$\Gamma(x) := \int_0^\infty e^{-t} t^{x-1} \, dt \tag{10.1.5}$$

is due to Legendre (1809). Euler preferred the equivalent expression

$$\Gamma(x) = \int_0^1 (-\ln t)^{x-1} \, dt. \tag{10.1.6}$$

186

This integral appears in Gradshteyn and Ryzhik (1994) [G & R]: 4.215.1. The history of this function is presented by Davis (1959). Dunham (1999) offers more information about the mathematical work of Euler.

One of the most important properties of the gamma function is the **functional equation**. This is proved in the next proposition.

Proposition 10.1.1. *The Γ function satisfies the functional equation*

$$\Gamma(x+1) = x\,\Gamma(x). \qquad (10.1.7)$$

In the case of integer argument we have

$$\Gamma(k) = (k-1)! \qquad (10.1.8)$$

Proof. The functional equation is obtained by integration by parts. The value $\Gamma(1) = 1$ and (10.1.7) yield (10.1.8). □

Note 10.1.1. The gamma function definition in (10.1.5) is valid for $x \in \mathbb{R}^+$. The functional equation (10.1.7) provides an extension of $\Gamma(x)$ to $x \in \mathbb{R}$ except for the nonpositive integers. This gives the desired extension of the factorial. Compare with Exercise 5.8.2.

Corollary 10.1.1. *The Γ function satisfies*

$$\Gamma(x+k) = \Gamma(x)(x)_k \qquad (10.1.9)$$

where $(x)_k = x(x+1)\cdots(x+k-1)$ is the ascending factorial symbol.

Proof. Apply (10.1.7) k times. □

There are many alternative ways to characterize the Gamma function. For instance Bohr and Mollerup (1922) proved that $\Gamma(x)$ is the only function $f : (0, \infty) \to (0, \infty)$ with $f(1) = 1$ that for $x > 0$ satisfies a) $f(x) > 0$, b) $f(x+1) = xf(x)$ and c) f is log-convex, that is, $\ln f$ is convex.

Wielandt characterized the gamma function as the only analytic function that satsifies the functional equation $f(z+1) = zf(z)$, $z \in \mathbb{C}$ and is bounded on the strip $\{1 \le \operatorname{Re} z \le 2\}$. The reader should consult Remmert (1996), who proves classical properties of $\Gamma(z)$ from this point of view. A different characterization based on an approximation for $\ln n!$ is discussed by Laugwitz and Rodewald (1987).

We now follow Berndt (unpublished) and prove that the functional equation, the value $\Gamma(1) = 1$ and limiting behavior characterize Γ uniquely. The original definition of Euler is a consequence of the proof.

Theorem 10.1.1. *Let $x \in \mathbb{R} - \{0, -1, -2, \cdots\}$. Then there is a unique function $F(x)$ that satisfies*

$$F(1) = 1 \qquad (10.1.10)$$
$$F(x + 1) = xF(x) \qquad (10.1.11)$$
$$\lim_{n \to \infty} \frac{F(x + n)}{n^x F(n)} = 1. \qquad (10.1.12)$$

The function F is given by

$$F(x) = \lim_{n \to \infty} \frac{(n-1)! \, n^x}{(x)_n}. \qquad (10.1.13)$$

Proof. From (10.1.10) and (10.1.11) we obtain $F(n) = (n-1)!$ and

$$F(x + n) = (x + n - 1)(x + n - 2) \cdots (x + 1)xF(x) \quad (10.1.14)$$

so that

$$\frac{F(x + n)}{n^x F(n)} = \frac{(x)_n F(x)}{n^x F(n)} \qquad (10.1.15)$$

and the limiting behavior yields (10.1.13). □

The original expression for $\Gamma(x)$ discovered by Euler is a consequence of Theorem 10.1.1.

Corollary 10.1.2. *The gamma function is given by*

$$\Gamma(x) = \lim_{n \to \infty} \frac{(n-1)! \, n^x}{(x)_n}. \qquad (10.1.16)$$

Proof. It suffices to check that the right-hand side of (10.1.5) satisfies the condition of Theorem 10.1.1. The first two have already been established. We now verify (10.1.12) for $0 < x < 1$, which sufficient in view of the identity (10.1.11).
    Let

$$H(x) = \int_0^\infty t^{x-1} e^{-t} dt$$

so that $H(n) = (n-1)!$. Now let $t = ny$ to obtain

$$\frac{H(x + n)}{H(n)n^x} = \frac{n^n}{(n-1)!} \int_0^\infty y^{x+n-1} e^{-ny} dy. \qquad (10.1.17)$$

Now

$$\int_0^\infty y^{x+n-1} e^{-ny} dy = \int_0^1 y^{x+n-1} e^{-ny} dy + \int_1^\infty y^{x+n-1} e^{-ny} dy$$

$$> \int_0^1 y^n e^{-ny} dy + \int_1^\infty y^{n-1} e^{-ny} dy. \quad (10.1.18)$$

Now integrate

$$\frac{1}{n} \frac{d}{dy} \left(y^n e^{-ny} \right) = y^{n-1} e^{-ny} - y^n e^{-ny}$$

from $y = 0$ to 1 to obtain

$$\int_0^1 y^{n-1} e^{-ny} dy - \int_0^1 y^n e^{-ny} dy = \frac{e^{-n}}{n} \quad (10.1.19)$$

so that (10.1.18) yields

$$\int_0^\infty y^{x+n-1} e^{-ny} dy > \frac{(n-1)!}{n^n} - \frac{e^{-n}}{n}. \quad (10.1.20)$$

Now conclude that

$$\frac{(n-1)!}{n^n} - \frac{e^{-n}}{n} < \int_0^\infty y^{x+n-1} e^{-ny} dy < \frac{(n-1)!}{n^n} + \frac{e^{-n}}{n}.$$

The result now follows from Stirling's formula given in Theorem 5.7.1. □

Exercise 10.1.1. Check the details.

Proposition 10.1.2. *The Γ function satisfies the reflection rule*

$$\Gamma(x)\Gamma(1-x) = \frac{\pi}{\sin \pi x}. \quad (10.1.21)$$

Proof. Use the expression for $\Gamma(x)$ and $\Gamma(1-x)$ given in Corollary 10.1.2 to obtain

$$\Gamma(x)\Gamma(1-x) = \lim_{n\to\infty} \frac{(n-1)!n^x}{x(x+1)\cdots(x+n-1)} \times \frac{(n-1)!n^{1-x}}{(1-x)(2-x)\cdots(n-x)}$$

$$= \lim_{n\to\infty} \frac{(n-1)!^2 n}{x(1^2-x^2)(2^2-x^2)\cdots((n-1)^2-x^2)(n-x)}$$

$$= \lim_{n\to\infty} \left[x \left(1 - x^2/1^2\right) \left(1 - x^2/2^2\right) \cdots \left(1 - x^2/(n-1)^2\right) \right.$$
$$\left. \times (1 - x/n) \right]^{-1}$$

$$= \left[x \prod_{n=1}^\infty \left(1 - x^2/n^2\right) \right]^{-1}.$$

The result now follows from the factorization of $\sin \pi x$ in (6.8.1). □

Exercise 10.1.2. Prove that

$$\Gamma\left(\tfrac{1}{2} - x\right)\Gamma\left(\tfrac{1}{2} + x\right) = \frac{\pi}{\cos \pi x}.$$

Corollary 10.1.3. *The gamma function satisfies*

$$\Gamma(\tfrac{1}{2}) = \sqrt{\pi} \tag{10.1.22}$$

Proof. The value $\Gamma(\tfrac{1}{2})$ is positive, and by (10.1.21) it satisfies $\Gamma(\tfrac{1}{2})^2 = \pi$.
$\square$

Note 10.1.2. This corollary confirms the evaluation of the normal integral given in (8.2.8).

Exercise 10.1.3. Prove that for $m \in \mathbb{N}$

$$\Gamma\left(m + \tfrac{1}{2}\right) = \frac{\sqrt{\pi}}{2^{2m}}\frac{(2m)!}{m!}. \tag{10.1.23}$$

Hint. Use induction and $\Gamma(x + 1) = x\Gamma(x)$.

Exercise 10.1.4. Check [G & R] 3.371. For $n \in \mathbb{N}$ and $\mu > 0$,

$$\int_0^\infty x^{n-1/2} e^{-\mu x}\, dx = \frac{\sqrt{\pi}\,(2n-1)!!}{2^n\,\mu^{n+1/2}} = \frac{\Gamma(n + 1/2)}{\mu^{n+1/2}}.$$

Exercise 10.1.5. Check [G & R] 4.215.2:

$$\int_0^1 (-\ln x)^{-\mu}\, dx = \frac{\pi}{\Gamma(\mu)\sin \mu\pi}.$$

and obtain the special cases [G & R] 4.215.3:

$$\int_0^1 \sqrt{-\ln x}\, dx = \frac{\sqrt{\pi}}{2}.$$

and [G & R] 4.215.4:

$$\int_0^1 \frac{dx}{\sqrt{-\ln x}} = \sqrt{\pi}.$$

Exercise 10.1.6. Check the identities (3.5.7) and (3.5.8).

Exercise 10.1.7. Prove that $\Gamma'(1) = -\gamma$. **Hint.** Differentiate (10.1.5) and use (9.3.3).

Exercise 10.1.8. In this exercise we collect some of the many integrals appearing in [G & R] that can be reduced to values of the gamma function.
a) Check that, for μ, $\nu > 0$,

$$\int_0^\infty x^{\nu-1} e^{-\mu x} \, dx = \frac{\Gamma(\nu)}{\mu^\nu}. \tag{10.1.24}$$

This appears in [G & R]: 3.381.4.
b) Check that, for $\mu > 0$, $\nu > -1$, $u > 0$,

$$\int_u^\infty (x - u)^\nu e^{-\mu x} dx = \mu^{-\nu-1} e^{-u\mu} \Gamma(\nu + 1). \tag{10.1.25}$$

This appears in [G & R]: 3.382.2.
c) Check [G & R]: 3.326. For $\mu > 0$,

$$\int_0^\infty \exp(-x^\mu) \, dx = \frac{1}{\mu} \Gamma\left(\frac{1}{\mu}\right). \tag{10.1.26}$$

d) Check [G & R]: 3.338. For $\mu > 0$,

$$\int_{-\infty}^\infty \exp(-e^x) e^{\mu x} \, dx = \Gamma(\mu). \tag{10.1.27}$$

Exercise 10.1.9. Determine the values of a, b, $c \in \mathbb{R}$ for which

$$\int_0^\infty x^a e^{-bx^c} dx \tag{10.1.28}$$

is convergent. For those values express the integral in terms of the gamma function. Discuss the special cases a, b, $c \in \mathbb{N}$.

Extra 10.1.1. The gamma function satisfies many interesting inequalities. For instance, Gautschi (1974) showed that the harmonic mean of $\Gamma(x)$ and $\Gamma(1/x)$ is greater than or equal to 1,

$$\frac{2}{1/\Gamma(x) + 1/\Gamma(1/x)} \geq 1, \quad x > 0 \tag{10.1.29}$$

and Alzer (1999) proved the analogous result:

$$\frac{2}{1/\Gamma^2(x) + 1/\Gamma^2(1/x)} \geq 1, \quad x > 0. \tag{10.1.30}$$

10.2. The Beta Function

The **beta function** given by

$$B(x, y) = \int_0^1 t^{x-1}(1 - t)^{y-1}\, dt \tag{10.2.1}$$

is considered an essential companion to $\Gamma(x)$. The first result provides a fundamental relation between these two functions. The proof presented here is given by Brown (1961) and employs the notion of **convolution**: given two functions f and g define the convolution $f * g$ by

$$(f * g)(t) = \int_0^t f(\tau)g(t - \tau)\, d\tau. \tag{10.2.2}$$

The reader will recall that the **Laplace transform**, defined by

$$\mathcal{L}(f)(s) = \int_0^\infty e^{-st} f(t)\, dt \tag{10.2.3}$$

satisfies

$$\mathcal{L}(f * g) = \mathcal{L}(f) \cdot \mathcal{L}(g). \tag{10.2.4}$$

Exercise 10.2.1. The gamma function appears in the evaluation of one the simplest Laplace transforms. Confirm this by establishing the identity

$$\mathcal{L}\left(t^{x-1}\right) = \Gamma(x)s^{-x}.$$

Proposition 10.2.1. *The functions beta and gamma are related by the functional equation*

$$B(x, y) = \frac{\Gamma(x)\Gamma(y)}{\Gamma(x + y)}. \tag{10.2.5}$$

Proof. Form

$$\Gamma(x)s^{-x}\Gamma(y)s^{-y} = \mathcal{L}\left(\int_0^t \tau^{x-1}(t - \tau)^{y-1}\, d\tau\right)$$

$$= \mathcal{L}\left(t^{x+y-1}\int_0^1 \rho^{x-1}(1 - \rho)^{y-1}\, d\rho\right)$$

$$= \mathcal{L}\left(t^{x+y-1}B(x, y)\right).$$

Exercise 10.2.1, used in the first line is now used again to complete the proof.   □

Exercise 10.2.2. Check that for $n, m \in \mathbb{N}$

$$B(m, n) = \left(\frac{1}{m} + \frac{1}{n}\right) \left(\frac{m+n}{m}\right)^{-1}.$$

In particular

$$B(m, m) = \frac{2}{m} \left(\frac{2m}{m}\right)^{-1}.$$

10.3. Integral Representations for Gamma and Beta

In this section we consider some elementary properties of the gamma and beta functions. The section consists mostly of definite integrals that can be expressed in terms of these functions.

The next exercise establishes the symmetry of B.

Exercise 10.3.1. The beta function is symmetric in x and y, that is

$$B(x, y) = B(y, x). \tag{10.3.1}$$

Hint. The identity is equivalent to

$$\int_0^1 x^{\nu-1}(1-x)^{\mu-1}dx = \int_0^1 x^{\mu-1}(1-x)^{\nu-1}dx \tag{10.3.2}$$

that follows by a simple change of variable. This appears in [G & R]: 3.191.3.

Exercise 10.3.2. The beta function is represented by

$$B(x, y) = \int_0^\infty \frac{t^{x-1}}{(1+t)^{x+y}} dt. \tag{10.3.3}$$

Hint. Find a change of variables that maps $[0, \infty)$ to $[0, 1]$.

Exercise 10.3.3. Check [G & R] 3.251.6:

$$\int_0^\infty \frac{x^{\mu+1}}{(1+x^2)^2} dx = \frac{\mu\pi}{4\sin(\mu\pi/2)}. \tag{10.3.4}$$

Hint. Reduce to (10.3.3).

Exercise 10.3.4. Evaluate

$$I_{m,n} = \int_0^\infty \frac{dx}{(x^n + 1)^{m+1}} \tag{10.3.5}$$

in terms of the beta function.

Exercise 10.3.5. Check [G & R] 3.166.16

$$\int_0^1 \frac{dx}{\sqrt{1 - x^4}} = \frac{1}{4\sqrt{2\pi}} \Gamma^2 \left(\tfrac{1}{4}\right) \tag{10.3.6}$$

and [G & R] 3.166.18

$$\int_0^1 \frac{x^2 \, dx}{\sqrt{1 - x^4}} = \frac{1}{\sqrt{2\pi}} \Gamma^2 \left(\tfrac{3}{4}\right). \tag{10.3.7}$$

Evaluate also

$$\int_0^1 \frac{x \, dx}{\sqrt{1 - x^4}} \quad \text{and} \quad \int_0^1 \frac{x^3 \, dx}{\sqrt{1 - x^4}}. \tag{10.3.8}$$

Exercise 10.3.6. Prove that

$$\int_0^\infty \left((x^n + 1)^{1/n} - x\right) dx = \frac{1}{2n} B(1 - 2/n, 1/n). \tag{10.3.9}$$

This was proposed by Spiegel and Rosenbaum (1955).

Extra 10.3.1. The results of Exercise 10.3.5 lead to the relation

$$\int_0^1 \frac{dx}{\sqrt{1 - x^4}} \times \int_0^1 \frac{x^2 \, dx}{\sqrt{1 - x^4}} = \frac{\pi}{4}. \tag{10.3.10}$$

This is the lemniscatic identity of Euler (1781). The formula is a special case of an important identity of Legendre among the periods of an elliptic integral. See McKean and Moll (1997), page 69, for details.

The next exercise establishes an integral representation for the beta function in terms of trigonometric functions.

Exercise 10.3.7. The beta function is given by

$$B(x, y) = 2 \int_0^{\pi/2} \cos^{2x-1} \theta \sin^{2y-1} \theta \, d\theta. \tag{10.3.11}$$

Thus

$$\int_0^{\pi/2} \cos^p \theta \, \sin^q \theta \, d\theta = \frac{1}{2} B \left(\frac{p+1}{2}, \frac{q+1}{2} \right). \qquad (10.3.12)$$

In particular

$$\int_0^{\pi/2} \cos^p \theta \, d\theta = \int_0^{\pi/2} \sin^p \theta \, d\theta = \frac{1}{2} B \left(\frac{p+1}{2}, \frac{1}{2} \right) \qquad (10.3.13)$$

$$= \frac{\Gamma((p+1)/2) \sqrt{\pi}}{2 \, \Gamma(p/2 + 1)}.$$

The identity (10.3.12) appears in [G & R] 3.621.5. Use this to confirm [G & R] 3.621.2:

$$\int_0^{\pi/2} \sqrt{\sin x} \, dx = \frac{\sqrt{2}}{\sqrt{\pi}} \Gamma^2 \left(\tfrac{3}{4} \right),$$

$$\int_0^{\pi/2} \sin^{3/2} x \, dx = \frac{1}{6\sqrt{2\pi}} \Gamma^2 \left(\tfrac{1}{4} \right).$$

Exercise 10.3.8. Give a proof of Wallis' formula (6.4.5) using (10.3.12).

Exercise 10.3.9. This exercise outlines a new proof of the functional equation for the gamma and beta functions given in Proposition 10.2.1. First check that

$$\Gamma(x) = 2 \int_0^\infty s^{2x-1} e^{-s^2} \, ds$$

and now compute the product $\Gamma(x)\Gamma(y)$ in polar coordinates to obtain

$$\Gamma(x)\Gamma(y) = 4 \int_0^{\pi/2} \cos^{2x-1} \theta \, \sin^{2y-1} \theta \, d\theta \times \int_0^\infty r^{2x+2y-1} e^{-r^2} \, dr.$$

Now identify the last integral as $\frac{1}{2}\Gamma(x+y)$ and use (10.3.11) to obtain the result.

10.4. Legendre's Duplication Formula

The trigonometric functions satisfy an addition theorem: the relation

$$\sin(x+y) = \sin x \, \cos y + \sin y \, \cos x \qquad (10.4.1)$$

and the special case $\sin 2x = 2 \sin x \cos x$ are familiar to the reader. This last result can be written as

$$\frac{\sin(2\pi x)}{\sin(\pi x)} = 2 \sin(\pi(x + \tfrac{1}{2})). \qquad (10.4.2)$$

In this section we establish a duplication formula of Legendre for Γ that is reminiscent of (10.4.2).

In order to motivate the final result, we try to evaluate the integral $J_{2,m}$ defined in (6.4.5) directly by a symbolic language. Such an attempt yields

$$J_{2,m} = \frac{\sqrt{\pi}}{2} \frac{\Gamma(m+1/2)}{\Gamma(m+1)}, \tag{10.4.3}$$

and using the value for $J_{2,m}$ established in (6.4.5) we obtain

$$\Gamma\left(m + \tfrac{1}{2}\right) = \frac{\sqrt{\pi}}{2^{2m}} \frac{(2m)!}{m!}. \tag{10.4.4}$$

This is Exercise 10.1.3. Legendre's relation extends (10.4.4) for $m \notin \mathbb{N}$. The proof presented here is due to S. K. Lakshmana Rao (1955). It employs the **Mellin transform** defined by

$$\mathfrak{M}\left(f(x)\right)(s) := \int_0^\infty f(x)x^{s-1}dx. \tag{10.4.5}$$

This transform satisfies a convolution rule analog to (10.2.4):

$$\mathfrak{M}f_1(x) \cdot \mathfrak{M}f_2(x) = \mathfrak{M}g(x) \tag{10.4.6}$$

where

$$g(x) = \int_0^\infty f_1\left(\frac{x}{u}\right) f_2(u)\frac{du}{u}$$

is the convolution of f_1 and f_2. Paris and Kaminski (2001) provide more information about this transform and its uses.

Theorem 10.4.1. *Let $x \in \mathbb{R}$. Then*

$$\Gamma(x + \tfrac{1}{2}) = \frac{\Gamma(2x)\Gamma(\tfrac{1}{2})}{\Gamma(x)2^{2x-1}}. \tag{10.4.7}$$

In particular, if $m \in \mathbb{N}$ we recover (10.4.4).

Proof. Consider the functions

$$f_1(x) = e^{-x} \quad \text{and} \quad f_2(x) = e^{-x}x^{1/2}.$$

Then

$$\mathfrak{M}f_1(x) = \Gamma(s) \quad \text{and} \quad \mathfrak{M}f_2(x) = \Gamma(s+1/2)$$

and the convolution of f_1 and f_2 is

$$g(x) = \int_0^\infty e^{-(x/u+u)} \frac{du}{u^{1/2}} = \sqrt{\pi} e^{-2\sqrt{x}} \tag{10.4.8}$$

in view of Exercise 8.4.1. This gives

$$\mathfrak{M}g(x) = \sqrt{\pi} \int_0^\infty e^{-2x^{1/2}} x^{s-1} dx$$

$$= \sqrt{\pi} \int_0^\infty e^{-y} (y/2)^{2s-1} dy$$

$$= \frac{\sqrt{\pi}\,\Gamma(2s)}{2^{2s-1}}.$$

The convolution rule (10.4.6) concludes the proof. $\qquad\square$

Exercise 10.4.1. This exercise reproduces Serret's proof of Legendre's identity. Compute

$$B(x, x) = \int_0^1 (u - u^2)^{x-1} du$$

$$= \int_0^1 (\tfrac{1}{4} - (\tfrac{1}{2} - u)^2)^{x-1} du$$

$$= 2 \int_0^{1/2} (\tfrac{1}{4} - (\tfrac{1}{2} - u)^2)^{x-1} du.$$

Change variables $u \mapsto (1 - \sqrt{v})/2$ to evaluate the last integral as $2^{1-2x} B(\tfrac{1}{2}, x)$.

Exercise 10.4.2. This exercise outlines Liouville's proof of Legendre's duplication formula (10.4.7) for the gamma function. **Hint.** Use (8.4.3) in the form

$$\int_0^\infty e^{-(x+k^2/x)} \frac{dx}{\sqrt{x}} = \sqrt{\pi} e^{-2k} \tag{10.4.9}$$

multiply by $k^{\mu-1}$ and integrate from $k = 0$ to ∞. Evaluate the resulting integrals to produce (10.4.7).

Exercise 10.4.3. Let $x \in \mathbb{R}^+$. Prove that

$$B(x, \tfrac{1}{2}) = \frac{\Gamma^2(x)\, 2^{2x-1}}{\Gamma(2x)} \tag{10.4.10}$$

$$B(x + \tfrac{1}{2}, \tfrac{1}{2}) = \frac{\pi}{x\, 2^{2x-1}} \cdot \frac{\Gamma(2x)}{\Gamma^2(x)} \tag{10.4.11}$$

In the case $x = n \in \mathbb{N}$ this can be written as

$$B(n, \tfrac{1}{2}) = \frac{2^{2n}}{n\binom{2n}{n}}$$

and

$$B(n + \tfrac{1}{2}, \tfrac{1}{2}) = \frac{\pi}{2^{2n}}\binom{2n}{n}.$$

Find also an expression for $B(n, m + \tfrac{1}{2})$ and $B(n + \tfrac{1}{2}, m + \tfrac{1}{2})$.

Exercise 10.4.4. Derive the value of Wallis' integral (6.4.5) from Legendre's duplication formula (10.4.7). **Hint.** Use the change of variables $u = x^2$ to express $J_{2,m}$ in terms of the beta function.

Exercise 10.4.5. Establish the dimidiation and duplication formulas for the ascending factorial symbol given in (1.5.4) and (1.5.5) by using Legendre's formula.

Exercise 10.4.6. Prove Legendre's duplication formula by applying the Ramanujan Master Theorem 7.8.2 to the function $1/\sqrt{1+4x}$. The required Taylor expansion is given in (4.2.6).

10.5. An Example of Degree 4

In Chapter 7 we have shown that

$$N_{0,4}(a; m) = \int_0^\infty \frac{dx}{(x^4 + 2ax^2 + 1)^{m+1}}$$
$$= \frac{\pi}{2^{m+3/2}}(a + 1)^{-m-1/2} P_m(a)$$

where $P_m(a)$ is a polynomial. In this section we evaluate some special cases in terms of Γ and B. The first example is a quartic version of Wallis' formula.

Theorem 10.5.1. *Let* $m \in \mathbb{N}$. *Then*

$$J_{4,m} := \int_0^\infty \frac{dx}{(x^4 + 1)^{m+1}} = \frac{\pi}{m!2^{2m+3/2}} \prod_{k=1}^m (4k - 1). \quad (10.5.1)$$

Proof. The change of variables $t = x^4$ and (10.3.3) produce

$$J_{4,m} = \frac{1}{4}\int_0^\infty \frac{t^{-3/4}\,dt}{(1 + t)^{m+1}} = \frac{1}{4}B\left(\tfrac{1}{4}, m + \tfrac{3}{4}\right).$$

Using (10.2.5) and (10.1.7) we obtain

$$J_{4,m} = \frac{1}{4m!}\Gamma\left(\tfrac{1}{4}\right)\Gamma(m+\tfrac{3}{4})$$

$$= \frac{1}{m!\,2^{2m+2}}\Gamma\left(\tfrac{1}{4}\right)\Gamma\left(\tfrac{3}{4}\right) \times \prod_{k=1}^{m}(4k-1).$$

Finally, the symmetry formula (10.1.21) gives $\Gamma(1/4)\Gamma(3/4) = \pi\sqrt{2}$. This yields (10.5.1). □

Exercise 10.5.1. Check that

$$\frac{J_{4,m}}{\pi\sqrt{2}} \in \mathbb{Q} \tag{10.5.2}$$

and

$$\mu_2\left(\frac{J_{4,m}}{\pi\sqrt{2}}\right) \geq -3m-1 \tag{10.5.3}$$

with equality if and only if m is a power of 2.

Exercise 10.5.2. Prove the identity

$$\prod_{k=1}^{m}(4k-1) = 2^{2m}\left(\tfrac{3}{4}\right)_m \tag{10.5.4}$$

and conclude that

$$J_{4,m} = \frac{\pi}{2\sqrt{2}\,m!}\left(\tfrac{3}{4}\right)_m. \tag{10.5.5}$$

Exercise 10.5.3. Establish the special values

$$N_{0,4}(0;m) = \frac{\pi}{m!}2^{-2m-3/2} \times \prod_{l=1}^{m}(4l-1)$$

$$N_{0,4}(1;m) = \frac{\pi}{2^{4m+2}}\binom{4m+1}{2m}$$

$$N'_{0,4}(0;m) = \frac{-\pi}{2^{2m+5/2}m!}\prod_{l=1}^{m}(4l+1)$$

$$N'_{0,4}(1;m) = -\frac{4m+3}{2m+3}\frac{\pi}{2^{4(m+1)}}\binom{4m+1}{2m} \tag{10.5.6}$$

and deduce the special values for P_m:

$$P_m(0) = \frac{1}{m!2^m} \prod_{l=1}^{m}(4l - 1)$$

$$P_m(1) = 2^{-2m} \binom{4m + 1}{2m}$$

$$P'_m(0) = \frac{1}{m!2^{m+1}} \left[-\prod_{l=1}^{m}(4l + 1) + (2m + 1)\prod_{l=1}^{m}(4l - 1) \right]$$

$$P'_m(1) = \frac{m(m + 1)}{2m + 3}P_m(1). \tag{10.5.7}$$

These values were used in Exercise 7.6.2. **Hint**. Use a method similar to the proof of Theorem 10.5.1.

Note 10.5.1. The values of $d_0(m)$ produced in (7.3.2) and (10.5.7) yield the (uninteresting) identity:

$$\sum_{s=0}^{m}\sum_{k=s}^{m}(-1)^{k-s}2^{-3k}\binom{2k}{k}\binom{2m + 1}{2s}\binom{m - s}{m - k} = \frac{1}{m!2^m}\prod_{l=1}^{m}(4l - 1).$$

In particular, the sum on the left-hand side is nonnegative.

Exercise 10.5.4. The previous identity shows that the odd part of $m!$ divides the product of the first m numbers congruent to 3 modulo 4. Give a direct proof. See Exercise 1.2.5 for a related problem.

Similarly, the two expressions for $d_1(m)$ yield

$$\sum_{s=0}^{m-1}\sum_{k=s+1}^{m}(-1)^{k-s-1}2^{-3k}\binom{2k}{k} \times (m - s)\binom{2m + 2}{2s + 1}\binom{m - s - 1}{m - k}$$

$$= \frac{1}{2^{m+1}m!}\left[(2m + 1)\prod_{l=1}^{m}(4l - 1) - \prod_{l=1}^{m}(4l + 1) \right].$$

Exercise 10.5.5. Prove that the odd part of $m!$ divides

$$A_1(m) := (2m + 1)\prod_{l=1}^{m}(4l - 1) - \prod_{l=1}^{m}(4l + 1). \tag{10.5.8}$$

Hint. Use the previous identity. Compare with Exercise 1.2.5.

10.6. The Expansion of the Loggamma Function

The values of the Riemann zeta function

$$\zeta(s) = \sum_{n=1}^{\infty} \frac{1}{n^s} \qquad (10.6.1)$$

at the even integers appeared in (6.9.6) in relation to the expansion of the cotangent function. In this section we show that all the values of $\zeta(k)$, for $k \in \mathbb{N}$, appear in the Taylor expansion of $\ln \Gamma(x + 1)$ and we postpone the study of $\zeta(s)$ to Chapter 11. This expansion also involves the Euler constant γ considered in Chapter 9.

Theorem 10.6.1. *The Taylor series expansion of* $\ln \Gamma(1 + x)$ *is given by*

$$\ln \Gamma(1 + x) = -\gamma x + \sum_{k=2}^{\infty} \frac{(-1)^k \zeta(k)}{k} x^k \qquad (10.6.2)$$

is valid for $|x| < 1$.

Proof. The expression

$$\lim_{\alpha \to 0^+} \left(\frac{1 - e^{-\alpha v}}{\alpha} \right)^x = v^x$$

is replaced in

$$\Gamma(1 + x) = \int_0^\infty v^x e^{-v} dv$$

to produce

$$\Gamma(1 + x) = \lim_{\alpha \to 0^+} \int_0^\infty \frac{e^{-v} (1 - e^{-\alpha v})^x}{\alpha^x} dv.$$

Let $\alpha = 1/b$ and use the change of variable $y = e^{-v/b}$ to obtain

$$\Gamma(1 + x) = \lim_{b \to \infty} b^{x+1} \int_0^1 y^{b-1} (1 - y)^x \, dy$$

$$= \lim_{b \to \infty} b^{x+1} B(b, x + 1).$$

We conclude that

$$\lim_{b \to \infty} b^{x+1} \frac{\Gamma(b)}{\Gamma(b + x + 1)} = 1,$$

and thus

$$\lim_{b\to\infty} \frac{(x+b)(x+b-1)\cdots(x+1)\Gamma(x+1)}{(b-1)! \times b^{x+1}} = 1.$$

Therefore

$$\ln\Gamma(x+1) = \lim_{b\to\infty} x\ln b - \ln(1+x) - \ln(1+x/2) - \cdots - \ln(1+x/b).$$
(10.6.3)

Expanding the logarithms in (10.6.3) and using the notation

$$H_n^{(k)} = \sum_{k=1}^{n} \frac{1}{j^k}$$
(10.6.4)

for the **harmonic numbers of order** k, we obtain

$$\ln\Gamma(x+1) = \lim_{b\to\infty} -(H_b - \ln b)x + \tfrac{1}{2}H_b^{(2)}x^2 - \cdots$$

and this is (10.6.2).   □

Exercise 10.6.1. Use the relation

$$\Gamma(1+x/\pi)\Gamma(1-x/\pi) = \frac{x}{\sin x}$$
(10.6.5)

to derive an expansion for $\ln\sin x$.

Exercise 10.6.2. Derive Legendre's formulas

$$\ln\left(\frac{\Gamma(1+x)}{\Gamma(1-x)}\right) = -\ln\left(\frac{1+x}{1-x}\right) - (\gamma-1)x$$
$$- \sum_{k=1}^{\infty} \frac{\zeta(2k+1)-1}{2k+1} x^{2k+1}$$

and

$$\ln\Gamma(x+1) = \frac{1}{2}\ln\left(\frac{\pi x}{\sin\pi x}\right) - \frac{1}{2}\ln\left(\frac{1+x}{1-x}\right) - (\gamma-1)x$$
$$- \sum_{k=1}^{\infty} \frac{\zeta(2k+1)-1}{2k+1} x^{2k+1}.$$

Exercise 10.6.3. Prove Euler's formula

$$\gamma = \sum_{k=2}^{\infty} \frac{(-1)^k\zeta(k)}{k}.$$
(10.6.6)

Establish also the identities

$$\gamma = \ln\left(\frac{4}{\pi}\right) + 2\sum_{k=2}^{\infty}(-1)^k\frac{\zeta(k)}{2^k k}$$

and

$$\gamma = 1 - \ln\left(\frac{3}{2}\right) - \sum_{k=1}^{\infty} \frac{\zeta(2k+1) - 1}{(2k+1)4^k}.$$

Note 10.6.1. The text by H. Srivastava and J. Choi (2001) contains an over-whelming number of series representations for the Euler constant and other constants of analysis. For instance

$$\sum_{k=2}^{\infty} (-1)^k \frac{\zeta(k) - 1}{k} \left(\frac{3}{2}\right)^k = \frac{3\gamma}{2} - \frac{3}{2} + \ln\left(\frac{15\sqrt{\pi}}{8}\right) \quad (10.6.7)$$

appears on page 174, formula (154).

Extra 10.6.1. The value

$$\int_0^1 \ln \Gamma(x)\,dx = \ln \sqrt{2\pi} \quad (10.6.8)$$

is due to Euler. The example

$$\int_0^1 \ln^2 \Gamma(x)\,dx = \frac{\gamma^2}{12} + \frac{\pi^2}{48} + \frac{1}{3}\gamma \ln \sqrt{2\pi}$$
$$+ \frac{4}{3}\ln^2 \sqrt{2\pi} - (\gamma + 2\ln\sqrt{2\pi})\frac{\zeta'(2)}{\pi^2} + \frac{\zeta''(2)}{2\pi^2}$$

was obtained by Espinosa and Moll (2002). These two are examples of the family

$$L_n = \int_0^1 \ln^n \Gamma(x)\,dx \quad (10.6.9)$$

which is the subject of current research. See Espinosa and Moll (2004) for details.

We now define a **weight** to some real numbers according to the rules:

- $w(r) = 0$ if $r \in \mathbb{Q}$.
- $w(\zeta^{(j)}(k)) = k + j$ for $k,\ j \in \mathbb{N}$. For example $w(\zeta''(3)) = 5$.
- $w(\pi) = 1$.
- $w(\gamma) = 1$. This is consistent with the heuritiscs $\zeta(1) = \gamma$.
- $w(xy) = w(x) + w(y)$ for $x,\ y \in \mathbb{R}$.
- The weight is invariant under ln or radicals. For example

$$w(\ln \sqrt{2\pi}) = w(2\pi) = w(2) + w(\pi) = 1. \quad (10.6.10)$$

Under these assumptions we observe that for $n = 1$ and 2 the integral L_n is a homogeneous form[1] of weight n.

---

[1] Every term has the same weight n.

10.7. The Product Representation for $\Gamma(x)$

The infinite product

$$\sin \pi x = \pi x \prod_{k=1}^{\infty} \left(1 - \frac{x^2}{k^2}\right) \tag{10.7.1}$$

given in (6.8.1) makes it explicit that $\sin \pi x$ is a function that vanishes precisely at $x = k \in \mathbb{Z}$. It turns out that the problem of finding an analytic function that vanishes precisely at the negative integers leads to the reciprocal of the gamma function. The naive candidate

$$f(x) = x \prod_{k=1}^{\infty} \left(1 + \frac{x}{k}\right) \tag{10.7.2}$$

has to be modified due to lack of convergence of the infinite product in (10.7.2). See Extra 6.8.1. This leads to the product

$$\frac{1}{\Gamma(x)} = x e^{\gamma x} \prod_{k=1}^{\infty} \left(1 + \frac{x}{k}\right) e^{-x/k}, \tag{10.7.3}$$

where γ is the Euler constant discussed in Chapter 9.

Exercise 10.7.1. Check the identity by showing that the reciprocal of the product satisfies Theorem 10.1.1. **Hint**. The identities

$$e^{-\gamma} = \prod_{k=1}^{\infty} \left(\frac{k+1}{k}\right) e^{-1/k}, \tag{10.7.4}$$

$$\frac{1}{x+1} = \prod_{k=1}^{\infty} \frac{k+x+1}{k+x} \times \frac{k}{k+1}$$

$$e^{-\gamma x} = \lim_{n \to \infty} n^x \prod_{k=1}^{n} e^{-x/k}$$

might be helpful.

Exercise 10.7.2. Derive

$$\ln \Gamma(x) = -\ln x - \gamma x - \sum_{n=1}^{\infty} \left\{\ln(1 + x/n) - \frac{x}{n}\right\}. \tag{10.7.5}$$

Compute $(\ln \Gamma(x))''$ and check that Γ is log-convex so it satisfies the hypothesis of the Bohr–Mollerup theorem.

Extra 10.7.1. The **second logarithmic derivative** operation in Exercises 6.9.4 and 10.7.2 reappears in the context of elliptic functions. The two competing theories are due to Weierstrass with his function

$$\wp(x) = \frac{1}{x^2} + \sum \left(\frac{1}{(x - n\omega_1 - m\omega_2)^2} - \frac{1}{(n\omega_1 + m\omega_2)^2} \right) \quad (10.7.6)$$

(where the sum extends over $(m, n) \in \mathbb{Z}^2 - (0, 0)$) and to Jacobi with his **theta function**

$$\vartheta_1(x, \omega) = i \sum_{n \in \mathbb{Z}} (-1)^n e^{(2n-1)\pi i x + (n-1/2)^2 \pi i \omega}. \quad (10.7.7)$$

The two are related by

$$\wp(x) = -(\log \vartheta_1(x))'' \quad (10.7.8)$$

up to a constant. See McKean and Moll (1997) for details.

Exercise 10.7.3. Derive Legendre's duplication formula (10.4.7) from the product representation (10.7.3). **Hint.** Use the identities

$$\left(1 + \frac{x + 1/2}{k}\right) \left(1 + \frac{x}{k}\right) = \left(\frac{2k + 1}{2k}\right) \left(1 + \frac{x}{k}\right) \left(1 + \frac{2x}{2k + 1}\right),$$

and

$$\frac{2x}{k} = \frac{x}{k} + \frac{2x}{2k + 1} + \frac{x}{k(2k + 1)}. \quad (10.7.9)$$

Project 10.7.1. In this project we present a proof of **Gauss' multiplicative formula**

$$\prod_{k=0}^{m-1} \Gamma\left(z + \frac{k}{m}\right) = (2\pi)^{(m-1)/2} m^{\frac{1}{2} - mz} \Gamma(mz).$$

The special case $m = 2$ gives Legendre's duplication formula (10.4.7).
a) Check that

$$(z)_n \left(z + \frac{1}{m}\right)_n \cdots \left(z + \frac{m - 1}{m}\right)_n = m^{-mn}(mz)_{mn}. \quad (10.7.10)$$

b) Let $G(z)$ be the left-hand side of the formula. Confirm that

$$\frac{\Gamma(mz)}{m^{mz} G(z)} = \lim_{n \to \infty} \frac{m^{-mn} (mn - 1)!}{(n - 1)!^m n^{(m-1)/2}}$$

is independent of z.
c) Use Stirling's asymptotic formula for $n!$ to check that the constant in part b) is $(2\pi)^{-(m-1)/2} m^{-1/2}$.

10.8. Formulas from Gradshteyn and Rhyzik (G & R)

The goal of this section is to confirm some of the integrals appearing in [G & R] by using the Eulerian functions gamma and beta introduced in this chapter.

Exercise 10.8.1. Check that

$$\int_0^\infty \frac{x^{\mu-1}\, dx}{\sqrt{1+x^\nu}} = \frac{1}{\nu} B\left(\frac{\mu}{\nu}, \frac{1}{2} - \frac{\mu}{\nu}\right). \tag{10.8.1}$$

This appears in [G & R]: 3.248.1.

Exercise 10.8.2. Check that

$$\int_0^1 \frac{x^{2n+1}\, dx}{\sqrt{1-x^2}} = \frac{(2n)!!}{(2n+1)!!} \tag{10.8.2}$$

and

$$\int_0^1 \frac{x^{2n}\, dx}{\sqrt{1-x^2}} = \frac{(2n-1)!!}{(2n)!!} \frac{\pi}{2}. \tag{10.8.3}$$

These appear in [G & R]: 3.248.2 and 3.248.3 respectively.

Exercise 10.8.3. Check that

$$\int_1^\infty x^{\mu-1}(x^p - 1)^{\nu-1} dx = \frac{1}{p} B\left(1 - \nu - \frac{\mu}{p}, \nu\right). \tag{10.8.4}$$

This appears in [G & R]: 3.251.3.

Exercise 10.8.4. Check that for $p > 0$

$$\int_0^1 (1 - \sqrt{x})^{p-1}\, dx = \frac{2}{p(p+1)}. \tag{10.8.5}$$

This appears in [G & R]: 3.249.6.

Exercise 10.8.5. Check that for $n \in \mathbb{N}$, $n > 1$

$$\int_{-\infty}^\infty \left(1 + \frac{x^2}{n-1}\right)^{-n/2} dx = \frac{\sqrt{\pi(n-1)}}{\Gamma(n/2)} \Gamma\left(\frac{n-1}{2}\right). \tag{10.8.6}$$

This appears in [G & R]: 3.249.8.

Exercise 10.8.6. Check that for $\mu > \nu > 0$

$$\int_{-\infty}^\infty \frac{e^{-\mu x}}{(1 + e^{-x})^\nu}\, dx = B(\mu, \nu - \mu). \tag{10.8.7}$$

This appears in [G & R]: 3.313.2.

Exercise 10.8.7. This exercise gives the evaluation of an integral that will be used in the next section. Prove that

$$\int_0^\infty \frac{x^{2r}\,dx}{(x^4+1)^{m+1}} = \tfrac{1}{4}B\left(\tfrac{r}{2}+\tfrac{1}{4}, m-\tfrac{r}{2}+\tfrac{3}{4}\right). \qquad (10.8.8)$$

10.9. An Expression for the Coefficients $d_l(m)$

In this section we prove the existence of polynomials $\alpha_l(x)$ and $\beta_l(x)$ with positive integer coefficients such that

$$d_l(m) = \frac{1}{l!\,m!\,2^{m+l}}\left(\alpha_l(m)\prod_{k=1}^m (4k-1) - \beta_l(m)\prod_{k=1}^m (4k+1)\right),$$

where $d_l(m)$ are the coefficients of the polynomial $P_m(a)$ introduced in Section 7.2. These polynomials are efficient for the calculation of $d_l(m)$ if l is small relative to m, so they complement the results of Corollary 7.9.1.
For example

$$\begin{aligned}
\alpha_0(m) &= 1 \\
\alpha_1(m) &= 2m+1 \\
\alpha_2(m) &= 2(2m^2+2m+1) \\
\alpha_3(m) &= 4(2m+1)(m^2+m+3) \\
\alpha_4(m) &= 8(2m^4+4m^3+26m^2+24m+9).
\end{aligned}$$

and

$$\begin{aligned}
\beta_0(m) &= 0 \\
\beta_1(m) &= 1 \\
\beta_2(m) &= 2(2m+1) \\
\beta_3(m) &= 12(m^2+m+1) \\
\beta_4(m) &= 8(2m+1)(2m^2+2m+9).
\end{aligned}$$

The terms α_1 and β_1 appeared in Exercise 1.2.5.
The proof of the next theorem will use the next two exercises.

Exercise 10.9.1. Prove that for $m, j \in \mathbb{N}$,

$$\prod_{v=1}^m (4v-1+2j) = \prod_{v=1}^m (4v+1)\left(\prod_{v=m+1}^{m+t-1}(4v+1)\Big/\prod_{v=1}^{t-1}(4v+1)\right)$$

with $j = 2t - 1$ and

$$\prod_{v=1}^{m}(4v - 1 + 2j) = \prod_{v=1}^{m}(4v - 1) \left(\prod_{v=m+1}^{m+t} (4v - 1) \bigg/ \prod_{v=1}^{t}(4v - 1) \right)$$

for $j = 2t$.

Exercise 10.9.2. Check that for $m, \; r \in \mathbb{N}$

$$\frac{(m + 1)!}{(m + 1 - r)!} = \prod_{j=1}^{r}(j + m + 1 - r)$$

and

$$\frac{(2m + 2)!}{(2m + 2 - 2r)!} = 2^r \left(\prod_{i=1}^{r}(i + m + 1 - r) \right) \left(\prod_{i=1}^{r}(2i + 2m + 1 - 2r) \right).$$

Exercise 10.9.3. Check the identity

$$\prod_{v=1}^{t}(4v - 1) \prod_{v=1}^{t-1}(4v + 1) = \frac{(4t)!}{2^{2t}(2t)!}. \tag{10.9.1}$$

Theorem 10.9.1. *There exist polynomials $\alpha_l(x)$ and $\beta_l(x)$ with integer coefficients such that*

$$d_l(m) = \frac{1}{l! \, m! \, 2^{m+l}} \left(\alpha_l(m) \prod_{k=1}^{m}(4k - 1) - \beta_l(m) \prod_{k=1}^{m}(4k + 1) \right).$$

Proof. The proof consists in computing the expansion of $P_m(a)$ via the Leibnitz rule:

$$P_m(a) = \frac{2^{m+3/2}}{\pi} \sum_{j=0}^{l} \binom{l}{j} \left(\frac{d}{da} \right)^{l-j} (a + 1)^{m+1/2} \bigg|_{a=0} \left(\frac{d}{da} \right)^{j} N_{0,4}(a; m) \bigg|_{a=0}.$$

We have

$$\left(\frac{d}{da} \right)^{r} (a + 1)^{m+1/2} \bigg|_{a=0} = 2^{-2r} \frac{(2m + 2)!}{(m + 1)!} \frac{(m - r + 1)!}{(2m - 2r + 2)!}$$

and

$$\left(\frac{d}{da} \right)^{r} N_{0,4}(a; m) \bigg|_{a=0} = (-1)^r \frac{(m + r)!}{m!} 2^r \int_0^{\infty} \frac{x^{2r} \, dx}{(x^4 + 1)^{m+r+1}}.$$

The integral is evaluated in Exercise 10.8.7. The beta expression can now be simplified using (10.1.9), (10.2.5) and (10.1.21). The final result is

$$\left(\frac{d}{da}\right)^r N_{0,4}(a;m)\Bigg|_{a=0} = \frac{(-1)^r(2r)!}{2^{2r+2m+3/2}}\frac{\pi}{m!r!}\prod_{l=1}^{m}(4l-1+2r).$$

Therefore

$$P_m^{(l)}(0) = \frac{l!(2m+2)!}{2^{m+2l}m!(m+1)!}\sum_{j=0}^{l}\frac{(-1)^j(m-l+j+1)!(2j)!}{j!^2(l-j)!(2m-2l+2j+2)!}$$

$$\times \prod_{v=1}^{m}(4v-1+2j).$$

We now split the sum according to the parity of j and use the result of Exercise 10.9.1 to obtain

$$d_l(m) = X(m,l)\prod_{v=1}^{m}(4v-1) - Y(m,l)\prod_{v=1}^{m}(4v+1)$$

with

$$X(m,l) = \frac{(2m+2)!}{2^{m+2l}m!(m+1)!}\sum_{t=0}^{\lfloor l/2\rfloor}\frac{(m-l+2t+1)!(4t)!}{(2t)!^2(l-2t)!(2m-2l+4t+2)!}$$

$$\times \frac{\prod_{v=m+1}^{m+t}(4v-1)}{\prod_{v=1}^{t}(4v-1)}$$

and

$$Y(m,l) = \frac{(2m+2)!}{2^{m+2l}m!(m+1)!}\sum_{t=1}^{\lfloor(l+1)/2\rfloor}\frac{(m-l+2t)!(4t-2)!}{(2t-1)!^2(l-2t+1)!(2m-2l+4t)!}$$

$$\times \frac{\prod_{v=m+1}^{m+t-1}(4v+1)}{\prod_{v=1}^{t-1}(4v+1)}.$$

The quotients of factorials appearing above can be simplified via the results of Exercise 10.9.2 to obtain

$$d_l(m) = \frac{1}{l!m!2^{m+l}}\left(\alpha_l(m)\prod_{v=1}^{m}(4v-1) - \beta_l(m)\prod_{v=1}^{m}(4v+1)\right)$$

with

$$\alpha_l(m) = l!\sum_{t=0}^{\lfloor l/2\rfloor}\frac{\binom{4t}{2t}}{2^{2t}(l-2t)!}\frac{\prod_{v=m+1}^{m+t}(4v-1)}{\prod_{v=1}^{t}(4v-1)}\left(\prod_{v=m-(l-2t-1)}^{m}(2v+1)\right)$$

and

$$\beta_l(m) = l! \sum_{t=1}^{\lfloor (l+1)/2 \rfloor} \frac{\binom{4t-2}{2t-1}}{2^{2t-1}(l-2t+1)!} \left(\frac{\prod_{v=m+1}^{m+t-1}(4v+1)}{\prod_{v=1}^{t-1}(4v+1)} \right)$$
$$\times \left(\prod_{v=m-(l-2t)}^{m} (2v+1) \right).$$

The identity (10.9.1) is now employed to produce

$$\alpha_l(m) = \sum_{t=0}^{\lfloor l/2 \rfloor} \binom{l}{2t} \prod_{v=m+1}^{m+t} (4v-1) \prod_{v=m-(l-2t-1)}^{m} (2v+1) \prod_{v=1}^{t-1}(4v+1)$$

and

$$\beta_l(m) = \sum_{t=1}^{\lfloor (l+1)/2 \rfloor} \binom{l}{2t-1} \prod_{v=m+1}^{m+t-1} (4v+1) \prod_{v=m-(l-2t)}^{m} (2v+1) \prod_{v=1}^{t-1}(4v-1).$$

$\square$

Exercise 10.9.4. Express the polynomials α_l and β_l in terms of the ascending factorial symbol. **Hint**. See Exercise 10.5.2.

Extra 10.9.1. The polynomials $\alpha_l(m)$ and $\beta_l(m)$ have all their roots on the line $\mathrm{Re}(m) = -1/2$. This remarkable fact was proved by John Little (2004). The proof employs the auxiliary polynomials

$$C_l(t) := (-i)^l \alpha_l \left(\tfrac{it-1}{2} \right) \tag{10.9.2}$$

that satisfy the three-term recurrence

$$C_{l+1}(t) = 2t C_l(t) - (t^2 + (2l-1)^2) C_{l-1}(t) \tag{10.9.3}$$

and that the same holds for

$$D_l(t) := (-i)^{l-1} \beta_l \left(\tfrac{it-1}{2} \right). \tag{10.9.4}$$

10.10. Holder's Theorem for the Gamma Function

The functions $f(x) = \ln x$ and $g(x) = \tan^{-1} x$ have been shown to be nonrational. On the other hand it is relatively simple to produce differential equations that they satisfy

$$xf'(x) - 1 = 0 \tag{10.10.1}$$

and

$$(1 + x^2)g'(x) - 1 = 0. \tag{10.10.2}$$

The goal of this section is to prove that $\Gamma(x)$ does not satisfy a differential equation with polynomial coefficients. The proof presented here is given by Totik (1993).

Theorem 10.10.1. *There is no polynomial* $P = P(x, y_0, y_1, \cdots, y_n)$ *such that*

$$P(x, \Gamma(x), \Gamma'(x), \cdots, \Gamma^{(n)}(x)) \equiv 0. \tag{10.10.3}$$

Proof. Let

$$P = \sum q_k(x) y_0^{a_0} y_1^{a_1} \cdots y_n^{a_n} \tag{10.10.4}$$

where $k = (a_0, a_1, \cdots, a_n)$ be the polynomial of minimal degree that satisfies (10.10.3). The leading term of P is the one with largest $(a_0, a_1, \cdots, a_n)$.
  The relation $\Gamma(x + 1) = x\Gamma(x)$ shows that

$$Q(x, y_0, y_1, \cdots, y_n) = P(x + 1, xy_0, xy_1 + y_0, \cdots, xy_n + ny_{n-1})$$

is also a counterexample and its leading term is

$$LT(Q) = q_k(x + 1)x^{a_0 + a_1 + \cdots + a_n} y_0^{a_0} \cdots y_n^{a_n}. \tag{10.10.5}$$

Applying the euclidean algorithm to the leading terms of Q and P shows that P must divide Q. Any nonzero remainder would violate the minimality of the degree of P. Thus

$$P(x + 1, xy_0, xy_1 + y_0, \cdots, xy_n + ny_{n-1}) = R(x)P(x, y_0, y_1, \cdots, y_n) \tag{10.10.6}$$

with $\deg(R) \geq 1$. Considering the leading terms we obtain

$$q_k(x)R(x) = q_k(x + 1)x^{a_0 + a_1 + \cdots + a_n} \tag{10.10.7}$$

Now replace $x = x_0$, a zero of R, in (10.10.6) to obtain

$$P(x_0 + 1, x_0 y_0, \cdots, x_0 y_n + ny_{n-1}) = 0.$$

In the case $x_0 \neq 0$, we conclude that $x - (x_0 + 1)$ must divide P, contradicting the minimality of its degree. Therefore $x_0 = 0$ and

$$P(1, 0, z_1, \cdots, z_n) = 0. \tag{10.10.8}$$

The relation (10.10.6) now yields

$$P(m, 0, z_1, \cdots, z_n) = 0 \qquad (10.10.9)$$

for any $m \in \mathbb{N}$. Therefore $P(x, 0, z_1, \cdots, z_n) \equiv 0$. This shows that P is divisible by y_0 and we obtain a final contradiction to the minimality of P. □

10.11. The Psi Function

In this section we consider the logarithmic derivative of the Γ function

$$\psi(x) := \frac{\Gamma'(x)}{\Gamma(x)}. \qquad (10.11.1)$$

This function is the analog of $\cot x$ studied in Chapter 6.

The first representation of $\psi(x)$ is by a series.

Proposition 10.11.1. *The ψ function is given by*

$$\psi(x) = -\gamma - \frac{1}{x} - \sum_{k=1}^{\infty} \left(\frac{1}{k+x} - \frac{1}{k} \right). \qquad (10.11.2)$$

Proof. This follows directly by differentiating the series for $\ln \Gamma(x)$ given in (10.7.5). □

Exercise 10.11.1. Check that $\psi(1) = -\gamma$. Conclude that $\Gamma'(1) = -\gamma$.

Proposition 10.11.2. *Let H_k be the harmonic number and $H_0 = 0$. The function ψ satisfies*

$$\psi(x + k) = \psi(x) + \sum_{j=1}^{k} \frac{1}{x+j-1},$$
$$\psi(k) = -\gamma + H_{k-1},$$
$$\psi(k + 1/2) = -\gamma - 2\ln 2 + 2H_{2k} - H_k,$$
$$\psi(x) - \psi(1 - x) = -\pi \cot(\pi x). \qquad (10.11.3)$$

Proof. The logarithmic derivative of $\Gamma(x)\Gamma(1 - x) = \pi/\sin \pi x$ yields the first property. The second one follows from letting $x = 1$ in (10.11.3) and the proof of the third one is similar. Finally, the last property is obtained from the logarithmic derivative of $\Gamma(x)\Gamma(1 - x) = \pi/\sin \pi x$. □

Theorem 10.11.1. *The function $\psi(x)$ is given by*

$$\psi(x) = -\frac{1}{x} - \gamma + \sum_{k=2}^{\infty}(-1)^k \zeta(k)x^{k-1}. \qquad (10.11.4)$$

Proof. Differentiate the result of Theorem 10.6.1. □

Exercise 10.11.2. Prove that the function ψ satisfies

$$\psi(x + \tfrac{1}{2}) = 2\psi(2x) - \psi(x) - 2\ln 2, \qquad (10.11.5)$$
$$= -(\gamma + 2\ln 2) + \sum_{k=2}^{\infty}(-1)^k \zeta(k)\left[2^{k-1} - 1\right]x^{k-1}.$$

Exercise 10.11.3. Use Theorem 10.6.1 to obtain the values

$$\psi(1) = -\gamma, \quad \text{and} \quad \psi'(1) = \frac{\pi^2}{6}. \qquad (10.11.6)$$

Extra 10.11.1. The values of ψ for a rational argument were given by Gauss. For $p, q \in \mathbb{N}$, $0 < p < q$ we have:

$$\psi\left(\frac{p}{q}\right) = -\gamma - \frac{\pi}{2}\cot\left(\frac{\pi p}{q}\right)$$
$$- \ln(2q) + \sum_{k=1}^{q-1}\cos\left(\frac{2\pi kp}{q}\right)\ln\left(\sin\left(\frac{\pi k}{q}\right)\right).$$

See Andrews et al. (1999) for a proof.

Project 10.11.1. Use the expansion of ψ and Γ at $x = 1$ to produce

$\Gamma(1) = 1$

$\Gamma'(1) = -\gamma$

$\Gamma''(1) = \zeta(2) + \gamma^2$

$\Gamma^{(3)}(1) = -\left(2\zeta(3) + 3\gamma\zeta(2) + \gamma^3\right)$

$\Gamma^{(4)}(1) = 6\zeta(4) + 3\zeta^2(2) + 8\gamma\zeta(3) + 6\gamma^2\zeta(2) + \gamma^4$

$\Gamma^{(5)}(1) = -\left(24\zeta(5) + 20\zeta(2)\zeta(3) + 15\gamma\zeta^2(2) + 30\gamma\zeta(4) + 20\gamma^2\zeta(3)\right.$
$\left. + 10\gamma^3\zeta(2) + \gamma^5\right).$

Use the notion of weight defined in Extra 10.6.1 to prove that $\Gamma^{(n)}(1)$ is a homogeneous polynomial of degree n. **Hint.** Establish the recursion

$$\Gamma^{(n+1)}(1) = -\gamma \Gamma^{(n)}(1) + n! \sum_{k=1}^{n} \frac{(-1)^{k+1}}{(n-k)!} \zeta(k+1)\Gamma^{(n-k)}(1).$$

Shrivastava and Choi's text (2001) contains the values of $\Gamma^{(n)}(1)$ for $1 \leq n \leq 10$.

Exercise 10.11.4. Check the special values

$$\psi\left(\tfrac{1}{2}\right) = -(\gamma + 2\ln 2),$$
$$\Gamma'\left(\tfrac{1}{2}\right) = -\sqrt{\pi}\,(\gamma + 2\ln 2).$$

Exercise 10.11.5. Let $S(k) = \zeta(k) - 1$. Check that

$$\sum_{k=2}^{\infty} S(k)x^k = (1 - \gamma)x - x\psi(2 - x). \qquad (10.11.7)$$

Establish the identities

$$\sum_{k=1}^{\infty} S(2k)x^{2k} = \frac{x}{2}\left[\psi(2 + x) - \psi(2 - x)\right],$$

$$\sum_{k=2}^{\infty} \frac{S(k)}{k}x^k = (1 - \gamma)x + \ln \Gamma(2 - x),$$

$$\sum_{k=1}^{\infty} \frac{S(2k)}{k}x^{2k} = \ln\left(\Gamma(2 - x)\Gamma(2 + x)\right),$$

$$\sum_{k=1}^{\infty} \frac{S(2k + 1)}{2k + 1}x^{2k+1} = (1 - \gamma)x + \frac{1}{2}\ln\left(\frac{\Gamma(2 - x)}{\Gamma(2 + x)}\right).$$

The following special values appear in Bromwich (1926) in the section on miscellaneous examples, page 526, statement 6. A direct evaluation of these cases is provided by Johnson (1906).

Exercise 10.11.6. Confirm the evaluations

$$\sum_{k=2}^{\infty} S(k) = 1, \quad \sum_{k=2}^{\infty} \frac{S(k)}{k} = 1 - \gamma, \quad \sum_{k=1}^{\infty} S(2k) = \frac{3}{4}, \quad \sum_{k=2}^{\infty} \frac{S(2k)}{k} = \ln 2.$$

Extra 10.11.2. The ψ function is the first element of the family of **polygamma functions** defined by

$$\text{PolyGamma}[m, x] = \left(\frac{d}{dx}\right)^m \psi(x) \qquad (10.11.8)$$

so that PolyGamma[0, x] = $\psi(x)$. These functions are related to the Hurwitz zeta function

$$\zeta(s, q) = \sum_{n=0}^{\infty} \frac{1}{(n+q)^s} \tag{10.11.9}$$

by PolyGamma[m, q] = $(-1)^{m+1} m! \zeta(m+1, q)$. Gosper (1997) and Adamchik (1998) have generalized these functions to the case $m < 0$. These are so-called **negapolygamma functions**. An extension to $m \in \mathbb{C}$ is given the expression

$$\psi(z, q) = e^{-\gamma z} \frac{\partial}{\partial z} \left[e^{\gamma z} \frac{\zeta(z, q)}{\Gamma(1-z)} \right]_{z=1-m}. \tag{10.11.10}$$

See Espinosa and Moll (2004) for details.

10.12. Integral Representations for $\psi(x)$

In this section we consider several integrals associated with the ψ function. Differentiating

$$\Gamma(x) = \int_0^{\infty} t^{x-1} e^{-t} dt \tag{10.12.1}$$

yields [G & R] 4.352.4:

$$\Gamma'(x) = \int_0^{\infty} e^{-t} t^{x-1} \ln t \, dt. \tag{10.12.2}$$

In particular, for $x = k + 1$, using the values in Proposition 10.11.2 we have

$$\int_0^{\infty} e^{-t} t^k \ln t \, dt = k!(-\gamma + H_k). \tag{10.12.3}$$

This can be written as

$$\gamma = H_k - \frac{1}{k!} \int_0^{\infty} e^{-t} t^k \ln t \, dt \tag{10.12.4}$$

a one-parameter family of integral representations for the Euler constant.

Exercise 10.12.1. Write the evaluation of an integral that corresponds to the value of $\psi(k + 1/2)$.

Exercise 10.12.2. Check that the change of variables $t \to \mu t$ yields

$$\int_0^{\infty} t^{x-1} e^{-\mu t} \ln t \, dt = \frac{\Gamma(x)}{\mu^x} (\psi(x) - \ln \mu)$$

which appears in [G & R] 4.352.4. Confirm also the special values

$$\int_0^\infty x^n e^{-\mu x} \ln x \, dx = \frac{n!}{\mu^{n+1}} \left(H_n - \gamma - \ln \mu \right)$$

and

$$\int_0^\infty x^{n-1/2} e^{-\mu x} \ln x \, dx = \frac{\sqrt{\pi}(2n)!}{2^{2n} n! \mu^{n+1/2}} \left(2H_{2n} - H_n - \gamma - \ln 4\mu \right)$$

that appear in [G & R] 4.352.2 and 4.352.3 respectively.

The next result is an integral representation of $\psi(x)$ that generalizes (9.3.2).

Proposition 10.12.1. *The function ψ is given by*

$$\psi(x) = \int_0^\infty \left(\frac{e^{-t}}{t} - \frac{e^{-xt}}{1 - e^{-t}} \right) dt \qquad (10.12.5)$$

Proof. Start with the series (10.11.2) and write

$$\frac{1}{x+n} = \int_0^\infty e^{-t(x+n)} dt.$$

The result follows by summing the geometric series. □

Exercise 10.12.3. Prove the representation

$$\psi(x) = \int_0^\infty \frac{e^{-t} - e^{-xt}}{1 - e^{-t}} dt - \gamma. \qquad (10.12.6)$$

Exercise 10.12.4. The ψ function is given by

$$\psi(x) = \int_0^\infty \left(e^{-t} - (1+t)^{-x} \right) \frac{dt}{t} \qquad (10.12.7)$$

Hint. In the expression (10.12.2) replace $\ln t$ by its representation as a Frullani integral given in (5.8.8).

Extra 10.12.1. The ψ functions admit many other integral representations. For example

$$\psi(x) = -\int_0^1 \left(\frac{1}{\ln t} + \frac{t^{x-1}}{1-t}\right) dt.$$

$$= \int_0^\infty \left(\frac{1}{1+t} - \frac{1}{(1+t)^x}\right) \frac{dt}{t} - \gamma$$

$$= \int_0^1 \frac{t^{x-1} - 1}{t-1} dt - \gamma,$$

$$= \ln x + \int_0^\infty e^{-xt} \left(\frac{1}{t} - \frac{1}{1-e^{-t}}\right) dt.$$

See Whittaker and Watson (1961) for details.

10.13. Some Explicit Evaluations

In this section we evaluate several integrals that are direct consecuences of the integral representations of $\Gamma(x)$ and $\psi(x)$ described in the previous sections.

Proposition 10.13.1. *Let $a, x, y > 0$. Then*

$$\int_0^1 s^{x-1}(1 - s^a)^{y-1} ds = \frac{\Gamma(x/a)\Gamma(y)}{a\,\Gamma(x/a + y)}. \qquad (10.13.1)$$

Proof. In the representation

$$B(x, y) = \frac{\Gamma(x)\Gamma(y)}{\Gamma(x+y)} = \int_0^1 t^{x-1}(1-t)^{y-1} dt$$

let $t = s^a$ and then replace x by x/a. $\qquad \square$

Exercise 10.13.1. Check that

$$\int_0^1 s^{ax-1}(1 - s^a)^{y-1} \ln s \, ds = \frac{\Gamma(y)(\psi(x)\Gamma(x+y) - \Gamma(x)\psi(x+y))}{a^2\,\Gamma^2(x+y)}$$

Confirm the special case

$$\int_0^1 s^{x-1} \ln s \, ds = x^{-2}(\psi(1) - \psi(2)) = -\frac{1}{x^2}.$$

Hint. Differentiate (10.13.1) with respect to the parameter x.

Example 10.13.1. Let $r, q \in \mathbb{R}^+$ and define $p = \frac{1}{2}(q + 4r - 1)$ and $s = \frac{1}{2}(q + 1)$. Then

$$\int_0^{\pi/2} \sin^q u \, \cos^r u \ln \sin u \, du = \frac{\Gamma(2r + 1)}{4\Gamma^2(p)} [\psi(s)\Gamma(p) - \psi(p)\Gamma(s)]$$

Proof. The change of variables $s \mapsto \sin u$ in Proposition 10.13.1 with $a = 2$ gives the result. $\qquad \square$

As a special case we obtain

$$\int_0^{\pi/2} \sin(2u) \ln \sin u \, du = -\frac{1}{2}. \tag{10.13.2}$$

Project 10.13.1. Let

$$L(k) = \int_0^{\pi/2} \sin(ku) \ln \sin u \, du. \tag{10.13.3}$$

Check that

$$L(1) = -1 + \ln 2$$
$$L(2) = -\frac{1}{2}$$
$$L(3) = -\frac{7}{9} + \frac{1}{3} \ln 2$$
$$L(4) = -\frac{1}{2}.$$

Prove that for $k \in \mathbb{N}$ we have $L(2k) \in \mathbb{Q}$ and $L(2k + 1) - \frac{\ln 2}{2k+1} \in \mathbb{Q}$. Obtain closed forms for these integrals.

11

The Riemann Zeta Function

11.1. Introduction

The **Riemann zeta function** defined by

$$\zeta(s) := \sum_{k=1}^{\infty} \frac{1}{k^s} \tag{11.1.1}$$

is one of the fundamental functions of number theory. It appeared in Euler's work on prime numbers in the form

$$\zeta(s) = \prod_{p}(1 - p^{-s})^{-1} \tag{11.1.2}$$

where the product extends over all the primes.

This function plays a remarkable role in the study of the distribution of prime numbers. Riemann proposed to study the associated function $\sum_{n \leq x} \Lambda(n)$, where $\Lambda(n)$ is the von Mangoldt function which appeared in connection with iterates of primitives of $\ln(1 + x)$ in Project 5.3.2. Davenport (2000) gives the identity

$$\sum_{n \leq x} \Lambda(n) = x - \sum_{\rho} \frac{x^{\rho}}{\rho} - \frac{\zeta'(0)}{\zeta(0)} - \tfrac{1}{2}\log(1 - x^{-2}), \tag{11.1.3}$$

where the sum in ρ is over all the zeros of the Riemann zeta function in the critical strip $0 < \mathrm{Re}\,(\rho) < 1$. (The formula has some technical details that we will suppress). The famous **Riemann hypothesis** states that all these zeros are on the vertical line $\mathrm{Re}\,\rho = \tfrac{1}{2}$. It turns out that the asymptotic behavior of the sum on the left-hand side of (11.1.3) determines the behavior of the function

$$\pi(x) = \text{Number of primes less or equal than } x. \tag{11.1.4}$$

219

The famous **prime number theorem** states that

$$\pi(x) \sim \frac{x}{\ln x} \tag{11.1.5}$$

and Newman (1998) offers a simple analytic proof. The errors in this approximation are controlled by the zeros of the Riemann zeta function. Riemann proposed the approximation

$$\pi(x) \sim \mathrm{Li}(x) \tag{11.1.6}$$

where $\mathrm{Li}(x)$ is the logarithmic integral defined in (5.8.6). Koch proved in 1901 that the estimate

$$\left| \pi(x) - \mathrm{Li}(x) \right| \le C\sqrt{x}\,\ln x \tag{11.1.7}$$

is equivalent to the Riemann hypothesis. The last chapter of Havil (2003) has interesting accessible information about these issues.

In this book, the Riemann zeta function appeared in Theorem 6.9.2 where we establish the value of the Bernoulli number

$$B_{2n} = (-1)^{n-1}\frac{\zeta(2n)}{\pi^{2n}} \times \frac{(2n)!}{2^{2n-1}}, \tag{11.1.8}$$

for example

$$\zeta(2) = \frac{\pi^2}{6} \tag{11.1.9}$$

and

$$\zeta(4) = \frac{\pi^4}{90} \tag{11.1.10}$$

in view of the values $B_2 = 1/6$ and $B_4 = 1/30$. In particular we see that $\zeta(2n)$ is a rational multiple of π^{2n} and questions about $\zeta(2n)$ can be reduced to those about Bernoulli numbers. The next exercise illustrates this point.

Exercise 11.1.1. The values of the Riemann zeta function at the even integers satisfy the recursion

$$\zeta(2n) = 2\sum_{r=1}^{n-1}\frac{2^{2r}-1}{2^{2n}-1}\zeta(2r)\zeta(2n-2r). \tag{11.1.11}$$

Hint. Replace (11.1.8) in (5.9.5).

Note 11.1.1. Extra 5.9.2 can now be restated as follows: the optimal values of α and β for which

$$2^\alpha \leq 2^{2n} \left(1 - \zeta(2n)^{-1}\right) \leq 2^\beta$$

holds for all $n \in \mathbb{N}$ are $\alpha = 0$ and $\beta = 2 + \ln(1 - 1/\zeta(2))/\ln 2$.

Exercise 11.1.2. Prove the identities

$$\sum_{k=1}^{\infty} \frac{1}{(2k-1)^s} = \frac{(2^s - 1)}{2^s} \zeta(s) \qquad (11.1.12)$$

$$\sum_{k=1}^{\infty} \frac{(-1)^k}{k^s} = -\frac{(2^{s-1} - 1)}{2^{s-1}} \zeta(s). \qquad (11.1.13)$$

Note 11.1.2. The special case $s = 2$ yields

$$\sum_{k=1}^{\infty} \frac{1}{(2k-1)^2} = \frac{\pi^2}{8}. \qquad (11.1.14)$$

The alternating analogue of (11.1.14)

$$G := \sum_{k=1}^{\infty} \frac{(-1)^{k-1}}{(2k-1)^2} \qquad (11.1.15)$$

is the **Catalan's constant** discussed in Volume 2.

Project 11.1.1. This project comes from Elkies (2003). The goal is to prove that $S(n)$ is a rational multiple of π^n, where

$$S(n) = \sum_{k=-\infty}^{\infty} \frac{1}{(4k+1)^n}.$$

a) Prove that

$$S(n) = (1 - 2^{-n})\zeta(n) \qquad \text{if } n \text{ is even} \qquad (11.1.16)$$

and

$$S(n) = \sum_{k=0}^{\infty} \frac{(-1)^k}{(2k+1)^n} \qquad \text{if } n \text{ is odd.} \qquad (11.1.17)$$

b) Introduce the generating function

$$G(z) = \sum_{n=1}^{\infty} S(n)z^n$$

and prove that

$$G(z) = \frac{\pi z}{4}\left(\sec(\pi z/2) + \tan(\pi z/2)\right).$$

Hint. Compare the partial fraction decompositions of $zG(z)$ and $\frac{\pi z^2}{4}(\sec(\pi z/2) + \tan(\pi z/2))$.
c) Conclude that

$$S(2n) = \frac{(-1)^{n-1}(2^{2n} - 1)B_{2n}\pi^{2n}}{2(2n)!}$$

and

$$S(2n+1) = \frac{(-1)^n \pi^{2n+1} E_{2n}}{2^{2n+2}(2n)!}.$$

E_{2n} are the Euler numbers defined in (6.9.13).

11.2. An Integral Representation

This section describes a basic integral representation of $\zeta(s)$ in terms of the gamma function.

Theorem 11.2.1. *Let $s \in \mathbb{R}$ and $s > 1$. Then*

$$\zeta(s) = \frac{1}{\Gamma(s)}\int_0^{\infty} \frac{u^{s-1}}{e^u - 1}\, du. \tag{11.2.1}$$

This is [G & R] 3.411.1.

Proof. Observe that

$$\Gamma(s) = \int_0^{\infty} y^{s-1}e^{-y}\, dy = k^s \int_0^{\infty} u^{s-1}e^{-ku}\, du$$

so that

$$\frac{1}{k^s} = \frac{1}{\Gamma(s)}\int_0^{\infty} u^{s-1}e^{-ku}\, du. \tag{11.2.2}$$

Adding over k we obtain

$$\zeta(s) = \frac{1}{\Gamma(s)} \int_0^\infty u^{s-1} \sum_{k=1}^\infty e^{-ku} \, du$$

and (11.2.1) follows by summing the geometric series. □

Corollary 11.2.1. *Let $n \in \mathbb{N}$. Then*

$$\int_0^\infty \frac{u^n \, du}{e^u - 1} = n! \, \zeta(n+1). \tag{11.2.3}$$

Exercise 11.2.1. Obtain the integral representation of the Bernoulli number

$$B_{2n} = 4n \int_0^\infty \frac{v^{2n-1}}{e^{2\pi v} - 1} \, dv. \tag{11.2.4}$$

Exercise 11.2.2. Prove that

$$\int_0^\infty \frac{u^{s-1}}{e^u + 1} \, du = \frac{2^{s-1} - 1}{2^{s-1}} \Gamma(s) \zeta(s).$$

From the representation (11.2.1) we prove a relation between $\zeta(s)$ and Euler's constant γ.

Proposition 11.2.1. *The zeta function satisfies*

$$\lim_{s \to 1} \zeta(s) - \frac{1}{s-1} = \gamma. \tag{11.2.5}$$

Proof. From (11.2.1) and $\Gamma(s) = (s-1)\Gamma(s-1)$ it follows that

$$\left(\zeta(s) - \frac{1}{s-1}\right) \Gamma(s) = \int_0^\infty \frac{t^{s-1}}{e^t - 1} \, dt - \int_0^\infty e^{-t} t^{s-2} \, dt$$

$$= \int_0^\infty t^{s-1} \left(\frac{1}{e^t - 1} - \frac{1}{te^t}\right) dt.$$

Now let $s \to 1$ to produce

$$\lim_{s \to 1} \zeta(s) - \frac{1}{s-1} = \int_0^\infty e^{-t} \left(\frac{1}{1 - e^{-t}} - \frac{1}{t}\right) dt$$

and the integral is Euler's constant; see (9.3.5). □

Extra 11.2.1. The expansion of $\zeta(s)$ at the pole $s = 1$ is written as

$$\zeta(s) = \frac{1}{s-1} + \sum_{n=0}^{\infty} A_n(s-1)^n \tag{11.2.6}$$

where A_n are called **Stieltjes constants**. Little is known about them. Briggs and Chowla (1955) established that

$$A_n = \frac{(-1)^n}{n!} \left(\lim_{m \to \infty} \sum_{j=1}^{m} \frac{\ln^n j}{j} - \frac{\ln^{n+1} m}{n+1} \right). \tag{11.2.7}$$

Observe that (11.2.5) yields $A_0 = \gamma$ and (11.2.7) reduces to the definition of γ when $n = 0$.

Project 11.2.1. Prove that

$$\gamma = -\frac{1}{2} \sum_{k=2}^{\infty} \frac{\Lambda(k) - 1}{k} \tag{11.2.8}$$

where Λ is the von Mangoldt function defined in (5.3.18).
Hints. Prove first the identity

$$\sum_{d|n} \Lambda(d) = \ln n. \tag{11.2.9}$$

This follows directly from the prime decomposition of n. Then multiply the series $\sum \Lambda(n) n^{-s}$ and $\zeta(s)$ to obtain

$$-\frac{\zeta'(s)}{\zeta(s)} = \sum_{n=1}^{\infty} \frac{\Lambda(n)}{n^s}.$$

Conclude that

$$\zeta(s) + \frac{\zeta'(s)}{\zeta(s)} = -\sum_{k=1}^{\infty} \frac{\Lambda(k) - 1}{k^s}. \tag{11.2.10}$$

The left hand side approaches 2γ as $s \to 1$ by (11.2.6).

Chapter 11 in Apostol's text (1976) contains more information about the *Dirichlet series* $\sum f(n)/n^s$.

11.3. Several Evaluations for $\zeta(2)$

In this section we present several evaluations of the identity

$$\zeta(2) := \sum_{k=1}^{\infty} \frac{1}{k^2} = \frac{\pi^2}{6}.$$

Some of the proofs provide the equivalent form (11.1.14). An entertaining discussion of the many ways to evaluate this famous series is presented by Kalman (1993).

11.3.1. Euler's Proof

Euler's original proof is based on the representation (6.8.1) of $\sin x$ as an infinite product. That is

$$\frac{\sin x}{x} = \prod_{k=1}^{\infty} \left(1 - \frac{x^2}{(\pi k)^2} \right). \tag{11.3.1}$$

Exercise 11.3.1. a) Give a formal proof of $\zeta(2) = \pi^2/6$ by applying (2.4.8) to the function $\frac{\sin x}{x}$.
b) Obtain the value of $\zeta(4)$ by comparing coefficients in the expansion of (11.3.1).

11.3.2. Apostol's Proof

This appears in Apostol (1983). Start with

$$\frac{1}{k^2} = \int_0^1 \int_0^1 x^{k-1} y^{k-1} \, dx \, dy$$

and sum over k to produce

$$\zeta(2) = \int_0^1 \int_0^1 \frac{dx \, dy}{1 - xy}. \tag{11.3.2}$$

In order to evaluate the double integral let $u = (x+y)/2$ and $v = (y-x)/2$ to get

$$\zeta(2) = \int \int \frac{du \, dv}{1 - u^2 + v^2}$$

over the square of vertices $(0,0)$, $(1/2, -1/2)$, $(1,0)$, $(1/2, 1/2)$. Then using the symmetry of the integrand we have

$$\zeta(2) = 4 \int_0^{1/2} \int_0^u \frac{dv \, du}{1 - u^2 + v^2} + 4 \int_{1/2}^1 \int_0^{1-u} \frac{dv \, du}{1 - u^2 + v^2}$$

$$= 4 \int_0^{1/2} \frac{\tan^{-1}(u/\sqrt{1-u^2})}{\sqrt{1-u^2}} \, du + 4 \int_{1/2}^1 \frac{\tan^{-1}(1 - u/\sqrt{1-u^2})}{\sqrt{1-u^2}} \, du.$$

Now observe that in the first integral

$$\tan^{-1}\left(\frac{u}{\sqrt{1-u^2}}\right) = \sin^{-1} u$$

and in the second one

$$\tan^{-1}\left(\frac{1-u}{\sqrt{1-u^2}}\right) = \frac{\pi}{4} - \frac{1}{2} \sin^{-1} u$$

(this relation was proved in Exercise 6.2.3) so that

$$\zeta(2) = 4 \int_0^{1/2} \frac{\sin^{-1} u}{\sqrt{1-u^2}} \, du + 4 \int_{1/2}^1 \frac{\pi/4 - \sin^{-1} u}{\sqrt{1-u^2}} \, du$$

$$= 4 \int_0^{\pi/6} t \, dt + 4 \int_{\pi/6}^{\pi/2} \left(\frac{\pi}{4} - \frac{t}{2}\right) dt$$

$$= \frac{\pi^2}{6}.$$

11.3.3. Calabi's Proof

The next proof is due to E. Calabi. The story behind this proof and its gener-
alizations is provided by Elkies (2003). Start as in Apostol's proof with

$$\sum_{k=1}^{\infty} \frac{1}{(2k-1)^2} = \int_0^1 \int_0^1 \frac{dx \, dy}{1 - x^2 y^2}. \tag{11.3.3}$$

The change of variables

$$(x, y) = \left(\frac{\sin u}{\cos v}, \frac{\sin v}{\cos u}\right) \tag{11.3.4}$$

has Jacobian $1 - x^2 y^2$ and we obtain that $3\zeta(2)/4$ is the area of the image of
the unit square under (11.3.4).

11.3.4. Matsuoka's Proof

The next proof is given by Matsuoka (1961). Consider the integrals

$$I_n = \int_0^{\pi/2} \cos^{2n} x \, dx \quad \text{and} \quad J_n = \int_0^{\pi/2} x^2 \cos^{2n} x \, dx.$$

The integral I_n has been evaluated by Wallis' formula as

$$I_n = \frac{(2n)!}{2^{2n} \, n!^2} \frac{\pi}{2}.$$

We now prove an estimate on J_n that yields the value of $\zeta(2)$. Integrate by parts to produce

$$I_n = 2n \int_0^{\pi/2} x \sin x \, \cos^{2n-1} x \, dx$$

$$= -n \int_0^{\pi/2} x^2 \left(\cos^{2n} x - (2n-1)\sin^2 x \, \cos^{2n-2} x \right) \, dx$$

$$= n(2n-1)J_{n-1} - 2n^2 J_n.$$

It follows that

$$\frac{\pi}{4n^2} = \frac{2^{2n-2}\,(n-1)!^2}{(2n-2)!} J_{n-1} - \frac{2^{2n}n!^2}{(2n)!} J_n.$$

Now summing from $n = 1$ to N

$$\frac{\pi}{4} \sum_{n=1}^{N} \frac{1}{n^2} = J_0 - \frac{2^{2N} N!^2}{(2N)!} J_N. \tag{11.3.5}$$

The result follows from $J_0 = \pi^3/24$ and the inequality $x \leq \frac{\pi}{2} \sin x$ to obtain

$$J_N \leq \frac{\pi^2}{4} \int_0^{\pi/2} \sin^2 x \, \cos^{2N} x \, dx$$

$$= \frac{\pi^2}{4} (I_N - I_{N+1})$$

$$= \frac{\pi^2 I_N}{8(N+1)}.$$

Therefore

$$0 \leq \frac{2^{2N} N!^2}{(2N)!} J_N \leq \frac{\pi^3}{16(N+1)} \to 0$$

as $N \to \infty$. The value of $\zeta(2)$ follows from (11.3.5).

Corollary 11.3.1. *Let $n \in \mathbb{N}$. Then*

$$\int_0^{\pi/2} x^2 \cos^{2n} x \, dx = \frac{1}{2^{2n}} \binom{2n}{n} \frac{\pi}{4} \left(\zeta(2) - \sum_{k=1}^n \frac{1}{k^2} \right). \quad (11.3.6)$$

Note 11.3.1. The previous formula can be written as

$$\zeta(2) - \sum_{k=1}^n \frac{1}{k^2} = \left(\frac{1}{2^{2n}} \binom{2n}{n} \frac{\pi}{4} \right)^{-1} \int_0^{\pi/2} x^2 \cos^{2n} x \, dx \quad (11.3.7)$$

and the integral gives an expression for the error obtained in the approximation of the real number $\zeta(2) = \pi^2/6$ by the rational number obtained by cutting the series after n terms.

The next project discusses the integral

$$f(n, p) = \int_0^{\pi/2} x^p \cos^{2n} x \, dx. \quad (11.3.8)$$

In the previous proof we have employed $f(n, 0) = I_n$ and $f(n, 2) = J_n$.

Exercise 11.3.2. The goal of this exercise is to produce a closed form for

$$f(n, 1) = \int_0^{\pi/2} x \cos^{2n} x \, dx. \quad (11.3.9)$$

a) Prove that

$$f(n, 1) = f(n - 1, 1) - K_n \quad (11.3.10)$$

where

$$K_n = \int_0^{\pi/2} x \cos^{2n-2} x \sin^2 x. \quad (11.3.11)$$

b) Integrate by parts to check that

$$f(n, 1) = \frac{2n - 1}{2n} f(n - 1, 1) - \frac{1}{4n^2}. \quad (11.3.12)$$

c) Use Mathematica to evaluate the first few values of $f(n, 1)$ and conjecture that $f(n, 1) = a_n + b_n \pi^2$ with $a_n, b_n \in \mathbb{Q}$. Then use the recurrence in b) to obtain

$$a_n = \frac{2n - 1}{2n} a_{n-1} - \frac{1}{4n^2}$$

$$b_n = \frac{2n - 1}{2n} b_{n-1}$$

with initial conditions $a_0 = 0$ and $b_0 = 1/8$.

d) Prove that $b_n = 2^{-2n-3} \binom{2n}{n}$. **Hint.** Let

$$c_n = 2^{2n} \binom{2n}{n}^{-1} b_n \qquad (11.3.13)$$

and obtain a recurrence for c_n.

e) Similarly define

$$d_n = 2^{2n} \binom{2n}{n}^{-1} a_n \qquad (11.3.14)$$

and check that

$$d_n - d_{n-1} = -\frac{2^{2n-2}}{n^2 \binom{2n}{n}}.$$

Now sum from $n = 2$ to n to produce

$$a_n = -\sum_{j=1}^{n} \binom{2n}{n} \binom{2j}{j}^{-1} \frac{2^{-2(n-j)-2}}{j^2}. \qquad (11.3.15)$$

Exercise 11.3.3. This exercise establishes the recurrence

$$f(n, p) = \frac{2n-1}{2n} f(n-1, p) - \frac{p(p-1)}{4n^2} f(n, p-2). \quad (11.3.16)$$

Hint. Integrate by parts.

In particular, the values of $f(n, 0)$ given by Wallis' formula and $f(n, 1)$ given in Exercise 11.3.2 determine the value of $f(n, p)$.

Project 11.3.1. Use the results of Exercise 11.3.3 to determine a closed form formula for $f(n, p)$.

Exercise 11.3.4. In this exercise we obtain a closed form expression for

$$T(p) = \int_0^{\pi/2} x \tan^p x \, dx \qquad -2 < p < 1. \qquad (11.3.17)$$

This is given by Lossers (1985).
a) Check that

$$T(p) = \int_0^{\infty} \frac{y^p}{1+y^2} \tan^{-1} y \, dy.$$

b) Define

$$S(p, a) = \int_0^\infty \frac{y^p}{1 + y^2} \tan^{-1}(ay) \, dy$$

and confirm that

$$\frac{d}{da} S(p, a) = \frac{\pi}{2 \sin(\pi p/2)} \times \frac{a^{-p} - 1}{1 - a^2}.$$

Conclude that

$$T(p) = S(p, 1) = \frac{\pi}{2 \sin(\pi p/2)} \int_0^1 \frac{a^{-p} - 1}{1 - a^2} \, da.$$

c) Use the result in part b) and the integral representation (10.12.7) to conclude that

$$T(p) = \frac{\pi}{4 \sin(\pi p/2)} \left(\psi\left(\frac{1}{2}\right) - \psi\left(\frac{1 - p}{2}\right) \right).$$

In particular

$$T(0) = \lim_{p \to 0} T(p) = \frac{\pi^2}{8}. \qquad (11.3.18)$$

11.3.5. Boo Rim Choe's Proof

This proof appeared in Choe (1987). We use the expansion (6.6.6)

$$\sin^{-1} x = x + \sum_{k=1}^\infty \frac{c_k}{2k + 1} x^{2k+1}. \qquad (11.3.19)$$

with

$$c_k = 2^{-2k} \binom{2k}{k}. \qquad (11.3.20)$$

First, substitute $x \mapsto \sin t$, and then integrate (11.3.19) from 0 to $\pi/2$ and use Wallis' formula in the form

$$c_k \int_0^{\pi/2} \sin^{2k+1} x \, dx = \frac{1}{2k + 1} \qquad (11.3.21)$$

to produce the equivalent form (11.1.14).

11.3.6. The Proof of Yue and Williams (1994)

In this proof we start with the expansion (6.6.21):

$$\left(\sin^{-1} x\right)^2 = \frac{1}{2} \sum_{k=1}^{\infty} \frac{(2x)^{2k}}{k^2 C_k} \qquad (11.3.22)$$

with $C_k = \binom{2k}{k}$, and put $x = \sin t$ to get

$$t^2 = \frac{1}{2} \sum_{k=1}^{\infty} \frac{\sin^{2k} t}{k^2 C_k}. \qquad (11.3.23)$$

Integration from 0 to $\pi/2$ and Wallis' formula yield the result.

11.4. Apery's Constant: $\zeta(3)$

In this section we discuss several expressions for

$$\zeta(3) = \sum_{k=1}^{\infty} \frac{1}{k^3}. \qquad (11.4.1)$$

This constant is called **Apery's constant** honoring R. Apery's celebrated proof of its irrationality. The values $\zeta(2n)$ are rational multiples of π^{2n}, therefore they are all irrational numbers in view of the trascendence of π mentioned in Section 6.5.1. The corresponding result for the odd values of zeta has shown to be much more difficult. In 1978 Roger Apery in Luminy announced his remarkable result. van der Poorten (1979) gives a description of the reaction of mathematicians after Apery's lecture. The essence of his proof is the recurrence

$$(n+1)^3 y_{n+1} - (34n^3 + 51n^2 + 27n + 5)y_n + n^3 y_{n-1} = 0. \qquad (11.4.2)$$

The sequence a_n solves (11.4.2) with initial conditions $a_0 = 1$, $a_1 = 5$ and b_n solves the same recurrence with $b_0 = 0$, $b_1 = 6$. Apery shows that

$$a_n = \sum_{k=0}^{n} \binom{n}{k}^2 \binom{n+k}{k}^2$$

and

$$b_n = \sum_{k=0}^{n} \binom{n}{k}^2 \binom{n+k}{k}^2 \left(\sum_{m=1}^{n} \frac{1}{m^3} + \sum_{m=1}^{k} \frac{(-1)^{m-1}}{2m^3 \binom{n}{m} \binom{n+m}{m}} \right),$$

that $a_n/b_n \to \zeta(3)$ and then he concludes with a proof of the irrationality of $\zeta(3)$. Beukers (1979) used the triple integral

$$I_{\mathbb{R},n} = \int_0^1 \int_0^1 \int_0^1 \frac{u^n(1-u)^n v^n(1-v)^n w^n(1-w)^n}{((1-w)z+uvw)^{n+1}} \, du \, dv \, dw \quad (11.4.3)$$

to provide a new proof of Apery's result.

None of these proofs extend to $\zeta(5)$. The irrationality of this number is still an open question, but may be not for too long. In a series of papers T. Rivoal (2002) and W. Zudilin (2001) have managed to prove that at least one of the numbers $\zeta(5)$, $\zeta(7)$, $\zeta(9)$, $\zeta(11)$ is irrational. *There is hope.*

Huylebrouck (2001) presents Beukers' proof and provides unified irrationality proofs for π, $\ln 2$ and $\zeta(2)$ along these lines.

In this section we discuss several representations of Apery's constant $\zeta(3)$. Naturally one obtains a representation of $\zeta(3)$ as a special case of expressions for $\zeta(s)$. For instance, Exercise 11.1.2 yields

$$\zeta(3) = \frac{8}{7} \sum_{k=1}^{\infty} \frac{1}{(2k-1)^3} \quad (11.4.4)$$

and

$$\zeta(3) = \frac{4}{3} \sum_{k=1}^{\infty} \frac{(-1)^{k-1}}{k^3} \quad (11.4.5)$$

and the (11.2.1) gives

$$\zeta(3) = \frac{1}{2} \int_0^{\infty} \frac{u^2 \, du}{e^u - 1}.$$

11.4.1. A Formula of Ewell

The expression for $\zeta(3)$ in the next equation is offered by Ewell (1990).

Proposition 11.4.1. *The value of $\zeta(3)$ is given by*

$$\zeta(3) = \frac{\pi^2}{7} \left(1 - 4 \sum_{n=1}^{\infty} \frac{\zeta(2n)}{(2n+1)(2n+2)2^{2n}} \right). \quad (11.4.6)$$

Proof. Start with the expansion (6.6.6) and integrate to produce

$$\int_0^x \frac{\sin^{-1} t}{t} \, dt = x + \sum_{k=1}^{\infty} \frac{c_k}{(2k+1)^2} x^{2k+1},$$

with $c_k = 2^{-2k} \binom{2k}{k}$ as in (11.3.20). Let $u = \sin^{-1} t$ to obtain

$$\int_0^t \frac{u}{\tan u}\, du = \sum_{k=0}^{\infty} \frac{c_k}{(2k+1)^2} \sin^{2k+1} t,$$

then integrate the expansion of the cotangent (6.9.7) to get

$$t - 2\sum_{n=1}^{\infty} \frac{\zeta(2n)}{\pi^{2n}} \frac{t^{2n}}{2n+1} = \sum_{k=0}^{\infty} \frac{c_k}{(2k+1)^2} \sin^{2k+1} t.$$

Finally integrate from 0 to $\pi/2$ and use Wallis' formula in the form (11.3.21) to produce

$$\frac{\pi^2}{8} - 2\sum_{n=1}^{\infty} \frac{\zeta(2n)}{\pi^{2n}} \frac{(\pi/2)^{2n+2}}{(2n+1)(2n+2)} = \sum_{k=0}^{\infty} \frac{1}{(2k+1)^3}.$$

This result reduces to (11.4.6) by using (11.4.4). $\qquad\qquad\qquad$ $\square$

Note 11.4.1. The typical term in the series (11.4.6) is asymptotic to $n^{-2}2^{-2n}$ and so the new series converges much faster than the original.

11.4.2. A Formula of Yue and Williams

In this section we describe an expression for $\zeta(3)$ due to Yue and Williams (1993).

Theorem 11.4.1. *The Riemann zeta function satisfies*

$$\zeta(3) = -2\pi^2 \sum_{n=0}^{\infty} \frac{\zeta(2n)}{(2n+2)(2n+3)2^{2n}}. \qquad (11.4.7)$$

Proof. Start with the expansion (6.6.21)

$$\left(\sin^{-1} x\right)^2 = \sum_{n=1}^{\infty} \frac{2^{2n-1}}{n^2 \binom{2n}{n}} x^{2n}. \qquad (11.4.8)$$

and integrate to produce

$$\int_0^{\sin t} \frac{(\sin^{-1} x)^2}{x}\, dx = \sum_{n=1}^{\infty} \frac{2^{2n-2}}{n^3 \binom{2n}{n}} \sin^{2n} t$$

and with $x = \sin u$ we obtain

$$\int_0^t u^2 \cot u \, du = \sum_{n=1}^\infty \frac{2^{2n-2}}{n^3 \binom{2n}{n}} \sin^{2n} t. \qquad (11.4.9)$$

Now recall the expansion (6.9.7), integrate from 0 to $\pi/2$ and use Wallis' formula to produce the result. $\qquad\square$

Exercise 11.4.1. Establish

$$\zeta(3) - \sum_{k=1}^{n-1} \frac{1}{k^3} = \frac{1}{2} \int_0^\infty \frac{x^2 e^{-nx}}{1 - e^{-x}} \, dx \qquad (11.4.10)$$

which appears in [G & R] 3.411.14. This identity is in the same style as (11.3.7). There are similar expressions for other constants, for instance

$$\int_0^\infty \frac{x e^{-2nx} \, dx}{\sqrt{e^{2x} + 1}} = \frac{(2n-1)!!}{(2n)!!} \frac{\pi}{2} \left(\ln 2 + \sum_{k=1}^{2n} \frac{(-1)^k}{k} \right)$$

and

$$\int_0^\infty \frac{x e^{-(2n-1)x} \, dx}{\sqrt{e^{2x} - 1}} = -\frac{(2n-2)!!}{(2n-1)!!} \left(\ln 2 + \sum_{k=1}^{2n-1} \frac{(-1)^k}{k} \right)$$

provide representations for $\ln 2$. These appear in [G & R] 3.454.1 and 3.454.2 respectively. Check them.

Exercise 11.4.2. Prove the analogue of (11.3.2) due to Beukers (1979):

$$\zeta(3) = -\frac{1}{2} \int_0^1 \int_0^1 \frac{\ln xy}{1 - xy} \, dy \, dy. \qquad (11.4.11)$$

Extra 11.4.1. Sondow (2002) has established the representation

$$\gamma = - \int_0^1 \int_0^1 \frac{1 - x}{(1 - xy) \ln xy} \, dx \, dy$$

and the antisymmetric formula Sondow (1998)

$$\gamma = \lim_{x \to 1^+} \sum_{n=1}^\infty \left(\frac{1}{n^x} - \frac{1}{x^n} \right).$$

There are many other integral representations of special values of the Riemann zeta function. A remarkable one is given by Borwein and Borwein (1995):

$$\frac{1}{\pi} \int_0^{\pi/2} x^2 \ln^2 (2\cos x)\, dx = \frac{11}{16}\zeta(4).$$

11.5. Apery Type Formulae

The proof of the irrationality of $\zeta(3)$ given by Apery is based on the representation

$$\zeta(3) = \frac{5}{2} \sum_{k=1}^{\infty} \frac{(-1)^{k+1}}{k^3 \binom{2k}{k}}. \tag{11.5.1}$$

The analogous formulae

$$\zeta(2) = 3 \sum_{k=1}^{\infty} \frac{1}{k^2 \binom{2k}{k}} \tag{11.5.2}$$

and

$$\zeta(4) = \frac{36}{17} \sum_{k=1}^{\infty} \frac{1}{k^4 \binom{2k}{k}} \tag{11.5.3}$$

suggest the possibility of expressions of the form

$$\zeta(5) = \frac{a}{b} \sum_{k=1}^{\infty} \frac{(-1)^{k+1}}{k^5 \binom{2k}{k}} \tag{11.5.4}$$

and

$$\zeta(6) = \frac{c}{d} \sum_{k=1}^{\infty} \frac{1}{k^6 \binom{2k}{k}}. \tag{11.5.5}$$

Extensive computation by Borwein and Bradley (1997) ruled out the existence of such a formula when a, b, c, d are moderately sized integers.

There are beautiful representations for $\zeta(5)$ due to Koecher (1980).

$$\zeta(5) = 2 \sum_{k=1}^{\infty} \frac{(-1)^{k+1}}{k^5 \binom{2k}{k}} + \frac{5}{2} \sum_{k=1}^{\infty} \frac{(-1)^k}{k^3 \binom{2k}{k}} \sum_{j=1}^{k-1} \frac{1}{j^2}.$$

and for $\zeta(7)$ due to Borwein and Bradley:

$$\zeta(7) = \frac{5}{2} \sum_{k=1}^{\infty} \frac{(-1)^{k+1}}{k^7 \binom{2k}{k}} + \frac{25}{2} \sum_{k=1}^{\infty} \frac{(-1)^{k+1}}{k^3 \binom{2k}{k}} \sum_{j=1}^{k-1} \frac{1}{j^4}. \tag{11.5.6}$$

So far these formulae have not produced the desired proof of irrationality of $\zeta(5)$ and $\zeta(7)$. These cases have been extended by Borwein and Bradley (1997) and Almkvist and Granville (1999) to produce a generating function of $\zeta(4n + 3)$:

$$\sum_{k=0}^{\infty} \zeta(4k + 3)z^{4k} = \sum_{n=1}^{\infty} \frac{1}{n^3(1 - z^4/n^4)}$$

$$= \frac{5}{2} \sum_{n=1}^{\infty} \frac{(-1)^{n-1}}{n^3 \binom{2n}{n}} \frac{n^4}{n^4 - z^4} \prod_{m=1}^{n-1} \frac{m^4 + 4z^4}{m^4 - z^4}$$

and

$$\sum_{k=0}^{\infty} \zeta(2k + 3)\, z^{2k} = \sum_{n=1}^{\infty} \frac{1}{n^3(1 - z^2/n^2)}$$

$$= \sum_{n=1}^{\infty} \frac{(-1)^{n-1}}{n^3 \binom{2n}{n}} \left(\frac{1}{2} + \frac{2n^2}{n^2 - z^2} \right) \prod_{m=1}^{n-1} \left(1 - \frac{z^2}{m^2} \right).$$

There are similar identities that involve the constant $\tau := \ln((1 + \sqrt{5})/2)$:

$$\sum_{n=1}^{\infty} \frac{(-1)^{n-1}}{\binom{2n}{n}} = \frac{4\tau}{5\sqrt{5}} + \frac{1}{5}$$

$$\sum_{n=1}^{\infty} \frac{(-1)^{n-1}}{n \binom{2n}{n}} = \frac{2\tau}{\sqrt{5}}$$

$$\sum_{n=1}^{\infty} \frac{(-1)^{n-1}}{n^2 \binom{2n}{n}} = 2\tau^2.$$

Finally we mention Amdeberhan's beautiful formula (1996):

$$\zeta(3) = \frac{1}{4} \sum_{n=1}^{\infty} \frac{(-1)^{n-1}}{n^3 \binom{3n}{n} \binom{2n}{n}} \times \frac{(56n^2 - 32n + 5)}{(2n - 1)^2}$$

that has been used to compute the decimal expansion of $\zeta(3)$.

12

Logarithmic Integrals

The goal of this chapter is to explore the evaluation of integrals of the type

$$\int_a^b R_1(x) \ln R_2(x)\,dx$$

where R_1, R_2 are rational functions. The examples presented here are elementary and Chapter 13 contains some more advanced ones.
The evaluation

$$\int_0^{x^*} \frac{\ln(1-x)}{x}\,dx = \ln^2 x^* - \frac{\pi^2}{10}$$

with $x^* = (\sqrt{5}-1)/2$ appeared in the description of the dilogarithm (4.1.8) and it gives a measure of the complexity of this type of problems.
    Integrals of the form

$$\int_0^1 \frac{R(x)}{\ln x}\,dx$$

are much more complicated and are the subject of current research. Adamchik (1997) studied them in the equivalent form

$$\int_0^1 R(x) \ln\ln\left(\frac{1}{x}\right)dx$$

and evaluates some of them with the help of the Hurwitz zeta function. The evaluations lead to magnificent expressions. For instance,

$$\int_0^1 \frac{1}{1+x^2} \ln\ln\left(\frac{1}{x}\right)dx = \frac{\pi}{2} \ln\left(\frac{\sqrt{2\pi}\,\Gamma(\frac{3}{4})}{\Gamma(\frac{1}{4})}\right)$$

the example in Vardi's paper (1988) coming from [G & R] 4.229.7 which was

237

our original motivation, and

$$\int_0^1 \frac{x}{(1-x+x^2)^2} \ln \ln \left(\frac{1}{x}\right) dx = -\frac{\gamma}{3} - \frac{1}{3} \ln \left(\frac{6\sqrt{3}}{\pi}\right)$$

$$+ \frac{\pi\sqrt{3}}{27} \left(5 \ln 2\pi - 6 \ln \Gamma(\tfrac{1}{6})\right).$$

12.1. Polynomial Examples

In this section we describe integrals that are combinations of polynomials and powers of logarithms. They will be expressed in terms of

$$I_{n,k} = \int_0^1 x^n \ln^k x \, dx. \tag{12.1.1}$$

Starting with a polynomial

$$P(x) = \sum_{n=0}^{m} p_n x^n \tag{12.1.2}$$

we have

$$\int_a^b P(x) \ln^k x \, dx = \sum_{n=0}^{m} p_n \int_a^b x^n \ln^k x \, dx. \tag{12.1.3}$$

Exercise 5.3.1 shows that the integrand admits an elementary primitive. The change of variables $x = bt$ yields

$$\int_a^b x^n \ln^k x \, dx = b^{n+1} \sum_{j=0}^{k} \binom{k}{j} \ln^{k-j} b \int_{a/b}^1 t^n \ln^j t \, dt, \tag{12.1.4}$$

so we need to evaluate integrals of the form

$$L(c) = \int_c^1 x^n \ln^k x \, dx$$

$$= \int_0^1 x^n \ln^k x \, dx - \int_0^c x^n \ln^k x \, dx.$$

The second integral can be scaled as before to write it as a finite sum of integrals of the form $I_{n,k}$.

Exercise 12.1.1. Check this.

Proposition 12.1.1. *Let* k, $n \in \mathbb{N}$. *Then*

$$\int_0^1 x^n \ln^k x\, dx = (-1)^k \frac{k!}{(n+1)^{k+1}}. \qquad (12.1.5)$$

Proof. The change of variable $t = -\ln x$ gives

$$\int_0^1 x^n \ln^k x\, dx = (-1)^k \int_0^\infty t^k e^{-(n+1)t}\, dt$$

$$= \frac{(-1)^k}{(n+1)^{k+1}} \int_0^\infty t^k e^{-t}\, dt$$

and the integral is $\Gamma(k+1) = k!$. □

Exercise 12.1.2. Obtain the value

$$\int_2^3 x^3 \ln^2 x\, dx = \frac{1}{32}\left(65 + 64\ln 2 - 128\ln^2 2 - 324\ln 3 + 648\ln^2 3\right).$$

12.2. Linear Denominators

In this section we consider the closed-form evaluation of

$$I = \int_a^b \frac{\ln^k x}{q_1 x + q_0}\, dx \qquad (12.2.1)$$

in terms of the parameters a, b, p_0 and p_1. In general the combination of logarithms and rational functions produces integrals that are evaluated in terms of the **polylogarithm function** defined by

$$\mathrm{Li}_k(x) = \mathrm{PolyLog}[k, x] = \sum_{n=1}^\infty \frac{x^n}{n^k}. \qquad (12.2.2)$$

The special case $k = 2$ is the dilogarithm introduced by Euler and defined in (4.1.8). For instance, Mathematica gives

$$\int_a^b \frac{\ln x\, dx}{q_1 x + q_0} = \frac{1}{q_1}\left(\ln b \ln(1 + bq^*) - \ln a \ln(1 + aq^*)\right)$$

$$-\frac{1}{q_1}\left(\mathrm{PolyLog}[2, -aq^*] - \mathrm{PolyLog}[2, -bq^*]\right),$$

with $q^* = q_1/q_0$.

Simple expressions will exist only for special values of these parameters. Gosper (1996) has studied the Nielsen–Ramanujan constants

$$a_k = \int_1^2 \frac{\ln^k x}{x-1} dx, \quad \text{for } k \geq 1 \tag{12.2.3}$$

and proved that $a_1 = \zeta(2)/2$, $a_2 = \zeta(3)/4$ and

$$a_k = k!\zeta(k+1) - \frac{k}{k+1} \ln^{k+1} 2 - k! \sum_{j=0}^{k-1} \frac{\ln^j 2}{j!} \mathrm{Li}_{k+1-j}(\tfrac{1}{2}). \tag{12.2.4}$$

Example 12.2.1. Let $k \in \mathbb{N}$. Then

$$\int_0^1 \frac{\ln^k x}{1-x} dx = (-1)^k k! \zeta(k+1). \tag{12.2.5}$$

This appears in [G & R] 4.271.3. To check it, expand the denominator in a power series to get

$$\int_0^1 \frac{\ln^k x}{1-x} dx = \sum_{j=0}^{\infty} \int_0^1 x^j \ln^k x \, dx$$

$$= \sum_{j=0}^{\infty} (-1)^k \int_0^{\infty} t^k e^{-(j+1)t} \, dt$$

$$= k! \sum_{j=0}^{\infty} \frac{(-1)^j}{(j+1)^{k+1}}.$$

The result follows from Exercise 11.1.2.

Exercise 12.2.1. Prove the identities [G & R] 4.271.1, 4.271.2:

$$\int_0^1 \frac{\ln^k x}{x+1} dx = \frac{(-1)^k k! (2^k - 1)}{2^k} \times \zeta(k+1). \tag{12.2.6}$$

Exercise 12.2.2. Establish the value

$$\int_0^1 \ln\left(\frac{1+x}{1-x}\right) \frac{dx}{x} = \frac{\pi^2}{4}. \tag{12.2.7}$$

Exercise 12.2.3. Check [G & R] 0.241.4:

$$\int_1^p \frac{\ln(1-x)}{x} dx = \frac{\pi^2}{6} - \mathrm{PolyLog}[2, p]. \tag{12.2.8}$$

12.3. Some Quadratic Denominators

In this section we consider integrals of the form

$$Q_k(a, b, c) = \int_0^1 \frac{\ln^k x \, dx}{ax^2 + bx + c}. \tag{12.3.1}$$

Exercise 12.3.1. Confirm [G & R] 4.271.7:

$$\int_0^\infty \frac{\ln^{2n+1} x \, dx}{1 + bx + x^2} = 0$$

provided $|b| < 2$. **Hint.** Let $y = 1/x$. The special case $b = n = 0$ yields

$$\int_0^\infty \frac{\ln x \, dx}{1 + x^2} = 0 \tag{12.3.2}$$

a result due to Euler. This evaluation has been reproduced by Arora, Goel and Rodriguez (1988).

Exercise 12.3.2. Confirm [G & R] 4.271.9:

$$\int_0^\infty \frac{\ln^{2n} x \, dx}{1 - x^2} = 0.$$

Proposition 12.3.1. *Let $n \in \mathbb{N}$. Then*

$$\int_0^1 \frac{\ln^k x \, dx}{1 - x^2} = (-1)^k k! \zeta(k+1) \frac{2^{k+1} - 1}{2^{k+1}} \tag{12.3.3}$$

and

$$\int_0^1 \frac{x}{1 - x^2} \ln^k x \, dx = \frac{(-1)^k k! \zeta(k+1)}{2^{k+1}}. \tag{12.3.4}$$

The first integral appears in [G & R] 4.271.7.

Proof. The first one is the average of (12.2.5) and (12.2.6), the second comes from their difference.   □

Note 12.3.1. The previous evaluation can be written in terms of the Bernoulli numbers as

$$\int_0^1 \frac{\ln^{2n-1} x}{1 - x^2} dx = \frac{\pi^{2n}(2^{2n} - 1)}{4n} B_{2n}.$$

Some more complicated integrals appear from elementary manipulations of the one presented here, but a systematic evaluation requires more advanced functions. For example, the value

$$\int_0^1 \ln(1 - x^2) \ln^{k-1} x \frac{dx}{x} = \frac{(-1)^k (k-1)! \, \zeta(k+1)}{2^k}. \quad (12.3.5)$$

comes from integrating (12.3.4) by parts.

The integrals discussed above also have a trigonometric version. The next exercise appears as a problem in Linis and Grosswald (1957) and it was solved by E. Grosswald.

Exercise 12.3.3. Prove that

$$I = \int_0^{\pi/2} \frac{\ln \sec \theta}{\tan \theta} \, d\theta = \frac{\pi^2}{24}. \quad (12.3.6)$$

As before one obtains the evaluation of apparently more complicated definite integrals by introducing a parameter.

Example 12.3.1. Let $a \in \mathbb{R}^+$ and $x = y^a$ in (12.3.2) yields

$$\int_0^\infty \frac{y^{a-1}}{y^{2a} + 1} \ln y \, dy = 0. \quad (12.3.7)$$

Differentiating with respect to a yields

$$\int_0^\infty \frac{y^{a-1}(y^{2a} - 1)}{(y^{2a} + 1)^2} \ln^2 y \, dy = 0. \quad (12.3.8)$$

We conclude the existence of a sequence of polynomials $C_n(x)$ such that

$$\int_0^\infty \frac{y^{a-1} \ln^n y}{(y^{2a} + 1)^{n+1}} C_n(-y^{2a}) \, dy = 0. \quad (12.3.9)$$

The polynomials

$$C_n(x) = \sum_{j=0}^n a_{j,n} x^n \quad (12.3.10)$$

satisfy the differential–difference equation

$$C_{n+1}(x) = 2x(1-x)C_n'(x) + [1 + (2n+1)x] C_n(x), \quad (12.3.11)$$

and their coefficients satisfy

$$\begin{aligned}
a_{0,n+1} &= a_{0,n} \\
a_{1,n+1} &= 3a_{1,n} + (2n+1)a_{0,n} \\
a_{j,n+1} &= (2j+1)a_{j,n} + 2(n+1-j)a_{j-1,n} \\
a_{n+1,n+1} &= a_{n,n}.
\end{aligned} \quad (12.3.12)$$

Therefore $a_{j,n}$ are positive integers. Moreover, the polynomial is symmetric, that is

$$C_n(x) = x^n C_n(1/x). \tag{12.3.13}$$

The first few polynomials are

$$
\begin{aligned}
C_0(x) &= 1, \\
C_1(x) &= x + 1 \\
C_2(x) &= x^2 + 6x + 1 \\
C_3(x) &= x^3 + 23x^2 + 23x + 1 \\
C_4(x) &= x^4 + 76x^3 + 230x^2 + 76x + 1 \\
C_5(x) &= x^5 + 237x^4 + 1682x^3 + 1682x^2 + 237x + 1.
\end{aligned}
\tag{12.3.14}
$$

The list of coefficients

$$\{1, 1, 1, 1, 6, 1, 1, 23, 23, 1, 1, 76, 230, 76, 1\}$$

are identified by N. Sloane's *Handbook of Integer Sequences* (1973) as the **triangle numbers** $T(n, k)$.

Example 12.3.2. The evaluation of

$$\int_0^1 \frac{\ln(1+x)}{1+x^2}\, dx = \frac{\pi}{8}\ln 2 \tag{12.3.15}$$

is due to Serret (1844). The integral can be found in [G & R]: 4.291.8.

The change of variable $x = \tan t$ yields

$$
\begin{aligned}
\int_0^1 \frac{\ln(1+x)}{1+x^2}\, dx &= \int_0^{\pi/4} \ln(1+\tan t)\, dt \\
&= \int_0^{\pi/4} \ln\left(\frac{\sqrt{2}\,\cos(\pi/4 - t)}{\cos t} \right) dt \\
&= \frac{\pi}{8}\ln 2 + \int_0^{\pi/4} \ln\cos(\pi/4 - t)\, dt - \int_0^{\pi/4} \ln\cos t\, dt,
\end{aligned}
$$

and the last two are equal by symmetry about $t = \pi/8$.

Project 12.3.1. Let $n \in \mathbb{N}$. Then

$$\int_0^1 \frac{(\ln x)^{2n+1}}{1+x^2}\, dx = \frac{\pi^{2n+1}}{2^{2n+2}} E_{2n}, \tag{12.3.16}$$

where E_{2n} are the Euler numbers defined in (6.9.13).

12.4. Products of Logarithms

In this section we present one example of an integral of the form

$$\int_a^b \ln R_1(x) \ln R_2(x)\,dx$$

where R_1 and R_2 are rational functions.

Example 12.4.1.

$$\int_0^1 \ln(1+x)\ln(1-x)\,dx = \ln^2 2 - 2\ln 2 + 2 - \zeta(2). \quad (12.4.1)$$

The evaluation of this problem is given by Kerney and Stenger (1976). Observe that

$$\begin{aligned}
\int_0^1 \ln(1+x)\ln(1-x)\,dx &= \frac{1}{2}\int_{-1}^1 \ln(1+x)\ln(1-x)\,dx \\
&= \int_0^1 \ln(2t)\ln(2-2t)\,dt \\
&= \int_0^1 [\ln 2 + \ln t] \times [\ln 2 + \ln(1-t)] \\
&= \ln^2 2 + 2\ln 2 \cdot \int_0^1 \ln t\,dt + \int_0^1 \ln t \ln(1-t)\,dt.
\end{aligned}$$

Integrate by parts to check that the first integral is -1. To evaluate the second one expand $\ln(1-t)$ to obtain

$$\begin{aligned}
\int_0^1 \ln t \ln(1-t)\,dt &= -\sum_{k=1}^\infty \frac{1}{k}\int_0^1 t^k \ln t\,dt \\
&= \sum_{k=1}^\infty \frac{1}{k(k+1)^2} \\
&= \sum_{k=1}^\infty \left(\frac{1}{k(k+1)} - \frac{1}{(k+1)^2}\right) \\
&= 1 - (\zeta(2)-1) = 2 - \zeta(2).
\end{aligned}$$

Exercise 12.4.1. Use the expansion in (5.2.8) to obtain

$$\sum_{k=1}^\infty \frac{1}{k(2k+1)}\left(H_k - H_{2k} - \tfrac{1}{2k}\right) = \ln^2 2 - 2\ln 2 + 2 - \zeta(2). \quad (12.4.2)$$

Exercise 12.4.2. Check that

$$\sum_{k=1}^{\infty} \frac{1}{2^k k^2} = \frac{\pi^2}{12} - \frac{\ln^2 2}{2}. \tag{12.4.3}$$

Hint. Let $u = \ln(1 + x)$ in (12.4.1) and integrate by parts. The resulting integral can be expanded in a geometric series.

Extra 12.4.1. The series (12.4.3) is the special value $x = 1/2$ of the **dilogarithm function** defined in (4.1.8):

$$\text{DiLog}(1/2) = \frac{\pi^2}{12} - \frac{\ln^2 2}{2}. \tag{12.4.4}$$

According to Loxton (1984), the only *known* values of the function $L(z) = \text{DiLog}(z) + \frac{1}{2}\log z \cdot \log(1 - z)$ are Euler's results

$$L(1) = \frac{\pi^2}{6} \quad \text{and} \quad L(\tfrac{1}{2}) = \frac{\pi^2}{12} \tag{12.4.5}$$

and those given by Landen

$$L\left(\tfrac{1}{2}(\sqrt{5} - 1)\right) = \frac{\pi^2}{10} \quad \text{and} \quad L\left(\tfrac{1}{2}(3 - \sqrt{5})\right) = \frac{\pi^2}{15}. \tag{12.4.6}$$

Mathematica 12.4.1. A direct Mathematica computation provides the value and also gives the primitive

$$\int \ln(1 - x)\ln(1 + x)\,dx = -1 + 2x - (1 + x)\ln(1 + x)$$
$$+ (1 - x - \ln 4 + (1 + x)\ln(1 + x))\ln(1 - x)$$
$$+ 2\,\text{PolyLog}(2,\,(1 - x)/2).$$

12.5. The Logsine Function

In this section we consider the evaluation of

$$S_n := \int_0^{\pi} (\ln \sin x)^n \, dx. \tag{12.5.1}$$

This function has appeared in Exercise 10.6.1.

The first integral is due to Euler and has reappeared in Arora, Goel and Rodriguez (1988).

Proposition 12.5.1.

$$S_1 = \int_0^\pi \ln \sin x \, dx = -\pi \ln 2. \qquad (12.5.2)$$

Proof. Use $\sin x = 2 \sin(x/2) \cos(x/2)$ to obtain

$$S_1 = \ln 2 + \int_0^\pi \ln \sin \left(\tfrac{x}{2}\right) \, dx + \int_0^\pi \ln \cos \left(\tfrac{x}{2}\right) \, dx.$$

Replace $x/2$ bt t in the first integral and by $\pi/2 - t$ in the second, we find

$$\begin{aligned}
S_1 &= \pi \ln 2 + 4 \int_0^{\pi/2} \ln \sin t \, dt \\
&= \pi \ln 2 + 2S_1
\end{aligned}$$

and the result follows. $\square$

We now follow Beumer (1961) to prove a recursion for the integrals S_n.

Theorem 12.5.1. *The integrals* S_n *satisfy*

$$S_1 S_{2n-1} - S_2 S_{2n-2} + \cdots + S_{2n-1} S_1 = (-1)^{n-1} \frac{(2^{2n} - 1)}{(2n)!} \pi^{2n} B_n \qquad (12.5.3)$$

$$\ln 2 \, S_{n-1} + \sum_{k=1}^{n-1} (1 - 2^{-k}) \zeta(k+1) S_{n-k-1} = (n-1) S_n. \qquad (12.5.4)$$

Proof. Consider the Dirichlet series

$$X(s) = \sum_{n=0}^{\infty} 2^{-2n} \frac{\binom{2n}{n}}{(2n+1)^s}. \qquad (12.5.5)$$

and observe that

$$\frac{1}{(2n+1)^s} = \frac{1}{\Gamma(s)} \int_0^\infty x^{s-1} e^{-(2n-1)x} dx \qquad (12.5.6)$$

so by the binomial theorem we obtain

$$\int_0^\infty \frac{x^{s-1} \, dx}{\sqrt{e^{2x} - 1}} = \Gamma(s) X(s). \qquad (12.5.7)$$

The change of variable $e^{-x} \mapsto \sin \theta$ yields, for $s = n$,

$$S_n = \frac{(-1)^{n-1}}{(n-1)!} \int_0^{\pi/2} (\sin \theta)^{n-1} \, d\theta. \qquad (12.5.8)$$

Integrate

$$(\sin t)^x = \sum_{n=0}^{\infty} \frac{(\ln \sin \theta)^n}{n!} x^n \qquad (12.5.9)$$

from $x = 0$ to $\pi/2$ to obtain

$$\sum_{n=0}^{\infty} \frac{x^n}{n!} \int_0^{\pi/2} (\ln \sin \theta)^n \, d\theta = \int_0^{\pi/2} \sin^x t \, dt.$$

Now use (10.3.12) to obtain

$$\sum_{n=0}^{\infty} (-1)^n S_{n+1} x^n = \frac{\sqrt{\pi}}{2} \Gamma\left(\frac{x+1}{2}\right) \times \Gamma^{-1}\left(\frac{x}{2}+1\right). \quad (12.5.10)$$

Similarly

$$\sum_{n=0}^{\infty} S_{n+1} x^n = \frac{\sqrt{\pi}}{x} \tan\left(\frac{\pi x}{2}\right) \Gamma\left(\frac{x}{2}+1\right) \times \Gamma^{-1}\left(\frac{x+1}{2}\right).$$

The product of these last two series yields

$$\left(\sum_{n=0}^{\infty} (-1)^n S_{n+1} x^n\right) \times \left(\sum_{n=0}^{\infty} S_{n+1} x^n\right) = \frac{\pi}{2x} \tan\left(\frac{\pi x}{2}\right)$$

$$= \sum_{n=1}^{\infty} \frac{(2^{2n} - 1)(-1)^{n-1} B_{2n} \pi^{2n}}{(2n)!} x^{2n}.$$

Then (12.5.3) follows from here. The values

$$S_1 = \frac{\pi}{2}$$

$$S_2 = \frac{\pi}{2} \ln 2$$

we obtain

$$S_3 = \frac{\pi^3}{48} + \frac{\pi}{4} \ln^2 2.$$

In order to evaluate the remaining S_n we differentiate (12.5.10) to produce

$$\sum_{n=1}^{\infty} (-1)^n n S_{n+1} x^{n-1} = \frac{1}{2} [\psi(x/2 + 1/2) - \psi(x/2 + 1)] \times \sum_{n=0}^{\infty} (-1)^n S_{n+1} x^n.$$

Expanding ψ in its Taylor series produces

$$\psi(x/2 + 1/2) - \psi(x/2 + 1) = -2 \ln 2 + 2 \sum_{n=1}^{\infty} (-1)^{n+1} (1 - 2^{-n}) \zeta(n+1) x^n,$$

and we obtain (12.5.4). □

Exercise 12.5.1. a) Prove that

$$S_n = 2^n \int_0^{\pi/2} (\ln \sin x)^n dx$$

$$= 2^n \int_0^{\pi/2} (\ln \cos x)^n dx.$$

b) Check that

$$S_n = 2^n \int_0^1 \frac{\ln^n t \, dt}{\sqrt{1 - t^2}},$$

and

$$S_n = \frac{(-1)^n}{2} \int_0^\infty \frac{v^n e^{-v/2} \, dv}{\sqrt{1 - e^{-v}}}$$

and

$$S_n = \frac{1}{2} \int_0^1 \frac{\ln^n t \, dt}{\sqrt{t(1 - t)}}.$$

Exercise 12.5.2. This appears in Tyler and Chernhoff (1985). Prove that

$$\sum_{n=1}^{\infty} \frac{\zeta(2n)}{n(2n + 1)2^{2n}} = \ln \pi - 1. \qquad (12.5.11)$$

Hint. Use the infinite product for $\sin x$ to obtain

$$\ln \sin \pi x = \ln \pi x - \sum_{n=1}^{\infty} \frac{\zeta(2n)}{n} x^{2n}. \qquad (12.5.12)$$

Now integrate from 0 to 1/2. Integration from 0 to 1 yields the companion formula

$$\sum_{n=1}^{\infty} \frac{\zeta(2n)}{n(2n + 1)} = \ln 2\pi - 1. \qquad (12.5.13)$$

It was pointed out by Danese (1967) that this series is a particular case of

$$\sum_{k=1}^{\infty} \frac{\zeta(2k, z)}{2^{2k}(2k^2 + k)} = (2z - 1)\ln(z - 1/2) - 2z + 1 + \ln 2\pi - 2\ln \Gamma(z),$$

where

$$\zeta(s, z) = \sum_{n=0}^{\infty} \frac{1}{(z + n)^s} \qquad (12.5.14)$$

is the **Hurwitz zeta function** which will be studied in Volume 2.

Exercise 12.5.3. Gradshteyn and Ryzhik's compilation (1994) contains many other integrals involving products of logarithms. For instance, 4.315.1 and 4.315.3 are

$$\int_0^1 \ln(1+x)\ln^{n-1} x \, \frac{dx}{x} = (-1)^{n-1}(n-1)!\left(1-2^{-n}\right)\zeta(n+1)$$

and

$$\int_0^1 \ln(1-x)\ln^{n-1} x \, \frac{dx}{x} = (-1)^n(n-1)!\,\zeta(n+1).$$

Check them. The integrals 4.315.2 and 4.315.4 are the cases n odd for which $\zeta(n+1)$ can be expressed in terms of the Bernoulli numbers.

13

A Master Formula

13.1. Introduction

The goal of this chapter is to present a transformation due to Schlomilch that yields the evaluation of many definite integrals. The main application of this transformation is to present an evaluation of the integral

$$
\begin{aligned}
M_4(a; r, s) &= \int_0^\infty \left(\frac{x^2}{x^4 + 2ax^2 + 1} \right)^r \times \frac{x^2 + 1}{x^2(x^s + 1)} \, dx \\
&= \frac{B\left(r - \frac{1}{2}, \frac{1}{2}\right)}{2^{r+1/2}(a + 1)^{r-1/2}}
\end{aligned}
\tag{13.1.1}
$$

in terms of Euler's beta function

$$
B(p, q) = \int_0^1 x^{p-1}(1 - x)^{q-1} \, dx.
\tag{13.1.2}
$$

The evaluation described here is a *master formula*, provided by Boros and Moll (1998). Many different integrals can be derived from it through varying the parameters, changes of variable, differentiation and other more sophisticated transformations. In this form, (13.1.1) unifies large classes of integrals, and we illustrate its power through a number of examples. Some of these are well known, by which we mean that they can be computed by a symbolic language or can be found in a table of integrals. We have used Mathematica as a source for the former and Gradshteyn and Ryzhik (1994) [G & R] for the latter; others appear to be entirely new, as we have been unable to find anything resembling them in the literature. The variety of definite integrals that can be deduced from (13.1.1) is immense and we just show some examples. For example, in (13.6.10) we show that

$$
\int_0^\infty \left[(x^2 + 1)\sqrt[3]{x(x^2 + 1)} \right]^{-1} \times \ln \left(\frac{x}{x^2 + 1} \right) \, dx = \left(-\frac{1}{2}\Gamma(1/3) \right)^3.
$$

250

13.2. Schlomilch Transformation

We now present a result that connects the integral of a function f in two different scales. The conditions that f must satisfy are simply that the improper integrals appearing in the next result are finite. This will be easy to verify in the examples presented below.

Theorem 13.2.1. *Let a, $c > 0$. Then*

$$\int_0^\infty f\left((cx - a/x)^2\right) dx = \frac{1}{c} \int_0^\infty f(u^2)\, du. \qquad (13.2.1)$$

Proof. Transform the integral I on the left of (13.2.1) by $t = a/cx$ and add it to the original to obtain

$$2I = \frac{1}{c} \int_0^\infty f\left((cx - a/x)^2\right)\left(c + a/x^2\right) dx. \qquad (13.2.2)$$

The change of variables $u = cx - a/x$ maps the half line $[0, \infty)$ to the real line $(-\infty, \infty)$ and it yields (13.2.1). □

Example 13.2.1. Let $f(x) = e^{-x}$. Then (13.2.1) gives Laplace's integral (8.4.1).

Example 13.2.2. The details in this example were shown to the authors by R. Posey. It simplifies the original proof given by Boros and Moll (1998). Let $f(x) = 1/(1+x)^r$ with $r > 1/2$. Then (13.2.1) yields

$$\int_0^\infty \left(1 + (cx - a/x)^2\right)^{-r} dx = \frac{1}{c} \int_0^\infty \frac{du}{(1+u^2)^r}. \qquad (13.2.3)$$

Take $c = a$ and replace a by $\sqrt{2(a+1)}$, to obtain

$$[2(a+1)]^r \int_0^\infty \left(\frac{x^2}{x^4 + 2ax^2 + 1}\right)^r dx \qquad (13.2.4)$$

for the left-hand side. The right-hand side is

$$\sqrt{2(a+1)} \int_0^\infty \frac{du}{(1+u^2)^r} = \frac{1}{2}\sqrt{2(a+1)}\, B\left(r - \tfrac{1}{2}, \tfrac{1}{2}\right) \qquad (13.2.5)$$

where we have used (10.3.3). We conclude that

$$\int_0^\infty \left(\frac{x^2}{x^4 + 2ax^2 + 1}\right)^r dx = \frac{B\left(r - \tfrac{1}{2}, \tfrac{1}{2}\right)}{2^{1/2+r}(1+a)^{r-1/2}}.$$

This example can be written in many different ways by using appropriate changes of variables. For example $x \to 1/x$ yields

$$\int_0^\infty \left(\frac{x^2}{x^4 + 2ax^2 + 1} \right)^r \frac{dx}{x^2} = \frac{B\left(r - \frac{1}{2}, \frac{1}{2}\right)}{2^{1/2+r}(1+a)^{r-1/2}}.$$

In the next section we show how to expand this idea and how to introduce free parameters into this evaluation.

13.3. Derivation of the Master Formula

In this section we present an evaluation of the integral with three parameters.

Theorem 13.3.1. *Let*

$$M_4(a; r, s) = \int_0^\infty \left(\frac{x^2}{x^4 + 2ax^2 + 1} \right)^r \times \frac{x^2 + 1}{x^s + 1} \cdot \frac{dx}{x^2}. \quad (13.3.1)$$

Then $M_4(a; r, s)$ is independent of s and

$$M_4(a; r, s) = \int_0^\infty \left(\frac{x^2}{x^4 + 2ax^2 + 1} \right)^r \times \frac{dx}{x^2} \qquad (13.3.2)$$

$$= \int_0^\infty \left(\frac{x^2}{x^4 + 2ax^2 + 1} \right)^r dx \qquad (13.3.3)$$

$$= \frac{1}{2} \int_0^\infty \left(\frac{x^2}{x^4 + 2ax^2 + 1} \right)^r \times \frac{x^2 + 1}{x^2} dx \qquad (13.3.4)$$

$$= \int_0^1 \left(\frac{x^2}{x^4 + 2ax^2 + 1} \right)^r \times \frac{x^2 + 1}{x^2} dx, \qquad (13.3.5)$$

$$= \frac{B\left(r - \frac{1}{2}, \frac{1}{2}\right)}{2^{1/2+r}(1+a)^{r-1/2}}.$$

Proof. The equivalence of the four integrals is easy to establish. The second one follows from the first by the change of variable $x \to 1/x$. The third one is the average of the first two and the last one is obtained by splitting the original integral on $[0, 1]$ and $[1, \infty)$ and converting the integral over $[1, \infty)$ back to $[0, 1]$.

To obtain the common value of these integrals observe that the result has been established in Example 13.2.2 for $s = 2$, so the result follows from the next lemma with $g(x) = (x^2 + 2a + x^{-2})^{-r} \cdot (x + x^{-1})$. □

Lemma 13.3.1. *Suppose g satisfies the functional equation* $g(1/x) = g(x)$. *Then*

$$K(s) = \int_0^\infty \frac{g(x)}{x^s + 1} \frac{dx}{x} \tag{13.3.6}$$

is independent of s.

Proof. Split the integral into two pieces on $[0, 1]$ and $[1, \infty)$ and make the substitution $x \mapsto 1/x$ in the second one. Then

$$K(s) = \int_0^1 \frac{g(x)}{x^s + 1} \frac{dx}{x} + \int_0^1 \frac{x^s g(1/x)}{x^s + 1} \frac{dx}{x} = \int_0^1 g(x) \frac{dx}{x}$$

and the last expression is independent of s. □

13.4. Applications of the Master Formula

In this section we present several classical evaluations that are consequences of the master formula.

We begin with some numerical evaluations.

Example 13.4.1. Let $a = 1/2$, $r = 3$ in (13.3.2) to obtain

$$\int_0^\infty \frac{x^4}{(x^4 + x^2 + 1)^3} \, dx = \frac{\pi}{48\sqrt{3}}.$$

Then take $a = 7/2$, $r = 5/2$ in (13.3.2) to get

$$\int_0^\infty \frac{x^3}{(x^4 + 7x^2 + 1)^{5/2}} \, dx = \frac{2}{243}.$$

The third integral presented here corresponds to the values $a = 7$, $r = 5/4$ in (13.3.3):

$$\int_0^\infty \frac{\sqrt{x}}{(x^4 + 14x^2 + 1)^{5/4}} dx = \frac{\Gamma^2(3/4)}{4\sqrt{2\pi}}.$$

A more complicated example appears by taking $a = 1/2$, $r = 3/4$ in (13.3.3):

$$\int_0^\infty \frac{dx}{\sqrt{x}(x^4 + x^2 + 1)^{3/4}} = \frac{\pi^{3/2}}{\Gamma^2(3/4) \sqrt[4]{12}}.$$

Example 13.4.2. Using the fact that $M_4(a; r, s)$ is independent of s we can evaluate some strange integrals. For example: $s = 10$, $a = (1 + 2\sqrt{2})/2$,

$r = 1$ in (13.3.3) yield

$$\int_0^\infty \frac{dx}{x^{12} + 2\sqrt{2}x^{10} + (1 - 2\sqrt{2})x^8 + (-1 + 2\sqrt{2})x^6 + (1 - 2\sqrt{2})x^4 + 2\sqrt{2}x^2 + 1}$$
$$= \frac{\pi}{2(1 + \sqrt{2})}.$$

Some more artificial evaluations come from $s = 102$, $a = (1 + 2\sqrt{2})/2$, $r = 1$ in (13.3.3). As mentioned above, the integral is independent of s, so we get the same answer as in the previous case:

$$\int_0^\infty \frac{dx}{\left[x^4 + (1 + 2\sqrt{2})x^2 + 1\right]\left[x^{100} - x^{98} + \cdots + 1\right]} = \frac{\pi}{2(1 + \sqrt{2})}.$$

Example 13.4.3. The next evaluation is a classical one. Let $a = 1$, $r = 1$ in (13.3.2) to obtain

$$\int_0^\infty \frac{dx}{(x^2 + 1)(x^s + 1)} = \frac{\pi}{4}$$

(see, for instance, Edwards (1922), page 262). This can be transformed via $x = \tan\theta$ to the familiar form

$$\int_0^{\pi/2} \frac{d\theta}{1 + (\tan\theta)^s} = \frac{\pi}{4}.$$

The next example is also well known.

Example 13.4.4. The case $s = 0$, $a = 1$ and $r = 3/4$ in (13.3.3) produces

$$\frac{1}{2}\int_0^{\pi/2} \frac{d\theta}{\sqrt{\sin\theta \, \cos\theta}} = \frac{\Gamma^2(1/4)}{4\sqrt{\pi}},$$

so that, after the change of variable $2\theta \mapsto \theta$ we get

$$\int_0^{\pi/2} \frac{d\theta}{\sqrt{\sin\theta}} = \frac{\Gamma^2(1/4)}{2\sqrt{2\pi}}.$$

This can also be obtained by letting $r = -1/2$ in (13.4.3).

We now present a series of classical results of analysis that form part of the master formula.

Example 13.4.5. Wallis' integral formula. Let $r \in \mathbb{R}$. Then $a = 1$ and $s = 2$ in the master formula yield

$$\int_0^\infty (x + 1/x)^{-2r} \, dx = 2^{-2r} B \left(r - \tfrac{1}{2}, \tfrac{1}{2} \right) \tag{13.4.1}$$

and

$$\int_0^\infty x^{-2} (x + 1/x)^{-2r} \, dx = 2^{-2r} B \left(r - \tfrac{1}{2}, \tfrac{1}{2} \right) \tag{13.4.2}$$

The change of variables $x = \tan \theta$ in the expression (13.4.1) yields

$$\int_0^{\pi/2} \sin^{2r-2} \theta \cos^{2r-2} \theta \, d\theta = 2^{1-2r} B \left(r - \tfrac{1}{2}, \tfrac{1}{2} \right).$$

The substitution 2θ by θ and $2r - 2$ by r produces Wallis' integral:

$$\int_0^{\pi/2} \sin^r \theta \, d\theta = \frac{1}{2} B \left(\frac{r+1}{2}, \frac{1}{2} \right), \quad \text{for } r > -1. \tag{13.4.3}$$

This appeared in (10.3.13).

Example 13.4.6. The duplication formula of Legendre. We now use the formula

$$\int_0^\infty \left(\frac{x^2}{bx^4 + 2ax^2 + 1} \right)^r dx = \frac{B \left(r - \tfrac{1}{2}, \tfrac{1}{2} \right)}{2^{c+1/2} \sqrt{b} (a + \sqrt{b})^{r-1/2}} \tag{13.4.4}$$

to derive Legendre's duplication formula (10.4.7) for the gamma function. The expression (13.4.4) follows from (13.3.4) by a simple scaling.

Theorem 13.4.1. *Let $r \in \mathbb{R}$. Then*

$$\Gamma(r + 1/2) = \frac{\Gamma(2r)\Gamma(1/2)}{\Gamma(r)2^{2r-1}}. \tag{13.4.5}$$

Proof. Consider the special case of (13.4.4) with $b = a^2$; we then have

$$\int_0^\infty \frac{x^{2r}}{(ax^2 + 1)^{2r}} dx = \frac{B \left(r - \tfrac{1}{2}, \tfrac{1}{2} \right)}{2^{2r} a^{r+1/2}}.$$

The integral on the left-hand side can be evaluated using (10.3.3) to produce

$$B(r - 1/2, 1/2) = 2^{2r-1} B(r + 1/2, r - 1/2).$$

Now use (10.2.5) to complete the proof. $\qquad\square$

Example 13.4.7. The generating function of the central binomial coefficients. This is given by

$$\sum_{n=0}^{\infty} \binom{2n}{n} b^n = \frac{1}{\sqrt{1-4b}} \tag{13.4.6}$$

and appeared in Exercise 4.2.3.

Proof. The expression (13.3.3) yields

$$\int_0^{\infty} \left(\frac{x^2}{x^4 + 2ax^2 + 1} \right)^n dx = \frac{B\left(n + \frac{1}{2}, \frac{1}{2}\right) 1}{2^{1/2+n}(1+a)^{n-1/2}}$$

$$= \frac{\Gamma(n-1/2)\sqrt{\pi}}{\Gamma(n)} \times \frac{\sqrt{1+a}}{2^{1/2+n}(1+a)^n}$$

$$= 2^{3/2}\pi\sqrt{1+a} \binom{2n-2}{n-1} [8(1+a)]^{-n}.$$

Now sum from $n = 1$ to $n = \infty$ to obtain, on the left-hand side,

$$\int_0^{\infty} \frac{x^2 \, dx}{x^4 + (2a-1)x^2 + 1} = \frac{\pi}{2\sqrt{2a+1}},$$

where the resulting integral has been evaluated in Exercise 7.2.10. The proof is complete by choosing $b = [8(a+1)]^{-1}$. □

Example 13.4.8. A series involving central binomial coefficients. We now derive the value

$$\sum_{n=1}^{\infty} \frac{1}{n\binom{2n}{n}} = \frac{\pi}{3\sqrt{3}} \tag{13.4.7}$$

from the master formula. This appeared in (6.6.11).

Start with (13.3.3). Replacing n by $2n + 1$ and using Legendre's formula (13.4.5) produces

$$\int_0^{\infty} \left(\frac{2x}{x^2 + 1} \right)^{2n+1} \frac{dx}{x^2} = \frac{\Gamma^2(n)}{4\Gamma(2n)} = \left(2n\binom{2n}{n} \right)^{-1}.$$

The bound $2x/(x^2 + 1) \leq 1/2$ permits to sum from $n = 1$ to $n = \infty$ to produce

$$\int_0^{\infty} \frac{x \, dx}{(x^2 + 1)(x^4 + x^2 + 1)} = \frac{1}{2} \sum_{n=1}^{\infty} \frac{1}{n\binom{2n}{n}}.$$

The change of variables $y = x^2$ permits an elementary evaluation of the integral on the left by partial fractions. This gives (13.4.7).

13.5. Differentiation Results

The formula

$$M_4(a; r, s) = \int_0^\infty \left(\frac{x^2}{x^4 + 2ax^2 + 1} \right)^r \times \frac{x^2 + 1}{x^s + 1} \cdot \frac{dx}{x^2}$$

$$= \frac{B\left(r - \frac{1}{2}, \frac{1}{2}\right)}{2^{r + \frac{1}{2}}(1 + a)^{r - \frac{1}{2}}}$$

and other equivalent versions used depend on the three *independent* parameters a, r, s. We have seen that it is possible to derive the evaluation in closed form of a large number of integrals by appropriate manipulations of the parameters. One can obtain additional results by differentiation with respect to r. For example, with $a = -1/2$ in (13.3.3), we have

$$f(r) := \int_0^\infty \left(\frac{x^2}{x^4 - x^2 + 1} \right)^r \cdot \frac{dx}{x^2} = \frac{1}{2} B\left(r - \frac{1}{2}, \frac{1}{2}\right) \quad (13.5.1)$$

$$= \frac{\sqrt{\pi}}{2} \frac{\Gamma(r - \frac{1}{2})}{\Gamma(r)}.$$

Differentiation of (13.3.2) produces

$$f'(r) = \int_0^\infty \left(\frac{x^2}{x^4 - x^2 + 1} \right)^r \times \ln\left(\frac{x^2}{x^4 - x^2 + 1} \right) \frac{dx}{x^2},$$

and differentiating the expression in (13.3.3) gives

$$f'(r) = \frac{\sqrt{\pi}}{2} \times \frac{\Gamma(r)\Gamma'(r - 1/2) - \Gamma(r - 1/2)\Gamma'(r)}{\Gamma^2(r)}.$$

Several interesting evaluations can now be obtained by specifying the value of the parameter r. Such calculations produce *nice results* in terms of well-known constants, provided we know the values of the function Γ and its derivatives at the arguments r and $r - 1/2$ in terms of these constants. The following examples illustrate this.

Example 13.5.1. $r = 1$ produces

$$f'(1) = \int_0^\infty \ln\left(\frac{x^2}{x^4 - x^2 + 1}\right) \frac{dx}{x^4 - x^2 + 1}$$

$$= \frac{\sqrt{\pi}}{2} \frac{\Gamma(1)\Gamma'(1/2) - \Gamma(1/2)\Gamma'(1)}{\Gamma^2(1)}$$

$$= -\pi \ln 2.$$

Thus

$$\int_0^\infty \ln\left(\frac{x^2}{x^4 - x^2 + 1}\right) \frac{dx}{x^4 - x^2 + 1} = -\pi \ln 2.$$

In this calculation we used the value $\Gamma'(1/2) = -\sqrt{\pi}(\gamma + 2\ln 2)$.

Example 13.5.2. The value $r = 3/2$ produces

$$\int_0^\infty \frac{x}{(x^4 - x^2 + 1)^{3/2}} \times \ln\left(\frac{x}{x^4 - x^2 + 1}\right) dx = \ln 2 - 1.$$

and $r = 3/4$ yields

$$\int_0^\infty \left(\frac{x^2}{x^4 - x^2 + 1}\right)^{3/4} \times \ln\left(\frac{x^2}{x^4 - x^2 + 1}\right) dx = -\frac{\sqrt{\pi}}{2\sqrt{2}}\Gamma^2(1/4).$$

Extra 13.5.1. Further differentiation of the function $f(r)$ produces more examples of integrals that can be evaluated in closed form. These calculations now require the explicit knowledge of the values of Γ, Γ' and Γ'' at the arguments r and $r - 1/2$. For example, the calculation of $f''(1)$ gives the result

$$\int_0^\infty \ln^2\left(\frac{x^2}{x^4 - x^2 + 1}\right) \frac{dx}{x^4 - x^2 + 1} = \frac{\pi}{2}\left(\frac{\pi^2}{3} + 4\ln^2 2\right).$$

Many other similar integrals can be evaluated.

Note 13.5.1. Differentiation with respect to s gives zero, reflecting the fact that the integral is independent of s, and differentiation with respect to a merely returns an equivalent form of the original integral.

13.6. The case $a = 1$

In this section we discuss in more detail the special case $a = 1$. Theorem 13.3.1 yields

$$\int_0^\infty \left(\frac{x}{x^2 + 1}\right)^{2r} \frac{dx}{x^2} = 2^{-2r} B\left(r - \tfrac{1}{2}, \tfrac{1}{2}\right)$$

and the equivalent form

$$\int_0^\infty \left(\frac{x}{x^2+1}\right)^{2r} dx = 2^{-2r} B\left(r - \tfrac{1}{2}, \tfrac{1}{2}\right).$$

Define G as

$$G(r) = \frac{\Gamma(r - \tfrac{1}{2})}{\Gamma(r)2^{2r-1}} = \frac{1}{\sqrt{\pi}} B\left(r - \tfrac{1}{2}, \tfrac{1}{2}\right) 2^{1-2r},$$

so that

$$\frac{\sqrt{\pi}}{2} G(r) = \int_0^\infty \left(\frac{x}{x^2+1}\right)^{2r} \frac{dx}{x^2}.$$

Differentiation produces

$$\int_0^\infty \left(\frac{x}{x^2+1}\right)^{2r} \times \ln\left(\frac{x}{x^2+1}\right) \frac{dx}{x^2} = \frac{\sqrt{\pi}}{4} G'(r), \qquad (13.6.1)$$

and logarithmic differentiation of G gives

$$G'(r) = G(r)\left[\psi(r - 1/2) - \psi(r) - 2\ln 2\right],$$

from which

$$G''(r) = G(r) \times \left[\psi'(r - 1/2) - \psi'(r) + (\psi(r - 1/2) - \psi(r) - 2\ln 2)^2\right]$$

follows.

As before, certain values of r produce some interesting integrals. In order to obtain a clean-looking result, in addition to knowing the values of Γ, Γ' at the arguments r and $r - 1/2$, we also need to know the values of ψ and ψ'.

Example 13.6.1. Let $r = 1$. Then (13.6.1) yields

$$\int_0^\infty \frac{1}{(x^2+1)^2} \times \ln\left(\frac{x}{x^2+1}\right) dx = -\frac{\pi \ln 2}{2},$$

where we have used the appropriate values of ψ to compute $G'(1)$. We now use [G & R]'s formula 4.234.6 on page 566:

$$\int_0^\infty \frac{\ln x \, dx}{(a_1^2 + b_1^2 x^2)(1 + x^2)} = \frac{\pi b_1}{2a_1(b_1^2 - a_1^2)} \ln\left(\frac{a_1}{b_1}\right)$$

in the limiting case $a_1 \to 1$, $b_1 \to 1$ to obtain

$$\int_0^\infty \frac{\ln x \, dx}{(x^2+1)^2} = -\frac{\pi}{4}, \qquad (13.6.2)$$

and combining this with the previous result we get

$$\int_0^\infty \frac{\ln(x^2 + 1)\, dx}{(x^2 + 1)^2} = \frac{\pi}{4}(2\ln 2 - 1). \qquad (13.6.3)$$

Note 13.6.1. If we repeat the same procedure used to obtain (13.6.1) but use Theorem 13.3.1 with (13.3.2) in lieu of (13.3.3), we obtain

$$\int_0^\infty \left(\frac{x}{x^2 + 1}\right)^{2r} \times \ln\left(\frac{x}{x^2 + 1}\right) \times \frac{x^2 + 1}{x^2(x^s + 1)}\, dx = \frac{1}{4}\sqrt{\pi}\, G'(r). \quad (13.6.4)$$

Example 13.6.2. The value $r = 1$ and $s = 0$ in (13.6.4) yields

$$\int_0^\infty \left(\frac{1}{x^2 + 1}\right) \times \ln\left(\frac{x}{x^2 + 1}\right) dx = \frac{1}{2}\sqrt{\pi}\, G'(1) = -\pi \ln 2. \quad (13.6.5)$$

Using (12.3.2) we thus get

$$\int_0^\infty \frac{\ln(x^2 + 1)}{x^2 + 1}\, dx = \pi \ln 2. \qquad (13.6.6)$$

The change of variables $x = \tan\theta$ yields

$$\int_0^{\pi/2} \ln\cos\theta\, d\theta = -\frac{\pi}{2}\ln 2. \qquad (13.6.7)$$

Extra 13.6.1. The integral

$$I(n) := \int_0^\infty \frac{\ln(x^2 + 1)}{(x^2 + 1)^{n+1}}\, dx$$

can be evaluated as follows: the substitution $x \mapsto \tan\theta$ yields

$$I = -2 \int_0^{\pi/2} \cos^{2n}\theta \ln(\cos\theta)\, d\theta,$$

and this is given in [G & R] 4.387.9 as

$$I = B\left(n + \tfrac{1}{2}, \tfrac{1}{2}\right) \{\ln 2 + H_n - H_{2n}\}. \qquad (13.6.8)$$

The examples given in (13.6.3) and (13.6.6) are particular cases ($n = 1$ and $n = 0$ respectively) of this formula. It would be interesting to derive a proof of (13.6.8) from the master formula.

Example 13.6.3. Differentiation of (13.6.1) yields

$$G''(r) = \frac{8}{\sqrt{\pi}} \int_0^\infty \left(\frac{x}{x^2 + 1}\right)^{2r} \times \ln^2\left(\frac{x}{x^2 + 1}\right) \frac{dx}{x^2} \qquad (13.6.9)$$

and, as usual, special values of r produce some exact evaluations. We provide just one such example.

Let $r = 1$ in (13.6.9). Then

$$\int_0^\infty \frac{1}{(x^2+1)^2} \times \ln^2\left(\frac{x}{x^2+1}\right) dx = \frac{\pi}{48}\left(\pi^2 + 48\ln^2 2\right).$$

We conclude this section with two examples that we find aesthetically pleasing.

We have not been able to evaluate these examples symbolically.

Example 13.6.4. Let $r = 3/4$ in (13.6.1). We then obtain

$$\int_0^\infty \left[(x^2+1)\sqrt{x(x^2+1)}\right]^{-1} \times \ln\left(\frac{x}{x^2+1}\right) dx = -\frac{1}{8\sqrt{\pi}}(\pi + 2\ln 2)$$
$$\times \Gamma^2(1/4).$$

The last one is obtained with the value $c = 5/6$ in (13.6.1):

$$\int_0^\infty \left[(x^2+1)\sqrt[3]{x(x^2+1)}\right]^{-1} \times \ln\left(\frac{x}{x^2+1}\right) dx = \left(-\frac{1}{2}\Gamma(1/3)\right)^3.$$

$$(13.6.10)$$

Aren't they pretty?

Project 13.6.1. It would be interesting to relate the integrals in Example 13.6.4 to [G & R] 4.244.1 which states

$$\int_0^1 \frac{\ln x}{\sqrt[3]{x(1-x^2)^2}} dx = -\frac{1}{8}\Gamma^3(1/3), \qquad (13.6.11)$$

and 4.244.11:

$$\int_0^1 \frac{\ln x}{\sqrt{x(1-x^2)}} dx = \frac{-\sqrt{2\pi}}{8}\Gamma^2(1/4).$$

Perhaps one can use the master formula to prove

$$\int_0^1 \frac{\ln x\, dx}{\sqrt[n]{1-x^{2n}}} = \frac{-\pi B(1/2n, 1/2n)}{8n^2 \sin(\pi/2n)} \qquad n > 1$$

$$\int_0^1 \frac{\ln x\, dx}{\sqrt[n]{x^{n-1}(1-x^{2n})}} = \frac{-\pi B(1/2n, 1/2n)}{8 \sin(\pi/2n)} \qquad n > 1$$

These are 4.243.1 and 4.243.2 respectively.

Extra 13.6.2. The fact that two apparently different integrals produce the same answer, as we have observed in (13.6.10) and (13.6.11), might be nothing more than a coincidence since there is a large amount of flexibility in the representation of a real number such as $\frac{1}{8}\Gamma^3(1/3)$. Kontsevich and Zagier (2001) have proposed a remarkable conjecture for **periods**. A period is a complex number whose real and imaginary part are values of absolutely convergent integrals of rational functions with rational coefficients, over domains in $\mathbb{R}^n$ given by polynomial (in)equalities with rational coefficients. For example

$$\ln 2 = \int_1^2 \frac{dx}{x}$$

and

$$\pi = \int_{x^2+y^2\leq 1} dx\,dy$$

are periods.

Conjecture. If a period has two integral representations, then one can pass from one formula to the other using only the following rules:

1) *Additivity:*

$$\int_a^b (f(x) + g(x))\,dx = \int_a^b f(x)\,dx + \int_a^b g(x)\,dx,$$

$$\int_a^b f(x)\,dx = \int_a^c f(x)\,dx + \int_c^b f(x)\,dx.$$

2) *Change of variables:* If $y = f(x)$ is an invertible change of variables, then

$$\int_{f(a)}^{f(b)} F(y)\,dy = \int_a^b F(f(x))\,f'(x)\,dx.$$

3) *Newton–Leibnitz:*

$$\int_a^b f'(x)\,dx = f(b) - f(a)$$

and higher dimensional versions of these rules.

All the functions and domains of integrations allowed in the passage from one representation to another are algebraic with algebraic coefficients.

Exercise 13.6.1. Prove that $\Gamma(p/q)^q$ is a period. In particular $\Gamma(1/3)^3$ is a period.

13.7. A New Series of Examples

Starting with the expression

$$G(r) = \frac{\Gamma(r - 1/2)}{\Gamma(r)2^{2r-1}} = \frac{2}{\sqrt{\pi}} \int_0^\infty \left(\frac{x}{x^2+1}\right)^{2r} \cdot \frac{dx}{x^2}$$

and differentiating n times we obtain

$$\int_0^\infty \left[\frac{x}{x^2+1}\right]^{2r} \times \left[\ln\left(\frac{x}{x^2+1}\right)\right]^n \cdot \frac{dx}{x^2} = \frac{\sqrt{\pi}}{2^{n+1}} \frac{d^n}{dr^n}\left(\frac{\Gamma(r-1/2)}{\Gamma(r)2^{2r-1}}\right).$$

$$(13.7.1)$$

In particular, for $r = 3/2$ and $n = 3$ we obtain

$$\int_0^\infty \left[\left(\frac{x}{x^2+1}\right) \times \ln\left(\frac{x}{x^2+1}\right)\right]^3 \cdot \frac{dx}{x^2} = \frac{3}{8}[\zeta(3) + \zeta(2) - 4].$$

From here, using the substitution $t = x/(x^2+1)$, it follows that

$$\zeta(3) = 4 - \zeta(2) + \frac{8}{3} \int_0^{1/2} \frac{t \ln^3 t}{\sqrt{1-4t^2}}\, dt. \qquad (13.7.2)$$

Exercise 13.7.1. Check (13.7.2) directly. **Hint.** Use the expansion for $1/\sqrt{1-4t^2}$ and integrate term by term.

Now let $r = 3/2$ and $n = 4$ in (13.7.1):

$$\int_0^\infty \left[\frac{x}{x^2+1}\right]^3 \times \left[\ln\left(\frac{x}{x^2+1}\right)\right]^4 \cdot \frac{dx}{x^2}$$

$$= \frac{1}{160}\left[-3\pi^4 - 80\pi^2 + 1920 - 480\zeta(3)\right].$$

Note 13.7.1. The values of the previous integrals are rational combinations of the values of the Riemann zeta function $\zeta(s)$ at $s = 2, 3,$ and 4. Higher values of n produce similar algebraic combinations of the values of $\zeta(j)$. Define

$$H(r;n) := \sqrt{\pi}\left(\frac{d}{dr}\right)^n G(r). \qquad (13.7.3)$$

Then, for example,

$$H(3, 3/2) = \pi^2 + 6\zeta(3) - 24$$
$$H(4, 3/2) = -3\pi^4/10 - 8\pi^2 + 192 - 48\zeta(3)$$
$$H(5, 3/2) = 360\zeta(5) + 480\zeta(3) + 80\pi^2 + 3\pi^4 - 20\zeta(3)\pi^2 - 1920.$$

Naturally, the appearance of the values of the Riemann zeta function is due to the expression for the polygamma function

$$\text{PolyGamma}[n, x] = \frac{d^{n+1}}{dx^{n+1}} \ln \Gamma(x)$$

$$= (-1)^{n+1} n! \sum_{k=0}^{\infty} \frac{1}{(x+k)^{n+1}}$$

and the special values at $x = 1$ and $x = 3/2$:

$$\text{PolyGamma}[n, 1] = (-1)^{n+1} n! \zeta(n+1)$$
$$\text{PolyGamma}[n, 3/2] = (-1)^{n+1} n! \left[(2^{n+1} - 1)\zeta(n+1) - 2^{n+1} \right].$$

The series expansion for $\ln \Gamma(x)$ appears in Theorem 10.6.1.

Example 13.7.1. The formula (13.7.1) can be used to produce many more exact evaluations of definite integrals. For example, we now show that

$$\int_0^\infty \left(\frac{x}{x^2+1} \right)^{2j+1} \times \left[\ln \left(\frac{x}{x^2+1} \right) \right] \cdot \frac{dx}{x^2} = \frac{H_j - H_{2j} - 1/2j}{2j \binom{2j}{j}} \quad (13.7.4)$$

for any positive integer j, and a similar formula for even exponents. Here H_j is the harmonic number.

We start with (13.7.1):

$$I(r, n) := \int_0^\infty \left(\frac{x}{x^2+1} \right)^{2r} \times \left[\ln \left(\frac{x}{x^2+1} \right) \right]^n \cdot \frac{dx}{x^2}$$

$$= \frac{\sqrt{\pi}}{2^{n+1}} \left(\frac{d}{dr} \right)^{n+1} \left[\frac{\Gamma(r-1/2)}{\Gamma(r) 2^{2r-1}} \right].$$

Now split the domain of integration into $[0, 1]$ and $[1, \infty)$, and let $u = 1/x$ in the second part to obtain

$$I(r, n) = \int_0^1 \left(\frac{x}{x^2+1} \right)^{2r-1} \times \left[\ln \left(\frac{x}{x^2+1} \right) \right]^n \cdot \frac{dx}{x}.$$

The change of variable $v = x/(x^2 + 1)$ yields

$$I(r, n) = \int_0^{1/2} u^{2r-2} \ln^n u \cdot \frac{du}{\sqrt{1-4u^2}}.$$

Letting $v = 2u$ and expanding the term $[\ln v - \ln 2]^n$ then gives

$$I(r, n) = 2^{1-2r} \sum_{k=0}^n (-1)^{n-k} \binom{n}{k} (\ln 2)^{n-k} \int_0^1 v^{2r-2} (\ln v)^k \frac{dv}{\sqrt{1-v^2}}.$$

$$(13.7.5)$$

Now define

$$F(r) := \int_0^1 \frac{v^{2r-2}\, dv}{\sqrt{1-v^2}}, \tag{13.7.6}$$

so that

$$\int_0^1 v^{2r-2}(\ln v)^k \frac{dv}{\sqrt{1-v^2}} = 2^{-k}\left(\frac{d}{dr}\right)^k F(r). \tag{13.7.7}$$

The function $F(r)$ can be computed by the change of variables $v = \sin\theta$ resulting in

$$F(r) = \frac{1}{2} B\left(r - \tfrac{1}{2}, \tfrac{1}{2}\right).$$

We conclude that

$$I(r, n) = (\ln 2)^n 2^{-2r} \sum_{k=0}^n (-1)^{n-k} \binom{n}{k}(2\ln 2)^{-k}\left(\frac{d}{dr}\right)^k B\left(r - \tfrac{1}{2}, \tfrac{1}{2}\right),$$

and the special case $n = 1$ gives

$$I(r, 1) = 2^{-2r}\ln 2 \times \left[-B\left(r - \tfrac{1}{2}, \tfrac{1}{2}\right) + \frac{1}{2\ln 2}\frac{d}{dr}B\left(r - \tfrac{1}{2}, \tfrac{1}{2}\right)\right].$$

This can be simplified using the function

$$\psi(r) = \frac{d}{dr}\ln\Gamma(r), \tag{13.7.8}$$

the final result being

$$I(r, 1) = 2^{-(2r+1)} B\left(r - \tfrac{1}{2}, \tfrac{1}{2}\right)\left[-2\ln 2 + \psi(r - 1/2) - \psi(r)\right].$$

In the special case $r = j + 1/2$ with j an integer, the previous expression can be simplified using the values

$$\psi(j) = -\gamma + H_{j-1} \quad \text{and} \quad \psi(j + 1/2) = -\gamma - 4 + \sum_{r=1}^j \frac{2}{2r-1},$$

yielding

$$\psi(r - 1/2) - \psi(r) = 2\left[H_j - H_{2j} + \ln 2 - 1/(2j)\right].$$

This gives the evaluation (13.7.4).

13.8. New Integrals by Integration

In this section we describe some interesting examples that appear as consequences of the formula

$$\int_0^\infty \left[\frac{x}{x^2+1}\right]^{2r} \times \left[\ln\left(\frac{x}{x^2+1}\right)\right]^m \frac{dx}{x^2} = \frac{\sqrt{\pi}}{2^{m+1}}\left(\frac{d}{dr}\right)^m\left[\frac{\Gamma(r-1/2)}{\Gamma(r)2^{2r-1}}\right].$$

Using our general principle, we can multiply the integrand by $(x^2+1)/(x^s+1)$ and the result is *independent* of s. Define

$$I := \frac{\sqrt{\pi}}{2^{m+1}}\left(\frac{d}{dr}\right)^m\left[\frac{\Gamma(r-1/2)}{\Gamma(r)2^{2r-1}}\right].$$

We then have

$$I = \int_0^\infty \left[\frac{x}{x^2+1}\right]^{2r} \times \left[\ln\left(\frac{x}{x^2+1}\right)\right]^m \times \frac{x^2+1}{x^s+1} \times \frac{dx}{x^2}.$$

Now integrate this equation with respect to s from $s=b$ to $s=c$ and use the value

$$\int_b^c \frac{ds}{x^s+1} = \frac{1}{\ln x}\ln\left[\frac{x^c(1+x^b)}{x^b(1+x^c)}\right]$$

to conclude that

$$\int_0^\infty \left[\frac{x}{x^2+1}\right]^{2r-1}\left[\ln\left(\frac{x}{x^2+1}\right)\right]^m \ln\left[\frac{x^c(1+x^b)}{x^b(1+x^c)}\right]\frac{dx}{x\ln x}$$
$$= (c-b)\frac{\sqrt{\pi}}{2^{m+1}}\left(\frac{d}{dr}\right)^m\left[\frac{\Gamma(r-1/2)}{\Gamma(r)2^{2r-1}}\right].$$

Example 13.8.1. Take $m=0$, $c=1$, $b=2p+1$ with a positive integer p. Then

$$\frac{x^c(1+x^b)}{x^b(1+x^c)} = x^{-2p}\left[1-x+x^2-x^3+\cdots+x^{2p}\right],$$

and the integral becomes

$$\int_0^\infty \left[\frac{x}{x^2+1}\right]^{2r-1}\left\{\frac{\ln(1-x+\cdots+x^{2p})}{\ln x}-2p\right\}\frac{dx}{x}$$
$$= -\frac{p}{2^{2r-1}}B\left(r-\tfrac{1}{2},\tfrac{1}{2}\right).$$

Now observe that

$$\int_0^\infty \frac{x^{2r-2}\,dx}{(x^2+1)^{2r-1}} = \frac{1}{2}\int_0^\infty \frac{t^{r-3/2}}{(1+t)^{2r-1}}\,dt$$

$$= \frac{1}{2}B(r-1/2, r-1/2),$$

where we have used the representation (10.2.1). Using Legendre duplication formula (10.4.7) and replacing r by $r-1$, we conclude that

$$\int_0^\infty \left[\frac{x}{x^2+1}\right]^{2r+1} \left\{\frac{\ln(1-x+\cdots+x^{2p})}{\ln x}\right\} \frac{dx}{x} = \frac{pr\pi\Gamma(2r)}{2^{4r}\Gamma^2(r+1)}. \qquad (13.8.1)$$

In particular, when r is an integer n:

$$\int_0^\infty \left[\frac{x}{x^2+1}\right]^{2n+1} \left\{\frac{\ln(1-x+\cdots+x^{2p})}{\ln x}\right\} \frac{dx}{x} = \frac{p\pi}{2^{4n+1}}\binom{2n}{n}. \qquad (13.8.2)$$

For instance,

$$\int_0^\infty \frac{\ln(1-x+x^2)}{(x^2+1)\ln x}\,dx = \frac{\pi}{2}. \qquad (13.8.3)$$

Example 13.8.2. We now sum (13.8.2) from $n=0$ to $n=\infty$ to obtain

$$\int_0^\infty \frac{x^2+1}{x^4+x^2+1} \times \frac{\ln(1-x+x^2-\cdots+x^{2p})}{\ln x}\,dx = \frac{p\pi}{2}\sum_{n=0}^\infty 2^{-4n}\binom{2n}{n},$$

and the series can be summed via (13.4.6) to obtain

$$\int_0^\infty \frac{x^2+1}{x^4+x^2+1} \times \frac{\ln(1-x+x^2-\cdots+x^{2p})}{\ln x}\,dx = \frac{p\pi}{\sqrt{3}}.$$

The case $p=1$ produces

$$\int_0^\infty \frac{x^2+1}{x^4+x^2+1} \times \frac{\ln(1-x+x^2)}{\ln x}\,dx = \frac{\pi}{\sqrt{3}}$$

and $x=\tan\theta$ transforms this into

$$\int_0^\pi \frac{\ln(2-\sin\theta)-\ln(1+\cos\theta)}{\ln(1-\cos\theta)-\ln(1+\cos\theta)} \times \frac{d\theta}{4-\sin\theta} = \frac{\pi}{4\sqrt{3}}. \qquad (13.8.4)$$

Exercise 13.8.1. Prove that

$$\int_0^\infty \frac{x^{2(r+p)}\, dx}{(1+x^2)^{2r+1}(x^{2p}+1)} = \frac{\pi}{2^{4r+1}}\binom{2r}{r}.$$

Many more similar integrals can be derived by these methods.

13.9. New Integrals by Differentiation

Several interesting evaluations of definite integrals can be obtained from the special case $a = 1$ of (13.3.1) written as

$$\int_0^\infty \left(\frac{x}{x^2+1}\right)^{c+1} \cdot \frac{x^2+1}{x^b+1} \cdot \frac{dx}{x^2} = 2^{-(c+1)} B\left(\tfrac{c}{2}, \tfrac{1}{2}\right) \qquad (13.9.1)$$

where $c := 2r - 1$. Differentiating (13.9.1) with respect to b yields a two-parameter family of vanishing integrals:

$$H(b, c) := \int_0^\infty \frac{x^{b+c-1}\, \ln x}{(1+x^2)^c\, (1+x^b)^2} dx = 0. \qquad (13.9.2)$$

The vanishing of $H(b, c)$ can be established directly by the change of variables $x \mapsto 1/t$.

Project 13.9.1. Observe that the three-parameter integral

$$H_1(a, b, c) := \int_0^\infty \frac{x^a\, \ln x}{(1+x^2)^c\, (1+x^b)^2} dx \qquad (13.9.3)$$

is not identically zero. For instance Mathematica yields

$$\int_0^\infty \frac{x^2\, \ln x}{(1+x)^2(1+x^2)}\, dx = \frac{\pi^2}{16}$$

and

$$\int_0^\infty \frac{x\, \ln x}{(1+x^2)(1+x^3)^2}\, dx = \frac{2\pi}{81}(\pi - 3\sqrt{3}).$$

The project is to evaluate $H_1(a, b, c)$ as a function of its parameters.

Example 13.9.1. Differentiation of (13.9.2) with respect to the parameter b yields

$$\int_0^\infty \frac{x^{b+c-1}\, (x^b - 1)\, (\ln x)^2}{(1+x^2)^c\, (1+x^b)^3}\, dx = 0 \qquad (13.9.4)$$

that in the special case $b = 2$ and $c = 1$ produces

$$\int_0^\infty \frac{x^2(x^2 - 1) (\ln x)^2}{(1 + x^2)^4} \, dx = 0.$$

Mathematica confirms this evaluation.

Extra 13.9.1. Further examples can be obtained by forcing the parameter c to be a function of b, before differentiating. This leads to

$$\int_0^\infty \frac{x^{b+c-1} \ln x}{(1 + x^2)^c (1 + x^b)^3} \left\{ \left((1 + x^b)(\ln x - \ln(1 + x^2)) \right) c'(b) \right.$$
$$\left. - (x^b - 1) \ln x \right\} dx = 0.$$

To obtain a larger class of results we differentiate (13.9.4) n times and observe that

$$\left(\frac{d}{db} \right)^n \frac{x^{b+c-1} \ln x}{(1 + x^2)^c (1 + x^b)^2} = \frac{x^{b+c-1} \ln^{n+1} x}{(1 + x^b)^{n+2} (1 + x^2)^c} \times Q_n(-x^b). \quad (13.9.5)$$

where $Q_n(t)$ is a polynomial in t, of degree n, with positive integer coefficients. The first few are

$$Q_0(t) = 1 \qquad\qquad\qquad (13.9.6)$$
$$Q_1(t) = t + 1$$
$$Q_2(t) = t^2 + 4t + 1$$
$$Q_3(t) = t^3 + 11t^2 + 11t + 1$$

and we recognize the Eulerian polynomials $A_n(t)$.

Exercise 13.9.1. Prove that $t \, Q_n(t) = A_{n+1}(t)$.

Example 13.9.2. Now differentiate (13.9.2) with respect to c to obtain

$$\int_0^\infty \frac{x^{b+c-1} \ln^2 x}{(1 + x^2)^c (1 + x^b)^2} \, dx = \int_0^\infty \frac{x^{b+c-1} \ln x \, \ln(1 + x^2)}{(1 + x^2)^c (1 + x^b)^2} \, dx. \quad (13.9.7)$$

For example, in the case $b = c = 1$, Mathematica 4.0 evaluates

$$\int_0^\infty \frac{x \ln^2 x \, dx}{(1 + x)^2 (1 + x^2)} = \frac{\pi^2 (3\pi - 8)}{48}$$

but it cannot evaluate the right-hand side of (13.9.7).

Example 13.9.3. The case $a = 1$ of (13.3.1) yields

$$\int_0^\infty \left(\frac{x}{x^2+1}\right)^{2r} \cdot \frac{x^2+1}{x^b+1} \cdot \frac{dx}{x^2} = 2^{-2r} B\left(r - \tfrac{1}{2}, \tfrac{1}{2}\right). \quad (13.9.8)$$

Differentiate (13.9.8) m times with respect to r to obtain

$$\int_0^\infty \left(\frac{x}{x^2+1}\right)^{2r} \ln^m\left(\frac{x}{x^2+1}\right) \times \frac{x^2+1}{x^b+1}\frac{dx}{x^2} = \frac{\sqrt{\pi}}{2^{m+1}}\left(\frac{d}{dr}\right)^m \frac{\Gamma(r-\tfrac{1}{2})}{\Gamma(r)\,2^{2r-1}}.$$

In the special case $r = 1$ we obtain that the right-hand side is a linear combination of products of the constants π, $\ln 2$ and the values of the zeta function at odd integers. This is one more instance of the weight assignment described in Extra 10.6.1. The integral above is a *homogeneous* polynomial of weight m. For example

$$\frac{1}{\pi}\int_0^\infty \ln\left(\frac{x}{x^2+1}\right)\frac{dx}{(x^2+1)^2} = \frac{-\ln 2}{2}$$

$$\frac{1}{\pi}\int_0^\infty \ln^2\left(\frac{x}{x^2+1}\right)\frac{dx}{(x^2+1)^2} = \frac{1}{48}(48\ln^2 2 + \pi^2)$$

$$\frac{1}{\pi}\int_0^\infty \ln^3\left(\frac{x}{x^2+1}\right)\frac{dx}{(x^2+1)^2} = \frac{-1}{8}(16\ln^3 2 + 3\zeta(3) + \pi^2 \ln 2)$$

$$\frac{1}{\pi}\int_0^\infty \ln^4\left(\frac{x}{x^2+1}\right)\frac{dx}{(x^2+1)^2} = \frac{19}{960}\pi^4 + \frac{1}{2}\pi^2 \ln^2 2 + 4\ln^4 2 + 3\zeta(3)\ln 2.$$

There are many more integrals to evaluate. We pause here.

Appendix: The Revolutionary WZ Method

A.1. Introduction

The goal of this chapter is to give a very short introduction to a revolutionary method discovered by H. Wilf and D. Zeilberger to find closed-form expressions for a special kind of finite sums. The reader will find in

```
http://www.math.temple.edu/~akalu/html/pg1.html
```

Akalu Tefera's very nice review of the methods presented here.

The reader should read George E. Andrews' entertaining article (1994) responding to Doron Zeilberger's opinion (1993, 1994) about rigorous proofs, computer-generated proofs and *The Death of Proof?*

The problem of evaluating finite sums involving binomial coefficients is described in many elementary textbooks. For instance, the binomial theorem

$$(x + y)^n = \sum_{k=0}^{n} \binom{n}{k} x^{n-k} y^k \tag{A.1.1}$$

in the special case $x = y = 1$, yields the value

$$\sum_{k=0}^{n} \binom{n}{k} = 2^n. \tag{A.1.2}$$

Many other sums can be obtained by algebraic manipulations of (A.1.1). For example, differentiating (A.1.1) with respect to y and setting $x = y = 1$ yields

$$\sum_{k=0}^{n} k \binom{n}{k} = n2^{n-1}. \tag{A.1.3}$$

The evaluation of these sums was proposed in Exercise 1.4.6.

A.2. An Introduction to WZ Methods

The WZ method developed by H. Wilf and D. Zeilberger has produced an algorithm that evaluates a large number of these sums. Let $F(n, k)$ be the *summand* of the expression that we are trying to evaluate and suppose there is a function $G(n, k)$ such that

$$F(n, k) = G(n, k + 1) - G(n, k). \tag{A.2.1}$$

Then, summing over k gives

$$\sum_{k=1}^{n} F(n, k) = G(n, n + 1) - G(n, 1). \tag{A.2.2}$$

Example A.2.1. Let $F(n, k) = k \cdot k!$, then $G(n, k) = k!$ and

$$\sum_{k=1}^{n} k \cdot k! = (n + 1)! - 1. \tag{A.2.3}$$

In this case both F and G are independent of n.

The condition (A.2.1) is very rare, but the great discovery of Wilf and Zeilberger is that for a large class of summands $F(n, k)$ there exists a function $G(n, k)$ such that

$$\sum_{j=0}^{m} a_j(n) F(n - j, k) = G(n, k + 1) - G(n, k) \tag{A.2.4}$$

for some coefficients $a_j(n)$. The functions (F, G) are called a WZ pair.

Example A.2.2. Let

$$F(n, k) = \frac{(-1)^{n-k}}{m - k} \binom{n}{k}$$

and let

$$\mathrm{Sum}(n) := \sum_{k=1}^{n} F(n, k)$$

be the sum that we are trying to evaluate. The command

```
zeilpap ( (-1)^{n-k} * binomial(n,k)/
     (m-k), k ,n,YourName)
```

writes a short paper entitled "A *proof of the* Your Name *identity*" authored by Shalosh B. Ekhad (Doron Zeilberger's computer), with the proof that Sum(n) satisfies the linear recurrence equation

$$(n+1)\,\text{Sum}(n) + (n - m + 1)\,\text{Sum}(n+1) = 0.$$

The proof is the construction of the function

$$G(n, k) = \frac{k(-m + k)(-1)^{n-k}}{(n + 1 - k)(m - k)} \binom{n}{k}$$

so that

$$(n+1)F(n, k) + (n - m + 1)F(n + 1, k) = G(n, k + 1) - G(n, k). \tag{A.2.5}$$

Naturally once the computer has produced (A.2.5) one is free to verify it by hand. In order to complete the evaluation one needs to check that the right-hand side satisfies (A.2.5) and that they have the same initial conditions. This is routine.

Example A.2.3. The identity (3.4.6) can be written as

$$S(n) := \sum_{j=0}^{n} (m - n) \binom{m}{n} \frac{(-1)^{n-j}}{m - j} \binom{n}{j} = 1. \tag{A.2.6}$$

The algorithm now shows that $S(n)$ satisfies $S(n + 1) = S(n)$, and in view of $S(0) = 1$, it is identically 1.

A.3. A Proof of Wallis' Formula

In this section we present a proof of Wallis' formula (6.4.5) in the form

$$J_{2,m} := \int_{0}^{\pi/2} \cos^{2m} \theta \, d\theta = \frac{\pi}{2^{2m+1}} \binom{2m}{m}. \tag{A.3.1}$$

First observe that

$$J_{2,m} = \int_{0}^{\pi/2} \left(\frac{1 + \cos 2\theta}{2} \right)^m d\theta, \tag{A.3.2}$$

introduce $\psi = 2\theta$, expand the power, and simplify the result by using the fact that, by symmetry, the odd powers of the cosine integrate to zero. We conclude that $J_{2,m}$ satisfies

$$J_{2,m} = 2^{-m} \sum_{k=0}^{\lfloor m/2 \rfloor} \binom{m}{2k} J_{2,k}.$$ (A.3.3)

Note that $J_{2,m}$ is uniquely determined by (A.3.3) along with the initial value $J_{2,0} = \pi/2$.

Exercise A.3.1. Use the recurrence (A.3.3) to produce the first few values of $J_{2,m}$. Use the prime factorization of this data to guess the formula

$$J_{2,m} = \frac{\pi}{2^{2m+1}} \binom{2m}{m}.$$ (A.3.4)

We now prove (A.3.4) by induction. The recursion (A.3.3) shows that the inductive step amounts to proving the identity

$$f(m) := \sum_{k=0}^{\lfloor m/2 \rfloor} 2^{-2k} \binom{m}{2k} \binom{2k}{k} = 2^{-m} \binom{2m}{m}.$$ (A.3.5)

To prove this consider the summand

$$F(m, k) = 2^{-2k} \binom{m}{2k} \binom{2k}{k}$$ (A.3.6)

and introduce the rational function

$$R(m, k) = \frac{4k^2}{m + 1 - 2k}.$$ (A.3.7)

Then

$$G(m, k) := F(m, k)R(m, k) = 2^{-2k+2}(2k - 1) \binom{m}{2k - 1} \binom{2k - 2}{k - 1}$$

satisfies

$$G(m, k + 1) - G(m, k) = (2m + 1)F(m, k) - (m + 1)F(m + 1, k).$$ (A.3.8)

We now sum (A.3.8) from $k = 0$ to $k = m$ to produce the recurrence

$$f(m + 1) = \frac{2m + 1}{m + 1} f(m).$$ (A.3.9)

Exercise A.3.2. Finish the proof of Wallis' formula by checking that $2^{-m} \binom{2m}{m}$ satisfies the recurrence (A.3.9) with the same initial value as f.

Extra A.3.1. The mystery of this proof is the appearance of the rational function $R(m, k)$ defined in (A.3.7). This function is called a *rational certificate* for $F(m, k)$. The construction of these certificates is now an automatic process due to the theory developed by Wilf and Zeilberger. This is explained in Nemes et al. (1997) and Petkovsek et al. (1996). The sum (A.3.5) is the example used by Petkovsek (1996) (page 113) to illustrate their method.

The automatic proof of (A.3.5) can be achieved by downloading the symbolic package EKHAD from D. Zeilberger's web site

$$\texttt{http://www.math.rutgers.edu/\~{}zeilberg}$$

and asking EKHAD to sum the left-hand side directly. The command

```
ct(binomial(m,2k) binomial(2k,k)
        2^{-2k}, 1, k, m, N)
```

produces the recursion (A.3.8) and the rational certificate R.

Exercise A.3.3. Use WZ methods to discuss the sum

$$\sum_{j=0}^{n-1} \binom{n}{j} \frac{(-1)^{n-j-1}}{n-j} (2^n - 2^j) \tag{A.3.10}$$

that appeared in (3.7.6).

Exercise A.3.4. Prove the identity

$$m \binom{2m}{m} \sum_{j=0}^{m-1} \frac{(-1)^j (m-1)!}{j!(m-j-1)!(2j+1)} = 2^{2m-1} \tag{A.3.11}$$

that appeared in (6.6.10).

Exercise A.3.5. Use the WZ method to discuss the sums in (3.4.8).

Many more examples of this technique can be found in Nemes et al. (1997).

Bibliography

Abel, N.: Beweis der Unmöglichkeit algebraische Gleichungen von hoheren Graden als dem vierten allgemeinen aufzulosen. *J. reine angew. Math.* **1**, 1826, 67–84. Reprinted in *Oeuvres Completes*, vol. 1, 66–94. Grondahl, Christiania, Sweden, 1881.

Abramowitz, M., and Stegun, I. A.: *Handbook of Mathematical Functions with Formulas, Graphs, and Mathematical Tables*. Applied Mathematical Series **55**, National Bureau of Standards, Washington, DC; Repr. Dover, New York, 1965.

Adamchik, V.: A class of logarithmic integrals. Proceedings of ISSAC'97. Maui, USA, 1997.

Adamchik, V.: On Stirling numbers and Euler sums. *Jour. Comp. Appl. Math.*, **79**, 1997, 119–130.

Adamchik, V.: Polygamma functions of negative order. *Jour. Comp. Appl. Math.*, **100**, 1998, 191–199.

Adamchik, V.: Integrals and series representations for Catalan's constant [1997]. Available at http://www-2.cs.cmu.edu/~adamchik/articles/catalan.htm

Adamchik, V., and Srivastava, H.: Some series of the zeta and related functions. *Analysis*, **18**, 1998, 131–144.

Adamchik, V., and Wagon, S.: π: A 2000-year old search changes direction. *Mathematica in Education and Research* **5:1**, 1996, 11–19.

Adamchik, V., and Wagon, S.: A simple formula for π. *Amer. Math. Monthly*, **104**, 1997, 852–855.

Addison, A. W.: A series representation for Euler's constant. *Amer. Math. Monthly*, **74**, 1967, 823–824.

Agnew, R. P., and Walker, R. J.: A trigonometric identity. *Amer. Math. Monthly*, **54**, 1947, 206–211.

Ahmed, Z., Dale, K., and Lamb, G.: Definitely an integral. *Amer. Math. Monthly*, **109**, 2002, 670–671.

Allouche, J. P.: A remark on Apery's numbers. *Preprint.*

Almkvist, G.: Many correct digits of π, revisited. *Amer. Math. Monthly*, **104**, 1997, 351–353.

Almkvist, G., and Granville, A.: Borwein and Bradley's Apery-like formulae for $\zeta(4n + 3)$. *Experimental Math.* **8**, 1999, 197–203.

Almkvist, G., Krattenhaler, C., and Petersson, J.: Some new formulas for π. Preprint available at http://arXiv.org/PS_cache/math/pdf/0110/0110238.pdf [10 Oct 2002]

Alvarez, J., Amadis, M., Boros, G., Karp, D., Moll, V., and Rosales, L.: An extension of a criterion for unimodality. *Elec. J. Comb.*, **8**, 2001, Paper 30.

Alzer, H., and Brenner, J. L.: On a double-inequality of Schlomilch-Lemonnier. *Jour. Math. Anal. Appl.*, **168**, 1992, 319–328.

Alzer, H.: Some gamma function inequalities. *Math. Comp.*, **60**, 1993, 337–346.

Alzer, H.: A proof of the arithmetic-geometric mean inequality. *Amer. Math. Monthly*, **103**, 1996, 585.

Alzer, H.: Inequalities for the gamma function. *Proc. Amer. Math. Soc.*, **128**, 1999, 141–147.

Alzer, H.: Sharp bounds for the Bernoulli numbers. *Arch. Math.* **74**, 2000, 207–211.

Amdeberhan, T.: Faster and faster convergent series for $\zeta(3)$. *Elec. Journal of Comb.* **3**, 1996, # R13.

Amdeberhan, T., and Zeilberger, D.: Hypergeometric series acceleration via the WZ method. *The Wilf Festchrift*, PA, 1996. *Electron. J. Combin.*, **4**, 1997, Research paper 3.

Andrews, G. E.: A theorem on reciprocal polynomials with applications to permutations and compositions. *Amer. Math. Monthly*, **82**, 1975, 830–833.

Andrews, G. E.: *Number Theory*. Dover, New York, 1994.

Andrews, G. E.: The Death of Proof? Semi-Rigorous Mathematics? You've Got to Be Kidding! *Math. Intelligencer*, **16**, 1994, 16–18.

Andrews, G. E., Askey, R., and Roy, R.: Special functions. *Encyclopedia of Mathematics and its Applications*, **71**, 1999. Cambridge University Press.

Andrews, L.: *Special Functions for Engineers and Applied Mathematicians*. Macmillan, New York, 1985.

Apelblat, A.: *Tables of Integrals and Series*. Verlag Harri Deutsch, 1996.

Apery, R.: Irrationalité de $\zeta(2)$ et $\zeta(3)$. In *Journées Arithmetiques de Luminy* (Colloq. Internat. CNRS, Centre Univ. Luminy, Luminy, 1978), 11–13. *Asterisque* **61**, 1979, Soc. Math. France, Paris.

Apostol, T.: Some series involving the Riemann zeta function. *Proc. Amer. Math. Soc.* **5**, 1954, 239–243.

Apostol, T.: *Mathematical Analysis*. Addison-Wesley, Reading, Mass., 1957.

Apostol, T.: Another elementary proof of Euler's formula for $\zeta(2n)$. *Amer. Math. Monthly*, **80**, 1973, 425–431.

Apostol, T.: *Introduction to Analytic Number Theory*, Undergraduate Texts in Mathematics, Springer-Verlag, New York, 1976.

Apostol, T.: A proof that Euler missed: evaluating $\zeta(2)$ the easy way. *The Mathematical Intelligencer*, **5**, 1983, 59–60.

Apostol, T.: Formulas for higher derivatives of the Riemann zeta function. *Math. Comp.*, **44**, 1985, 223–232.

Apostol, T.: An elementary view of Euler's summation formula. *Amer. Math. Monthly*, **106**, 409–418, 1999.

Arndt, J., and Haenel, C.: π *unleashed*. Springer-Verlag, 2001.

Arora, A. K., Goel, S., and Rodriguez, D.: Special integration techniques for trigonometric integrals. *Amer. Math. Monthly*, **95**, 1988, 126–130.

Arbuzov, A. B.: Tables of convolution integrals. Preprint available at http://arXiv.org/ PS_cache/hep-ph/pdf/0304/0304063.pdf [7 Apr 2003]

Askey, R., and Wilson, J.: A recurrence relation generalizing those of Apery. *J. Austral. Math. Soc. (Series A)*, **36**, 1984, 267–278.

Atkinson, M. D.: How to compute the series expansions of sec *x* and tan *x*. *Amer. Math. Monthly*, **93**, 1986, 387–389.

Artin, E.: *The Gamma Function*. Holt, Rinehart and Winston, New York, 1964.

Artin, E.: *Collected Papers*, Addison-Wesley, Reading, MA, pages viii–x, 1965.

Assmus, E. F.: Pi. *Amer. Math. Monthly*, **92**, 1985, 213–214.

Ayoub, R.: Euler and the zeta function. *Amer. Math. Monthly*, **81**, 1974, 1067–1086.

Ayoub, R.: On the nonsolvability of the general polynomial. *Amer. Math. Monthly*, **89**, 1982, 397–401.

de Azevedo, W. P.: Laplace's integral, the gamma function, and beyond. *Amer. Math. Monthly*, **109**, 2002, 235–245.

Bailey, D. H.: Numerical results on the transcendence of constants involving pi, *e*, and Euler's constant. *Math. Comp.*, **50**, 1988, 275–281.

Bailey, D. H., Borwein, J. M., Borwein, P. B., and Plouffe, S.: The quest for Pi. *Math. Intelligencer*, **19**, 1997, 50–57.

Bak, J., and Newman, D. J.: *Complex Analysis*. Springer-Verlag, Undergraduate Texts in Mathematics, 1982.

Balakrishnan, U.: A series for $\zeta(s)$. *Proc. Edinburgh Math. Soc.* **31**, 1988, 205–210.

Ball, K., and Rivoal, T.: Irrationalité d'une infinité de valeurs de la fonction zeta aux entiers impairs. *Invent. Math.*, **140**, 2001, 193–207.

Bank, S. B., and Kaufman, R. P.: A note on Holder's theorem concerning the gamma function. *Math. Ann.*, **232**, 1978, 115–120.

Barbeau, E. J.: Euler subdues a very obstreperous series. *Amer. Math. Monthly*, **86**, 1979, 356–372.

Barnes, C. W.: Euler's constant and *e*. *Amer. Math. Monthly*, **91**, 1984, 428–430.

Barnes, E. W.: On the expression of Euler's constant as a definite integral. *Messenger*, **33**, 1903, 59–61.

Barnes, E. W.: The theory of the gamma function. *Messenger Math.*, (2), **29**, 1929, 64–128.

Barnes, E. R., and Kaufman, W. E.: The Euler-Mascheroni constant. *Amer. Math. Monthly*, **72**, 1965, 1023.

Barnett, M. P.: Implicit rule formation in symbolic computation. *Computers Math. Applic.*, **26**, 1993, 35–50.

Barnett, M. P.: Symbolic computation of integrals by recurrence. *ACM SIGSAM Bulletin*, **37**, June 2003, 49–63. http://portal.acm.org

Barshinger, R.: Calculus II and Euler also (with a nod to series integral remainder bounds). *Amer. Math. Monthly*, **101**, 1994, 244–249.

Bateman, H.: *Higher Transcendental Functions*, Vol. I. *Compiled by the* Staff of the Bateman Manuscript Project. McGraw-Hill, 1953.

Beatty, S.: Elementary proof that *e* is not quadratically irrational. *Amer. Math. Monthly*, **62**, 1955, 32–33.

Beckmann, P.: *A History of π*. 2nd ed., Golem Press, Boulder, CO, 1971.

Beesley, E. M.: An integral representation for the Euler numbers. *Amer. Math. Monthly*, **76**, 1969, 389–391.

Berggren, L., Borwein, J., and Borwein, P.: *Pi: a Source Book*. Springer-Verlag, 1997.

Berndt, B.: Elementary evaluation of $\zeta(2n)$. *Math. Mag.*, **48**, 1975, 148–153.

Berndt, B.: The gamma function and the Hurwitz zeta function. *Amer. Math. Monthly*, **92**, 1985, 126–130.

Berndt, B.: Rudiments of the theory of the gamma function. *Unpublished notes.*

Berndt, B.: *Ramanujan's Notebooks. Part I.* Springer-Verlag, 1985.

Berndt, B.: *Ramanujan's Notebooks.* Part IV. Springer-Verlag, 1994.

Berndt, B., and Bhargava, S.: Ramanujan for lowbrows. *Amer. Math. Monthly*, **100**, 1993, 644–656.

Berndt, B., and Bowman, D.: Ramanujan's short unpublished manuscript on integrals and series related to Euler's constant. *Canad. Math. Soc. Conference Proceedings*, **27**, 2000, 19–27.

Beukers, F.: A note on the irrationality of $\zeta(2)$ and $\zeta(3)$. *Bull. London Math. Soc.* **11**, 1979, 268–272.

Beukers, F.: A rational approach to π. *Nieuw Archief Wisk.*, **5**, 2000, 372–379.

Beukers, F., Kolk, J. A. C., and Calabi, E.: Sums of generalized harmonic series and volumes. *Nieuw Arch. Wisk.*, **11**, 1993, 217–224.

Beumer, M. G.: Some special integrals. *Amer. Math. Monthly*, **68**, 1961, 645–647.

Bicheng, Y., and Debnath, L.: Some inequalities involving the constant e, and an application to Carleman's inequality. *Jour. Math. Anal. Appl.*, **223**, 1998, 347–353.

Birkhoff, G. D.: Note on the gamma function. *Bull. Amer. Math. Soc.*, **20**, 1914, 1–10.

Blatner, D.: *The Joy of π.* Walker Publ., 1999.

Blyth, C., and Pathak, P.: A note on easy proofs of Stirling's theorem. *Amer. Math. Monthly*, **93**, 1986, 376–379.

Boas, R.: Partial sums of infinite series, and how they grow. *Amer. Math. Monthly*, **84**, 1977, 237–258.

Boas, R., and Wrench, J.: Partial sums of the harmonic series. *Amer. Math. Monthly*, **78**, 1971, 864–870.

Bohr, H., and Mollerup, J.: *Laereborg i Matematisk Analyse*, Vol. III, Copenhagen, 1922.

Boo Rim Choe: An elementary proof of $\sum_{n=1}^{\infty} = \frac{\pi^2}{6}$. *Amer. Math. Monthly*, **94**, 1987, 662–663.

Boros, G., Joyce, M., and Moll, V.: A transformation on the space of rational functions. *Elemente der Mathematik*, **58**, 2003, 73–83.

Boros, G., Little, J., Moll, V., Mosteig, E., and Stanley, R.: A map on the space of rational functions, to appear in *Rocky Mountain Math Journal*. Preprint available at: http//arXiv.org/PS_cache/math/pdf/0308/0308039.pdf [5 Aug 2003].

Boros, G., and Moll, V.: An integral with three parameters. *SIAM Review*, **40**, 972–980, 1998.

Boros, G., and Moll, V.: An integral hidden in Gradshteyn and Rhyzik. *Jour. Comp. Appl. Math.*, **106**, 361–368, 1999.

Boros, G., and Moll, V.: A sequence of unimodal polynomials. *Jour. Math. Anal. Appl.* **237**, 272–287, 1999.

Boros, G., and Moll, V.: A criterion for unimodality. *Electron. J. Comb.* **6**, 1999, R10.

Boros, G., and Moll, V.: A rational Landen transformation. The case of degree 6. *Contemporary Mathematics*, **251**, 2000, 83–91.

Boros, G., and Moll, V.: The double square root, Jacobi polynomials and Ramanujan's master theorem. *Journal of Comp. Applied Math.* **130**, 2001, 337–344.

Boros, G., and Moll, V.: Landen transformations and the integration of rational functions. *Math. of Comp.* **71**, 2001, 649–668.

Boros, G., Moll, V., and Nalam, R.: An integral with three parameters. Part 2. *Jour. Comp. Applied Math.* **134**, 2001, 113–126.

Boros, G., Moll, V., and Shallit, J.: The 2-adic valuation of the coefficients of a polynomial. *Revista Scientia*, **7**, 2000–2001, 47–60.

Borwein, D., and Borwein, J.: On an intriguing integral and some series related to $\zeta(4)$. *Proc. Amer. Math. Soc.* **123**, 1995, 1191–1198.

Borwein, J., and Borwein, P.: *Pi and the AGM*. Canadian Mathematical Society, Wiley-Interscience Publication, 1987.

Borwein, J., Borwein, P., and Bailey, D. H.: Ramanujan, modular equations, and approximations to Pi or how to compute one billion digits of Pi. *Amer. Math. Monthly*, **96**, 1989, 201–219.

Borwein, J., Borwein, P., and Dilcher, K.: Pi, Euler numbers, and asymptotic expansions. *Amer. Math. Monthly*, **96**, 1989, 681–687.

Borwein, J., Borwein, P., Girgensohn, R., and Parnes, S.: Making sense of experimental mathematics. *Mathematical Intelligencer*, **18**, 1996, 12–18.

Borwein, J., and Bradley, D.: Searching symbolically for Apery-like formulae for values of the Riemann zeta function. *SIGSAM Bull. Communic. Comput. Algeb.*, **30**, (116), 1996, 2–7.

Borwein, J., and Bradley, D.: Empirically determined Apery-like formulae for $\zeta(4n+3)$. *Experimental Mathematics* **6**:3, 1997, 181–194.

Borwein, P., and Dykshoorn, W.: An interesting infinite product. *Jour. Math. Anal. Appl.*, **179**, 1993, 203–207.

Borwein, P., and Erderly, T.: *Polynomials and polynomial inequalities*. Springer Verlag, Graduate Texts in Mathematics, **161**, 1995.

Borwein, J., and Lisonek, P.: Applications of integer relation algorithms. *Discrete Math.*, **217**, 2000, 65–82.

Bowman, F.: Note on the integral $\int_0^{\pi/2}(\log \sin \theta)^n\, d\theta$. *Jour. London Math. Soc.* **22**, 172–173.

Boyd, D. W.: A p-adic study of the partial sums of the harmonic series. *Experimental Math.*, **3**, 1994, 287–302.

Bracken, P.: Properties of certain sequences related to Stirling's approximation for the gamma function. *Expo. Math.*, **21**, 2003, 171–178.

Bradley, D.: The harmonic series and the n-th term test for divergence. *Amer. Math. Monthly*, **107**, 2000, 651.

Bradley, D.: Representations of Catalan's constant (1998). Available at http://germain.umemat.maine.edu/faculty/bradley/papers/pub.html

Bradley, D.: Ramanujan's formula for the logarithmic derivative of the gamma function. *Math. Proc. Cambridge Philos. Soc.*, **120**, 1996, 391–401.

Brent, R. P., and McMillan, E. M.: Some new algorithms for high-precision computation of Euler's constant. *Math. Comp.* **34**, 1980, 305–310.

Brent, R. P.: Computation of the regular continued fraction for Euler's constant. *Math. Comp.* **31**, 1977, 771–777.

Brent, R. P.: Ramanujan and Euler's constant. *Proc. Symp. Applied Math.*, **48**, American Mathematical Society, 1994, 541–545.

Brenti, F.: Log-concave and unimodal sequences in algebra, combinatorics and geometry: an update. *Contemporary Mathematics*, **178**, 71–84, 1994.

Bressoud, D., and Wagon, S.: *A Course in Computational Number Theory*. Key College and Springer-Verlag, 2000.

Breusch, R.: A proof of the irrationality of π. *Amer. Math. Monthly*, **61**, 1954, 631–632.

Briggs, W. E.: Some constants associated with the Riemann zeta function. *Michigan Math. J.*, **3**, 1955–1956, 117–121.

Briggs, W. E.: On series which arise from a continuation of the zeta function. *Amer. Math. Monthly*, **69**, 1962, 406–407.

Briggs, W. E., and Chowla, S.: The power series coefficients of $\zeta(s)$. *Amer. Math. Monthly* **62**, 1955, 323–325.

Bromwich, T. J.: *An Introduction to the Theory of Infinite Series*. 2nd ed., Macmillan, New York, 1926.

Bronstein, M.: *Symbolic Integration I. Transcendental functions*. Algorithms and Computation in Mathematics, **1**. Springer-Verlag, 1997.

Bronstein, M.: Symbolic integration tutorial. ISAAC'98, Rostock, August 12, 1998.

Brothers, H. J., and Knox, J. A.: New closed-form approximations to the logarithmic constant e. *Mathematical Intelligencer*, **20**, 1998, 25–29.

Brown, J. W.: The beta-gamma function identity. *Amer. Math. Monthly*, **68**, 1961, 165.

Buhler, W. K.: *Gauss: A Biographical Study*. Springer-Verlag, New York, 1981.

Bustoz, J., and Ismail, M.: On gamma function inequalities. *Math. Comp.*, **47**, 1986, 659–667.

Butler, R.: On the evaluation of $\int_0^\infty (\sin^m t)/t^m \, dt$ by the trapezoidal rule. *Amer. Math. Monthly*, **67**, 1960, 566–569.

Cardano, G.: *Artis Magnae sive de regvlis algebraicis*. English translation: *The Great Art, or the Rules of Algebra*. Trans. ed. T. R. Witmer. MIT Press, Cambridge, MA, 1968. (original work published 1545)

Carlitz, L.: The coefficients of the reciprocal of a series. *Duke Math. J.*, **8**, 1941, 689–700.

Carlitz, L.: A divisibility property of the Bernoulli polynomials. *Proc. Amer. Math. Soc.*, **3**, 1952, 604–607.

Carlitz, L.: Eulerian numbers and polynomials. *Math. Magazine*, **32**, 1959, 247–260.

Cartier, P.: An introduction to zeta functions. In *From Number Theory to Physics*, Springer-Verlag, 1992.

Castellanos, D.: The ubiquitous Pi. Part I. *Math. Magazine*, **61**, 1988, 67–98.

Castellanos, D.: The ubiquitous Pi. Part II. *Math. Magazine*, **61**, 1988, 148–163.

Chao-Ping Chen, and Feng Qi: The best bounds of harmonic sequence. Available at http:// arXiv.org/PS_cache/math/pdf/0306/0306233.pdf [16 Jun 2003]

Chapman, R.: Evaluating $\zeta(2)$. 30 April 1999/7 July 2003. Preprint available at http:// www.maths.ex.ac.uk/~rjc/etc/zeta2.pdf

Chaudhuri, J.: Some special integrals. *Amer. Math. Monthly*, **74**, 1967, 545–548.

Chen, K. W.: Algorithms for Bernoulli numbers and Euler numbers. *Journal of Integer Sequences*, **4**, 2001, Article 01.1.6.

Chen, M. P., and Srivastava, H. M.: Some families of series representations for the Riemann zeta function. *Resultate Math.*, **33**, 1998, 179–197.

Chen, X.: Recursive formulas for $\zeta(2k)$ and $L(2k-1)$. *College Math. J.*, **26**, 1995, 372–376.

Cherry, G.: Integration in finite terms with special functions: the error function. *J. Symb. Comput.*, **1**, 1985, 283–302.

Cherry, G.: Integration in finite terms with special functions: the logarithmic integral. *SIAM J. Comput.*, **15**, 1986, 1–21.

Chong, Kong-Ming: An inductive proof of the arithmetic geometric mean inequality. *Amer. Math. Monthly*, **83**, 1976, 369.

Chowla, S.: The Riemann zeta and allied functions. *Bull. Amer. Math. Soc.*, **58**, 1952, 287–305.

Chu, J. T.: A modified Wallis product and some applications. *Amer. Math. Monthly*, **69**, 1962, 402–404.

Cohen, H.: Généralisation d'une construction de R. Apery. *Bull. Soc. Math. France*, **109**, 1981, 269–281.

Cohen, H.: On the 2-adic valuations of the truncated polylogarithm series. *Fibon. Quart.*, **37**, 1999, 117–121.

Cohen, H.: 2-adic behavior of numbers of domino tilings. *Preprint.*

Coleman, A. J.: The probability integral. *Amer. Math. Monthly*, **61**, 1954, 710–711.

Comtet, L.: Calcul practique des coefficients de Taylor d'une fonction algébrique. *L'Enseig. Math.*, **10**, 1964, 267–270.

Comtet, L.: Fonctions génératrices et calcul de certaines integrales. *Publ. Fac. Elec. Belgrade*, **197**, 1967, 77–87.

Conrey, J. B.: The Riemann hypothesis. *Notices Amer. Math. Soc.*, March 2003, 341–353.

Coolidge, J.: The number *e*. *Amer. Math. Monthly*, **57**, 1950, 591–602.

Coppo, M. A.: Nouvelle expressions des constantes de Stieltjes. *Exp. Math.*, **17**, 1999, 349–358.

Cox, D.: The arithmetic-geometric mean of Gauss. *L'Enseig. Math.* **30**, 1984, 275–330.

Cox, D., Little, J., and O'Shea, D.: *Using Algebraic Geometry*. Springer Verlag, Graduate Texts in Mathematics 185, 1998.

Coxeter, H. S. M., and Ringenberg, L. A.: A quotient of infinite series. *Amer. Math. Monthly*, **63**, 1956, 48.

Cusick, T. W.: Recurrences for sums of powers of binomial coefficients. *Jour. Comb. Theory*, Ser. A, **52**, 1989, 77–83.

Cvijovic, D., and Klinowski, J.: New formulae for the Bernoulli and Euler polynomials at rational arguments. *Proc. Amer. Math. Soc.*, **123**, 1995, 1527–1535.

Cvijovic, D., and Klinowski, J.: New rapidly convergent series representations for $\zeta(2n + 1)$. *Proc. Amer. Math. Soc.*, **125**, 1997, 1263–1271.

Cvijovic, D., and Klinowski, J.: Integral representations of the Riemann zeta function for odd-integer arguments. *Jour. Comp. Appl. Math.*, **142**, 2002, 435–439.

Dabrowski, A.: A note on the values of the Riemann zeta function at positive odd integers. *Nieuw Arch. Wisk.*, **14**, 1996, 199–207.

Dalzell, D. P.: Stirling's formula. *Jour. London Math. Soc.*, **5**, 1930, 145–148.

Danese, A. E.: Solution to Elementary problem 1801: A zeta-function identity. *Amer. Math. Monthly*, **74**, 1967, 80–81.

Davenport, H.: *Multiplicative number theory*. 3rd ed. Springer Verlag, 2000.

Davis, P. J.: Leonhard Euler's integral: a historical profile of the gamma function. *Amer. Math. Monthly*, **66**, 1959, 849–869.

De Doelder, P. J.: On some series containing $\psi(x) - \psi(y)$ and $(\psi(x) - \psi(y))^2$ for certain values of x and y. *Jour. Comp. Appl. Math.*, **37**, 1991, 125–141.

de Haan, B.: Nouvelles Tables d'Integrales Définies. Edition of 1867. Hafner Publ. Co., New York and London.

Desbrow, D.: On the irrationality of π^2. *Amer. Math. Monthly*, **97**, 1990, 903–906.

De Temple, D. W.: A quicker convergence to Euler's constant. *Amer. Math. Monthly*, **100**, 1993, 468–470.

De Temple, D. W., and Wang, S. H.: Half integer approximations for the partial sums of the harmonic series. *Jour. Math. Anal. Appl.*, **160**, 1991, 149–156.

Diaconis, P., and Freeman, D.: An elementary proof of Stirling's formula. *Amer. Math. Monthly*, **93**, 1986, 123–125.

Dilcher, K.: A bibliography of Bernoulli numbers. August 11, 2003. Available at http://www.mscs.dal.ca/~dilcher/bernoulli.html

Doetsch, G.: *Handbuch der Laplace Transformation*, Vol. I, Birkhauser, Basel, 1950.

Dunham, W.: *Euler. The master of us all*. Mathematical Association of America. Dolciani Mathematical Expositions No. 22, 1999.

Dutka, J.: The early history of the hypergeometric function. *Arch. Hist. Exact Sci.* **31**, 1984, 15–34.

Dvornicich, R., and Viola, C.: Some remarks on Beukers' integrals. *Colloq. Math. Soc. Janos Bolyai*, **51**, 1987, 637–657.

Ebbinghaus, H. et al: *Numbers*. Springer-Verlag, Reading in Mathematics, 1991.

Eberlein, W. F.: On Euler's infinite product for the sine. *J. Math. Anal. Appl.*, **58**, 1977, 147–151.

Edwards, H. M.: *Riemann's zeta function*. Academic Press, New York and London, 1974.

Edwards, J.: *A Treatise on the Integral Calculus*. Macmillan, 1922. Reprinted by Chelsea Publ. Co., New York.

Egecioglu, O., and Ryavec, C.: Polynomial families satisfying a Riemann hypothesis. *Preprint*.

Elizalde, E.: Zeta functions: formulas and applications. *Jour. Comp. Appl. Math.*, **118**, 2000, 125–142.

Elkies, N.: On the sums $\sum_{k=-\infty}^{\infty}(4k + 1)^{-n}$. *Amer. Math. Monthly*, **110**, 2003, 561–573.

Elsner, C.: On a sequence transformations with integral coefficients for Euler's constant. *Proc. Amer. Math. Soc.* **123**, 1995, 1537–1541.

English, B. J., and Rousseau, G.: Bounds for certain harmonic sums. *Jour. Math. Anal. Appl.*, **206**, 1997, 428–441.

Erdelyi, A., Magnus, W., Oberhettinger, F., and Tricomi, F. G.: Tables of integral transforms. Vol. I, McGraw-Hill, New York, 1954.

Erdelyi, A., Magnus, W., Oberhettinger, F., and Tricomi, F. G.: Tables of integral transforms. Vol. II, McGraw-Hill, New York, 1954.

Espinosa, O., and Moll, V.: On some definite integrals involving the Hurwitz zeta function. Part I. *Ramanujan Journal* **6**, 2002, 159–188.

Espinosa, O., and Moll, V.: A generalized polygamma function. *Integral Transforms and Special Functions*, 2004.

Espinosa, O., and Moll, V.: The evaluation of Tornheim double sums. *Preprint*.

Estermann, T.: Elementary evaluation of $\zeta(2k)$. *Jour. London Math. Soc.*, **22**, 1947, 10–13.

Estermann, T.: A theorem implying the irrationality of π^2. *Jour. London Math. Soc.*, **41**, 1966, 415–416.

Euler, L.: De summis serierum reciprocarum. *Commentarii Academiae Scientiarum Petropolitanae*, **7**, (1734–35), 1740, 123–134. Repr. in *Opera Omnia*, **14**, 73–86.

Euler, L.: *Introductio in Analysis Infinitorum*, 1748. Marcum-Michaelem Bousquet, Lausanne. English trans.: *Introduction to Analysis of the Infinite*. Trans. J. D. Blantan, Springer-Verlag, New York, 1988.

Euler, L.: De miris proprietatibus curvae elasticae sub aequatione $y = \int x \, x \, dx / \sqrt{(1 - x^4)}$ contentae. *Acta academiae scientiarum Petrop.* **2**, 1781, 34–61. Reprinted in *Opera Omnia*, ser. 1, vol. 21, 91–118.

Euler, L.: *Opera Omnia*, Leipzig-Berlin, 1924.

Everest, G., van der Poorten, A., Shparlinski, I., and Ward, T.: *Recurrence Sequences.* AMS Surveys and Monographs series, volume **104**, 2003.

Ewell, J.: A new series representation for $\zeta(3)$. *Amer. Math. Monthly*, **97**, 1990, 219–220.

Ewell, J.: On values of the Riemann zeta function at integral arguments. *Canad. Math. Bull.*, **34**, 1991, 60–66.

Ewell, J.: An eulerian method for representing π^2 by series. *Rocky Mount. Jour. Math.* **22**, 1992, 165–168.

Ewell, J.: On the zeta function values $\zeta(2k + 1)$, $k = 1, 2, \ldots$ *Rocky Mount. J. Math.*, **25**, 1995, 1003–1012.

Feller, W.: A direct proof of Stirling's formula. *Amer. Math. Monthly*, **74**, 1967, 1223–1225. Correction **75**, 1967, 518.

Finch, S.: *Mathematical Constants.* Encyclopedia of Mathematics and its Applications, **94**. Cambridge University Press, 2003.

Fine, B., and Rosenberger, G.: *The Fundamental Theorem of Algebra.* Springer Verlag, 1997.

Fischler, S.: Irrationalité de valeurs de zeta (d'après Apery, Rivoal, ...). *Seminaire Bourbaki*, no. 910, 55ème année, 2002–2003. Asterisque.

Galois, E.: Analyse d'un memoire sur la résolution algebrique des equations. *Ecrits Mem. Math.*, 1831, 163–165.

Gandhi, J. M.: Some integrals for Genocchi numbers. *Math. Mag.*, **33**, 1959, 21–23.

Gasper, G., and Rahman, M.: *Basic Hypergeometric Series.* Encyclopedia of Math. and Its Applications, **35**, Cambridge University Press, 1990.

Gauss, K. F.: *Arithmetische Geometrisches Mittel*, 1799. In *Werke*, **3**, 361–432. Königliche Gesellschaft der Wissenschaft, Göttingen. Reprinted by Olms, Hildescheim, 1981. (Original work published 1799).

Gautschi, W.: A harmonic mean inequality for the gamma function. *SIAM J. Math. Anal.*, **5**, 1974, 278–281.

Gautschi, W.: Some mean value inequalities for the gamma function. *SIAM J. Math. Anal.*, **5**, 1974, 282–292.

Geddes, K. O., Glasser, M. L., Moore, R. A., and Scott, T. C.: Evaluation of classes of definite integrals involving elementary functions via differentiation of special functions. *Applicable Algebra in Engineering, Communication and Computing*, **1**, 1990, 149–165.

Gerst, I.: Some series for Euler's constant. *Amer. Math. Monthly*, **76**, 1969, 273–275.

Giesy, D. P.: Still another elementary proof that $\sum 1/k^2 = \pi^2/6$. *Math. Magazine*, **45**, 1972, 148–149.

Glaisher, J. W. L.: On the history of Euler's constant. *Mess. of Math.*, **1**, 1872, 25–30 and **2**, 1873, 64.

Glaisher, J. W. L.: On Dr. Vacca's series for γ. *Quart. J. Pure Appl. Math.* **41**, 1909–10, 365–368.

Glaisher, J. W. L.: Summations of certain numerical series. *Messenger Math.*, **42**, 1912/13, 19–34.

Glaisher, J. W. L.: On the coefficients in the expansions of $\cos x/\cos 2x$ and $\sin x/\cos 2x$. *Quarterly J. Pure Appl. Math.* **45**, 1914, 187–222.

Glasser, M. L.: Some recursive formulas for evaluation of a class of definite integrals. *Amer. Math. Monthly*, **71**, 1964, 75–76.

Glasser, M. L.: Evaluation of some integrals involving the ψ-function. *Math. Comp.* **20**, 1966, 332–333.

Glasser, M. L.: Some integrals of the arctangent function. *Math. Comp.* **22**, 1968, 445–447.

Glasser, M. L.: A remarkable property of definite integrals. *Math. Comp.*, **40**, 1983, 561–563.

Goetgheluck, P.: Computing binomial coefficients. *Amer. Math. Monthly*, **94**, 1987, 360–365.

Goode, J.: A limit problem. Elementary problem 1187. *Amer. Math. Monthly*, **63**, 1956, 343–344.

Gordon, L.: A stochastic approach to the gamma function. *Amer. Math. Monthly*, **101**, 858–865, 1994.

Gosper, R. Wm., Jr.: Nielsen-Ramanujan constants. Private communication, 1996.

Gosper, R. Wm., Jr.: $\int_{n/4}^{m/6} \ln \Gamma(z)dz$. In *Special Functions, q-Series and Related Topics*, pages 71–76, M. Ismail, D. Masson, M. Rahman, eds. The Fields Institute Communications, AMS, 1997.

Gould, H. W.: Explicit formulas for Bernoulli numbers. *Amer. Math. Monthly*, **79**, 1972, 44–51.

Gourdon, X., and Sebah, P.: Numbers, constants and computation. (2002). Available at http://numbers.computation.free.fr/Constants/ constants.html

Gouvea, F.: *p-adic Numbers. An Introduction*. Universitytext, Springer-Verlag, 1997.

Gradshteyn, I. S., and Ryzhik, I. M.: *Table of Integrals, Series and Products*. 5th ed., ed. Alan Jeffrey. Academic Press, 1994.

Graham, R. L., Knuth, D. E., and Patashnik, O.: *Concrete Mathematics*, Addison-Wesley, Reading, MA, 1989.

Greenberg, R., Marsh, D., and Danese, A.: A zeta function summation. *Amer. Math. Monthly*, **74**, 1967, 80–81.

Greene, R., and Krantz, S.: *Function theory of one complex variable*. Graduate Studies in Mathematics, **40**, American Mathematical Society, 2002.

Greenstein, D. S.: A property of the logarithm. *Amer. Math. Monthly*, **72**, 1965, 767.

Griffiths, H. B., and Hirst, A. E.: Cubic equations, or where did the examination question come from? *Amer. Math. Monthly*, **101**, 1994, 151–161.

Grobner, W., and Hofreiter, N.: *Integraltafel*, Springer-Verlag, Vienna, 1973–1975.

Gurland, J.: On Wallis' formula. *Amer. Math. Monthly*, **63**, 1956, 643–645.

Hamming, R.: An elementary discussion of the transcendental nature of the elementary transcendental functions. *Amer. Math. Monthly*, **77**, 1970, 294–297.

Hancl, J.: A simple proof of the irrationality of π^4. *Amer. Math. Monthly*, **93**, 1986, 374–375.

Hansen, E. R.: *A Table of Series and Products*. Prentice-Hall, Englewood Cliffs, NJ, 1975.

Hardy, G. H.: A new proof of Kummer's series for $\log \Gamma(a)$. *Messenger Math.*, **31**, 1901–1902, 31–33.

Hardy, G. H.: Note on Dr. Vacca's series for γ. *Quart. J. Pure Appl. Math.*, **43**, 1912, 215–216.

Hardy, G. H.: *Orders of infinity*, 2nd ed., Cambridge University Press, 1924.

Hardy, G. H.: *The Integration of Functions of a Single Variable.* Cambridge Tracts in Mathematics and Mathematical Physics, **2**, 2nd ed., Cambridge University Press, 1958.

Hardy, G. H., and Wright, E.: *Introduction to Number Theory.* Oxford University Press, 5th ed., 1979.

Hata, M.: A note on Beukers' integral. *J. Austral. Math. Soc.* (series A), **58**, 1995, 143–153.

Hauss, M.: Fibonacci, Lucas and central factorial numbers, and π. *Fibonacci Quart.* **32**, 1994, 395–396.

Havil, J.: *Gamma: Exploring Euler's Constant.* Princeton University Press, 2003.

Hellman, M.: A unifying technique for the solution of the quadratic, cubic, and quartic. *Amer. Math. Monthly*, **65**, 1958, 274–276.

Hellman, M.: The insolvability of the quintic re-examined. *Amer. Math. Monthly*, **66**, 1959, 410.

Hermite, C.: Sur l'intégration des fractions rationelles. *Nouvelles Annales de Mathematiques* (2ème serie) **11**, 145–148, 1872.

Hermite, C.: Sur la fonction exponentielle. *C. R. Acad. des Sciences*, 1873, **77**, 18–24, 74–79, 226–233, 285–293.

Hijab, O.: *Introduction to Calculus and Classical Analysis.* Springer Verlag, New York, 1997.

Hjortnaes, M.: Overforing av rekken $\sum_{k=1}^{\infty}(1/k^3)$ til et bestemt integral. In *Proc. 12th Cong. Scand. Maths* (Lund 1953), Lund, 1954.

Hoffman, M. E.: Derivative polynomials for tangent and secant. *Amer. Math. Monthly*, **102**, 1995, 23–30.

Hoffman, M. E.: Derivative polynomials, Euler polynomials, and associated integer sequences. *Elec. J. Comb.*, **6**, 1999, Paper 21.

Holder, O.: Über die Eigenschaft der Γ-Funktion, keiner algebraischen Differentialgleichung zu genügen. *Math. Ann.*, **28**, 1887, 1–13.

Huylebrouck, D.: Similarities in irrationality proofs for π, $\ln 2$, $\zeta(2)$, and $\zeta(3)$. *Amer. Math. Monthly*, **108**, 2001, 222–231.

Johnson, W.: Note on the numerical transcendents S_n and $s_n = S_n - 1$. *Bull. Amer. Math. Soc.*, **12**, 1906, 477–482.

Johnsonbaugh, R. F.: Another proof of an estimate for e. *Amer. Math. Monthly*, **81**, 1974, 1011–1012.

Johnsonbaugh, R. F.: Summing alternating series. *Amer. Math. Monthly*, **86**, 1979, 637–648.

Johnsonbaugh, R. F.: The trapezoid rule, Stirling's formula, and Euler's constant. *Amer. Math. Monthly*, **88**, 1981, 696–698.

Jolley, L. B. W.: *Summation of Series.* 2nd ed., Dover, 1961.

Jordan, P. F.: Infinite sums of Psi functions. *Bull. Amer. Math. Soc.*, **79**, 1973, 681–683.

Kalman, D.: Six ways to sum a series. *College Math. J.*, **24**, 1993, 402–421.

Kaneko, M.: A recurrence formula for the Bernoulli numbers. *Proc. Japan Acad. Ser. A Math. Sci.*, **71**, 192–193.

Karatsuba, E. A.: Fast calculation of $\zeta(3)$. *Problemy Peredachi Informatsii*, **29**:1, 1993, 68–73. In Russsian; trans. in *Problems Inform. Transmission* **29**:1, 1993, 58–62.

Kaspar, T.: Integration in finite terms: The Liouville theory. *Math. Magazine*, **53**, 1980, 195–201.

Katsurada, M.: Rapidly convergent series representations for $\zeta(2n+1)$ and their χ-analogue. *Acta Arith.* **40**, 1999, 79–89.

Kazarinoff, N. D.: On Wallis' formula. *Edinburgh Math. Notes*, **40**, 1959, 19–21.

Kazarinoff, N. D.: *Analytic Inequalities*. Holt, Rinehart and Winston, New York, 1961.

Kenter, F. K.: A matrix representation for Euler's constant γ. *Amer. Math. Monthly*, **106**, 1999, 452–454.

Kerney, K., and Stenger, A.: A logarithmic integral. *Amer. Math. Monthly*, **83**, 1976, 384–385.

Khan, R. A.: A probabilistic proof of Stirling's formula. *Amer. Math. Monthly*, **81**, 1974, 366–369.

Klamkin, M. S.: A summation problem: Advanced Problem 4431. *Amer. Math. Monthly*, **58**, 1951, 195; *ibid.* **59**, 1952, 471–472.

Klamkin, M. S.: Advanced problem 4582. *Amer. Math. Monthly*, **61**, 1954, 199.

Klamkin, M. S.: Another summation. *Amer. Math. Monthly*, **62**, 1955, 129–130.

Kleinz, M., and Osler, T.: A child's garden of fractional derivatives. *College Math. Journal*, **31**, 2000, 82–88.

Knopp, K.: *Theory and Applications of Infinite Series*, 2nd English ed., Hafner, New York, 1951.

Knopp, M., and Robins, S.: Easy proofs of Riemann's functional equation for $\zeta(s)$ and of Lipschitz summation. *Proc. Amer. Math. Soc.*, **129**, 2001, 1915–1922.

Knuth, D. E.: Euler's constant to 1271 places. *Math. Comp.* **16**, 1962, 275–281.

Knuth, D. E., and Buckholtz, T. J.: Computation of tangent, Euler, and Bernoulli numbers. *Math. Comput.*, **21**, 1967, 663–688.

Koblitz, N.: *p-Adic Numbers, p-Adic Analysis, and Zeta Functions*, **58**, Graduate Texts in Mathematics, Springer-Verlag, 1984.

Koecher, M.: Letter to *Math. Intelligencer*, **2**, 1980, 62–64.

Koecher, M.: *Klassische Elementare Analysis*, Birkhauser Verlag, Basel, 1987.

Koepf, W.: *Hypergeometric Summation*. Advanced Lectures in Mathematics, Vieweg, 1998.

Kolasi, C., Egeralnd, W. O., and Hansen, C. E.: An integral of cosines. *Amer. Math. Monthly*, **95**, 1988, 354–356.

Kolbig, K. S.: Closed expressions for $\int_0^1 t^{-1} \log^{n-1} t \, \log^p(1-t)\,dt$. *Math. Comp.*, **39**, 1982, 647–654.

Kolbig, K. S.: On the integral $\int_0^\infty e^{-\mu t} t^{\nu-1} \log^m t \, dt$. Math. Comp. **41**, 1983, 171–182.

Kolbig, K. S.: On the integral $\int_0^{\pi/2} \log^n \cos x \log^p \sin x \, dx$. *Math. Comp.*, **40**, 1983, 565–570.

Kolbig, K. S.: Explicit evaluation of certain definite integrals involving powers of logarithms. *J. Symb. Comput.*, **1**, 1985, 109–114.

Kolbig, K. S.: On the integral $\int_0^\infty e^{-\mu t} t^{\nu-1} \log^m t \, dt$. *Math. Comp.*, **41**, 1985, 171–182.

Kolbig, K. S.: Nielsen's generalized polylogarithms. *SIAM J. Math. Anal.*, **17**, 1986, 1232–1258.

Kolbig, K. S.: On the integral $\int_0^\infty x^{\nu-1}(1 + \beta x)^{-\lambda} \ln^m x \, dx$. *J. Comp. Appl. Math.*, **14**, 1986, 319–344.

Kolbig, K. S.: *On Three Trigonometric Integrals of* $\ln \Gamma(x)$ *or Its Derivative*. CERN / Computing and Networks Division; CN/94/7, May 1994.

Kontsevich, M., and Zagier, D.: Periods. In *Mathematics Unlimited—2001 and Beyond*, pages 771–808. B. Engquist, W. Schmid, eds. Springer-Verlag, 2001.

Kortram, R. A.: Another computation of $\int_0^\infty e^{-u^2} du$. *Elem. Math.* **48**, 1993, 170–172.

Kortram, R. A.: Simple proofs for $\sum_{k=1}^\infty 1/k^2 = \pi^2/6$ and $\sin x = x \prod_{k=1}^\infty (1 - x^2/k^2\pi^2)$. *Math. Magazine*, **69**, 1996, 122–125.

Kummer, E. E.: Beitrag zur Theorie der Function $\Gamma(x) = \int_0^\infty e^{-v} v^{x-1} dv$. *J. Reine Angew. Math.*, **35**, 1847, 1–4. Repr. in *Collected Papers*, Vol. II, Springer-Verlag, Berlin, 1975, pages 325–328.

Lagarias, J.: An elementary problem equivalent to the Riemann hypothesis. *Amer. Math. Monthly*, **109**, 2002, 534–543.

Lambert J. H.: Mémoire sur quelques propriétés remarquables des quantités transcendentes circulaires et logarithmiques. *Histoire de l'Academie Royale des Sciences et des Belles-Lettres der Berlin*, 1761, 265–276.

Lammel, E.: Ein Beweis, dass die Riemannsche Zetafunktion $\zeta(s)$ in $|s - 1| \leq 1$ keine Nullstelle Besitzt. *Revista Univ. Nac. Tucuman, Ser. A*, **16**, 1966, 209–217.

Lan, Y.: A limit formula for $\zeta(2k + 1)$. *Jour. Number Theory*, **78**, 1999, 271–286.

Landen, J.: A disquisition concerning certain fluents, which are assignable by the arcs of the conic sections; wherein are investigated some new and useful theorems for computing such fluents. *Philos. Trans. Royal Soc. London*, **61**, 1771, 298–309.

Landen, J.: An investigation of a general theorem for finding the length of any arc of any conic hyperbola, by means of two elliptic arcs, with some other new and useful theorems deduced therefrom. *Philos. Trans. Royal Soc. London*, **65**, 1775, 283–289.

Lange, L. J.: An elegant continued fraction for π. *Amer. Math. Monthly*, **106**, 456–458, 1999.

Laplace, P. S.: *Théorie Analytique des Probabilités*. V. Courcier, Paris, 1812.

Larson, R., Edwards, B., and Heyd, D.: *Calculus of a Single Variable*, 6th ed. Houghton Mifflin, Boston–New York, 1998.

Laugwitz, D., and Rodewald, B.: A simple characterization of the gamma function. *Amer. Math. Monthly*, **94**, 1987, 534–536.

Legendre, A. M.: Mémoire de la classe des sciences mathèmatiques et physiques de l'Institute de France. Paris, 1809, 477, 485, 490.

Legendre, A. M.: *Théorie de Nombres*, Firmin Didot Frères, Paris, 1830.

Lehmer, D. H.: On the maxima and minima of Bernoulli polynomials. *Amer. Math. Monthly*, **47**, 1940, 533–538.

Lehmer, D. H.: Interesting series involving the central binomial coefficient. *Amer. Math. Monthly*, **92**, 1985, 449–457.

Lehmer, D. H.: A new approach to the Bernoulli polynomials. *Amer. Math. Monthly*, **95**, 1988, 905–911.

Leshchiner, D.: Some new identities for $\zeta(k)$. *Jour. Number Theory*, **13**, 1981, 355–362.

Levy, L. S.: Summation of the series $1^n + 2^n + \cdots + x^n$ using elementary calculus. *Amer. Math. Monthly*, **77**, 1970, 840–847.

Lewin, L.: *Polylogarithms and Associated Functions*. North Holland, New York, 1981.
Lewin, L. (editor): *Structural Properties of Polylogarithms*. Mathematical Surveys and Monographs, American Mathematical Society, **37**, 1991.
Liang, J. J. Y., and Todd, J.: The Stieltjes constants. *J. Res. Nat. Bur. Standards Sect. B Math. Sci.*, **76**, 1972, 161–178.
Lindemann, F.: Ueber die Zahl π. *Math. Annalen*, 1882, **20**, 213–225.
Linis, V., and Grosswald, E.: An improper integral. *Amer. Math. Monthly*, E 1260, 1957, 675.
Liouville, J.: Sur la détermination des integrales dont le valeur est algebrique. *J. École Polytech.*, **14**, 1833, 124–193.
Liouville, J.: Mémoire sur l'integration d'une classe de fonctions transcendentes. *Crelle J.*, **13**, 1835, 93–118.
Liouville, J.: Sur l'irrationalité du nombre $e = 2, 718 \cdots$. *Liouville Jour.* **5**, 1840, 192.
Liouville, J.: Sur l'integrale

$$\int_0^1 \frac{t^{\mu+1/2}(1-t)^{\mu-1/2}}{\left(a+bt-ct^2\right)^{\mu+1}}\, dt.$$

Journal des Mathématiques Pures et Appliquées. (2) **1**, 1856, 421–424.
Liouville, J.: Sur l'integrale

$$\int_0^1 \frac{t^{\mu+1/2}(1-t)^{\mu-1/2}}{\left(a+bt-ct^2\right)^{\mu+1}}\, dt.$$

Extrait d'une lettre de M. O. Schlomilch. Extrait d'une lettre de M. A. Cayley. Remarques de M. Liouville. *Journal des Mathématiques Pures et Appliquées.* (2) **2**, 1857, 47–55.
Little, J.: On the zeroes of two families of polynomials arising from certain rational integrals (2004) to appear in Rocky Mountain Mathematical Journal. Preprint available at http://mathes.holycross.edu/~little/pubs.html
Lodge, G., and Breusch, R.: Riemann zeta function. *Amer. Math. Monthly*, **71**, 1964, 446.
Loxton, J. H.: Special values of the dilogarithm function. *Acta Arith.*, **43**, 1984, 155–166.
Lossers, O. P., and Chico Problem Group, California State University: Evaluation of an integral. *Amer. Math. Monthly*, **92**, 1985, 516–517.
Lutzen, J.: *Joseph Liouville 1809–1882, Master of Pure and Applied Mathematics* (Studies in the History of Mathematics and Physical Sciences **15**), Springer-Verlag, New York, 1990.
Lutzsk, M.: Evaluation of some integrals by contour integration. *Amer. Math. Monthly*, **77**, 1970, 1080–1082.
Magnus, W., Oberhettinger, F., and Soni, R. P.: *Formulas and Theorems for the Special Functions of Mathematical Physics* (Die Grun. der Mathematischen Wiss. Ein. Anw., **52**), 1966. Springer-Verlag, New York.
Mallows, C. L.: A formula for expected value. *Amer. Math. Monthly*, **87**, 1980, 584.
Maor, E.: *e: The Story of a Number*. Princeton University Press, 1998.
Marchisotto, E. A., and Zakeri, G.: An invitation to integration in finite terms. *College Math. Jour.*, **25**, 1994, 295–308.

Marsaglia, G., and Marsaglia, J. C.: A new derivation of Stirling's approximation to $n!$. *Amer. Math. Monthly*, **97**, 1990, 826–829.

Massidda, V.: Analytical continuation of a class of integrals containing exponential and trigonometric functions. *Math. Comp.*, **41**, 1983, 555–557.

Matsuoka, Y. : An elementary proof of $\sum_{n=1}^{\infty} = \frac{\pi^2}{6}$. *Amer. Math. Monthly*, **68**, 1961, 485–487.

Mazia, V. G., Shaposhnikova, T. O., and Mazia, V. G.: Jacques Hadamard: A universal mathematician. *History of Mathematics*, **14**, American Mathematical Society, 1998.

McKean, H., and Moll, V.: *Elliptic Curves: Function Theory, Geometry, Arithmetic.* Cambridge University Press, 1997.

Mead, D. G.: Integration. *Amer. Math. Monthly*, **68**, 1961, 152–156.

Medina, H.: A sequence of Hermite interpolating-like polynomials for approximating arctangent (2003). Preprint available at http://myweb.lmu.edu/hmedina/Papers/arctan.pdf

Melzak, Z. A.: Infinite products for π e and π/e. *Amer. Math. Monthly*, **68**, 1961, 39–41.

Melzak, Z. A: *Companion to Concrete Mathematics*, Vol. I: *Mathematical Techniques and Various Applications.* John Wiley, New York, 1973.

Melzak, Z. A: *Companion to Concrete Mathematics*, Vol. II: *Mathematical Ideas, Modeling and Applications.* John Wiley, New York, 1976.

Mendelson, N. S.: An application of a famous inequality. *Amer. Math. Monthly*, **58**, 1951, 568.

Menon, P. K.: Some series involving the zeta function. *Math. Student*, **29**, 1961, 77–80.

Merkle, M.: Logarithmic convexity and inequalities for the gamma function. *Jour. Math. Anal. Appl.*, **203**, 1996, 369–380.

Mermin, N. D.: Stirling's formula! *Amer. J. Phys.*, **52**, 1984, 362.

Moll, V.: The evaluation of integrals: a personal story. *Notices Amer. Math. Soc.*, March 2002, 311–317.

Mordell, L. J.: The sign of the Bernoulli numbers. *Amer. Math. Monthly*, **80**, 1973, 547–548.

Murty, M. Ram: Artin's conjecture for primitive roots. *Mathematical Intelligencer*, **11**, 1988, 59–67.

Murty, M. Ram, and Reece, M.: A simple derivation of $\zeta(1-k) = -B_k/k$. Preprint.

Murty, M. Ram, and Srinivasan, S.: Some remarks on Artin's conjecture. *Canad. Math. Bull.*, **30**, 1987, 80–85.

Myerson, G., and van der Poorten, A. J.: Some problems concerning recurrence sequences. *Amer. Math. Monthly*, **102**, 1995, 698–705.

Nagaraja, K. S., and Verma, G. R.: Evaluation of the integral $\int_0^p u^n e^{-u^2} (u+x)^{-1} du$. *Math. Comp.*, **39**, 1982, 179–194.

Nahim, P.: *An imaginary tale. The story of* $\sqrt{-1}$. Princeton University Press, 1998.

Namias, V.: A simple derivation of Stirling's asymptotic series. *Amer. Math. Monthly*, **93**, 1986, 25–29.

Nathan, J. A.: The irrationality of e^x for nonzero rational x. *Amer. Math. Monthly*, **105**, 1998, 762–763.

Nemes, I., Petkovsek, M., Wilf, H., and Zeilberger, D.: How to do MONTHLY problems with your computer. *Amer. Math. Monthly*, **104**, 1997, 505–519.

Newman, D. J.: Advanced problem 4580. *Amer. Math. Monthly*, **61**, 1954, 199.

Newman, D. J.: Simple analytic proof of the prime number theorem. *Amer. Math. Monthly*, **87**, 1980, 693–696.

Newman, D. J.: A simplified version of the fast algorithm of Brent and Salamin. *Math. Comp.*, **44**, 1985, 207–210.

Newman, D. J.: *Analytic number theory*. Springer Verlag, 1998.

Newman, D. J., and Carlitz, L.: Series with negative coefficients. *Amer. Math. Monthly*, **66**, 1959, 430.

Nielsen, N.: *Traité Élémentaire des Nombres de Bernoulli*. Gauthier-Villars, Paris, 1923.

Nielsen, N.: *Die Gammafunktion*. Chelsea Publishing Company, Bronx and New York, 1965.

Ninham, B. W., Hughes, B. D., Frankel, N. E., and Glasser, M. L.: Mobius, Mellin, and mathematical physics. *Physica A*, **186**, 1992, 441–481.

Niven, I.: A simple proof that π is irrational. *Bull. Amer. Math. Soc.* **53**, 1947, 509.

Nuttall, A. H.: A conjectured definite integral. *SIAM Review*, **27**, 1985, 573.

Olds, C. D., and Davis, P.: Upper bound for an integral. *Amer. Math. Monthly*, **54**, 1947, 351–352.

Olds, C. D.: The simple continued fraction expansion of e. *Amer. Math. Monthly*, **77**, 1970, 968–974.

Ore, O.: *Cardano, the Gambling Scholar*. Dover, New York. Orig. Princeton University Press, Princeton, NJ, 1953.

Osler, T. J.: The union of Vieta's and Wallis's product for pi. *Amer. Math. Monthly*, **106**, 774–776, 1999.

Ostrowski, A.: Sur l'integrabilité élémentaire de quelques classes d'expressions. *Comm. Math. Helv.* **18**, 1946, 283–308.

Papadimitriou, I.: A simple proof of the formula $\sum_{k=1}^{\infty} k^{-2} = \pi^2/6$. *Amer. Math. Monthly*, **80**, 1973, 424–425.

Paris, R. B., and Kaminski, D.: *Asymptotics and Mellin-Barnes integrals*. Encyclopedia of Mathematics and its Applications, **85**, 2001. Cambridge University Press.

Parker, F. D.: Integrals of inverse functions. *Amer. Math. Monthly*, **62**, 1955, 439–440.

Parks, A. E.: π, e, and other irrational numbers. *Amer. Math. Monthly*, **93**, 1986, 722–723.

Patin, J. M.: A very short proof of Stirling's formula. *Amer. Math. Monthly*, **96**, 1989, 41–42.

Paule, P., and Schorn, M.: A Mathematica version of Zeilberger's algorithm for proving binomial coefficient identities. *J. Symbolic Computation*, **11**, 1994.

Pennisi, L. L.: Elementary proof that e is irrational. *Amer. Math. Monthly*, **60**, 1953, 474.

Perlstadt, M. A.: Some recurrences for sums of powers of binomial coefficients. *Jour. Number Theory*, **27**, 1987, 304–309.

Petkovsek, M., Wilf, H., and Zeilberger, D. : *A=B*. A. K. Peters, Wellesley, MA, 1996.

Philipp, S., Ismail, M., and Richberg, R.: Series involving the central binomial coefficient. *Amer. Math. Monthly*, **99**, 172–175.

Pippenger, N.: An infinite product for e. *Amer. Math. Monthly*, **87**, 1980, 391.

Plouffe, S.: On the computation of the n-th decimal digit of various transcendental constants. (March 2003). Available at http://www.labmath.uqam.ca/~plouffe/Simon/articlepi.html

Polya, G., and Szego, G.: *Problems and Theorems in Analysis*, I, Springer-Verlag, Berlin, 1972.

Potts, D. H.: Elementary integrals. *Amer. Math. Monthly*, **63**, 1956, 545–554.

Prevost, M.: A new proof of the irrationality of $\zeta(2)$ and $\zeta(3)$ using Pade approximants (2003) to appear in *J. Math. Pure Appl.*

Prudnikov, A. P., Brychkov, Yu. A., and Marichev, O. I.: *Integrals and Series*. Vol. 1. Trans. from the Russian. Gordon and Breach, New York, 1988.

Prudnikov, A. P., Brychkov, Yu. A., and Marichev, O. I.: *Integrals and Series*. Vol. 2: *Special functions*. Trans. from the Russian by G. G. Gould. Gordon and Breach, New York, 1988.

Prudnikov, A. P., Brychkov, Yu. A., and Marichev, O. I.: *Integrals and Series*. Vol. 3: *More Special Functions*. Trans. from the Russian by G. G. Gould. Gordon and Breach, New York, 1990.

Ramanujan, S.: A series for Euler's constant γ. *Messenger Math.*, 1916–1917, 73–80. Reprinted in *Collected Papers of Srinivasa Ramanujan*, (G. H. Hardy, P. V. Seshu Aiyar and B. M. Wilson, eds.), Cambridge University Press, 1927. Repr. by AMS-Chelsea, 2000.

Ramaswami, V.: Notes on Riemann ζ-function. *Jour. London Math. Soc.*, **9**, 1934, 165–169.

Rao, S. K.: A proof of Legendre's duplication formula. *Amer. Math. Monthly*, **62**, 1955, 120–121.

Rao, S. K.: On the sequence for Euler's constant. *Amer. Math. Monthly*, **63**, 1956, 572–573.

Raynor, G. E.: On Serret's integral formula. *Bull. Amer. Math. Soc.*, **45**, 1939, 911–917.

Remmert, R.: Wielandt's theorem about the Γ-function. *Amer. Math. Monthly*, **103**, 1996, 214–220.

Ribenbiom, P.: 13 *Lectures on Fermat's Last Theorem*. Springer-Verlag, 1979.

Ribenboim, P.: *The Book of Prime Number Records*. 2nd ed. Springer-Verlag, 1989.

Ribenboim, P.: *Lectures on Fermat's Last Theorem for Amateurs*. Springer-Verlag, 1999.

Risch, R. H.: The problem of integration in finite terms. *Trans. Amer. Math. Soc.*, **139**, 1969, 167–189.

Risch, R. H.: The solution of the problem of integration in finite terms. *Bull. Amer. Math. Soc.*, **76**, 1970, 605–608.

Ritt, J. F.: *Integration in Finite Terms: Liouville's Theory of Elementary Methods*, Columbia University Press, New York, 1948.

Rivoal, T.: La fonction zeta de Riemann prend une infinité de valeurs irrationnelles aux entiers impairs. *C. R. Acad. Sci. Paris*, **331**, Serie I, 2000, 267–270.

Rivoal, T.: Irrationalité d'au moins un des neuf nombres $\zeta(5)$, $\zeta(7)$, $\cdots$, $\zeta(21)$. *Acta Arith.* **103**, 2002, 157–167.

Rivoal, T.: Nombres d'Euler, approximants de Pade et constante de Catalan. *Preprint.*

Robbins, H.: A remark on Stirling's formula. *Amer. Math. Monthly*, **62**, 1955, 26–29.

Romik, D.: Stirling's approximation for $n!$: the ultimate short proof? *Amer. Math. Monthly*, **107**, 2000, 556–557.

Rosen, K.: *Discrete Mathematics and Its Applications*. 5th ed. McGraw Hill, 2003.

Rosen, M.: Niels Hendrik Abel and equations of the fifth degree. *Amer. Math. Monthly*, **102**, 1995, 495–505.

Rosenlicht, M.: Liouville's theorem on functions with elementary integrals. *Pacific Jour. Math.*, **24**, 1968, 153–161.

Rosenlicht, M.: Integration in finite terms. *Amer. Math. Monthly*, **79**, 1972, 963–972.

Roy, R.: The discovery of the series formula for π by Leibniz, Gregory and Nilakantha. *Math. Magazine*, **63**, 1990, 291–306.

Ruffini, P.: Teoria generale delle equazioni in cui si dimostra impossibile la soluzione algebraica delle equazioni generali di grado superiori al quarto. *Tomasso d'Aquino, Bologna.* Repr. in *Collected Works*, Vol. I, 159–186. Math. Circle of Palermo. Tipografia matematica, Palermo, Italy, 1950. (Original work published 1799)

Rutledge, G., and Douglas, R. D.: $\int_0^1 \frac{\log u}{u} \log^2(1 + u) \, du$ and related integrals. *Amer. Math. Monthly*, **41**, 1934, 29–36.

Salamin, E.: Computation of π using arithmetic-geometric mean. *Math. Comp.*, **30**, 1976, 565–570.

Sandham, H. F.: An approximate construction for *e*. *Amer. Math. Monthly*, **54**, 1947, 215–216.

Sandham, H. F., and Farnell, A. B.: A definite integral. *Amer. Math. Soc.*, **54**, 1947, 601–603.

Sandor, J.: Some integral inequalities. *Elem. Math.*, **43**, 1988, 177–180.

Sandor, J.: On certain limits related to the number *e*. *Libertas Math.*, **20**, 2000, 155–159.

Sandor, J., and Debnath, L.: On certain inequalities involving the constant *e* and their applications. *Jour. Math. Anal. Appl.*, **249**, 2000, 569–582.

Sasvari, Z.: An elementary proof of Binet's formula for the gamma function. *Amer. Math. Monthly*, **106**, 156–158, 1999.

Schmidt, A. L.: Legendre transforms and Apery's sequences. *J. Austral. Math. Soc.* (Series A), **58**, 1995, 358–375.

Serret, M. J. A.: Sur l'integrale $\int_0^1 \frac{\ln(1+x)}{1+x^2} \, dx$. *Jour. Math. Pure Appl. Math.* 9, 1844, 436.

Shafer, R. E., and Lossers, O. P.: Euler's constant. *Amer. Math. Monthly*, **76**, 1969, 1077–1079.

Shail, R.: A class of infinite sums and integrals. *Math. Comp.*, **70**, 2001, 788–799.

Sheldon, E. W.: Critical revision of de Haan's Tables of Definite Integrals. *Amer. J. Math.*, **34**, 1912, 88–114.

Shen, L. C.: Remarks on some integrals and series involving the Stirling numbers and $\zeta(n)$. *Trans. Amer. Math. Soc.*, **347**, 1995, 1391–1399.

Shurman, J.: *Geometry of the quintic*. John Wiley, New York, 1997.

Sierpinski, W.: *Elementary Theory of Numbers*. Ed. A. Schinzel. North Holland, Amsterdam, 1988.

Sloane, N. J. A.: *A Handbook of Integer Sequences*. Academic Press, New York, 1973.

Sondow, J.: Analytic continuation of Riemann's zeta function and values at negative integers via Euler's transformation of series. *Proc. Amer. Math. Soc.*, **120**, 1994, 421–424.

Sondow, J.: An antisymmetric formula for Euler's constant. *Math. Magazine*, **71**, 1998, 219–220.

Sondow, J.: Criteria for irrationality of Euler's constant. Proc. Amer. Math. Soc. **131**, 2003, 3335–3344.

Sondow, J.: A hypergeometric approach, via linear forms involving logarithms, to irrationality criteria for Euler's constant. Available at http://arXiv.org/PS_cache/math/pdf/0211/ 0211075.pdf [12 Nov 2002]

Sondow, J.: An infinite product for e^γ via hypergeometric formulas for Euler's constant γ. Preprint available at http://arXiv.org/PS_cache/math/pdf/0306/0306008.pdf

Sondow, J.: Double integrals for Euler's constant and $\ln(4/\pi)$. Available at http:// arXiv.org/ftp/ math/papers/0211/0211/48.pdf [20 Aug 2003]

Song, I. A recursive formula for even order harmonic series. *Jour. Comp. Appl. Math.*, **21**, 1988, 251–256.

Spanier, J., and Oldham, K.B.: *An Atlas of Functions*. Hemisphere Publishing, 1987.

Spiegel, M. R., and Rosenbaum, R. A.: An improper integral. *Amer. Math. Monthly*, **62**, 1955, 497.

Spiegel, M. R., and Stanaitis, O. E.: An improper integral. *Amer. Math. Monthly*, **62**, 1955, 262–263.

Spiegel, M. R.: Remarks concerning the probability integral. *Amer. Math.*, **63**, 1956, 35–37.

Spiegel, M. R.: *Calculus of Finite Differences and Difference Equations*. Schaum Outline Series, McGraw-Hill, New York, 1971.

Spivak, M.: *Calculus*. 2nd ed., Publish or Perish, 1980.

Srivastava, H. M.: Some infinite series associated with the Riemann zeta function. *Yokohama Math. Journal*, **35**, 1987, 47–50.

Srivastava, H. M.: A unified presentation of certain classes of series of the Riemann zeta function. *Riv. Mat. Univ. Parma* (4), **14**, 1988, 1–23.

Srivastava, H. M.: Sums of certain series of the Riemann zeta function. *Jour. Math. Anal. Appl.*, **134**, 1988, 129–140.

Srivastava, H. M.: Some rapidly convergent series for $\zeta(2n + 1)$. *Proc. Amer. Math. Soc.*, **127**, 1999, 385–396.

Srivastava, H. M.: Some simple algorithms for the evaluations and representations of the Riemann zeta function at positive integer arguments. *J. Math. Anal. Appl.*, **246**, 2000, 331–351.

Srivastava, H. M., and Choi, J.: *Series associated with the zeta and related functions*. Kluwer Academic Publishers, 2001.

Srivastava, H. M., Glasser, M. L., and Adamchik, V.: Some definite integrals associated with the Riemann zeta function. *Z. Anal. Anwendungen*, **19**, 2000, 831–846.

Staib, J. H.: The integration of inverse functions. *Math. Magazine*, **39**, 1966, 223–224.

Stanley, R.: Log-concave and unimodal sequences in algebra, combinatorics, and geometry. In *Graph Theory and its Applications: East and West* (Jinan, 1986), 500–535; *Ann. New York Acad. Sci.*, **576**, New York, 1989.

Stanley, R.: *Enumerative Combinatorics*, Vol. I. Cambridge Studies in Advanced Mathematics, Cambridge University Press, 1999.

Stark, E. L.: Another proof of the formula $\sum_{k=1}^{\infty} 1/k^2 = \pi^2/6$. *Amer. Math. Monthly*, **76**, 1969, 552–553.

Stark, E. L.: The series $\sum_{k=1}^{\infty} k^{-s}$, $s = 2, 3, 4, \cdots$, once more. *Math. Mag.*, **47**, 1974, 197–202.

Stein, S. K.: Formal integration: Dangers and suggestions. *Two-Year College Mathematics Journal*, **5**, 1974, 1–7.

Stieltjes, T. J.: Tables des valeurs des sommes $S_k = \sum_{n=1}^{\infty} n^{-k}$. *Acta Mathematica*, **10**, 1887, 299–302.

Stieltjes, T. J.: Sur le développement de $\log \Gamma(a)$. *Jour. Math. Pure Appl.*, (4)**5**, 1889, 425–444. Repr. in *Oeuvres*, **2**, Noordhoff 1918, 211–230; 2nd ed., Springer, 1993.

Stirling, J.: *Methodus Differentialis: Sive Tractatus de Summatione et Interpolatione Serierum Infinitarum*. London, 1730.

Stromberg, K. R.: *An Introduction to Classical Real Analysis*. Wadsworth, Belmont, CA, 1981.

Sweeney, D. W.: On the computation of Euler's constant. *Math. Comp.* **17**, 1963, 170–178.

Talvila, E.: Some divergent trigonometric integrals. *Amer. Math. Monthly*, **108**, 2001, 432–436.

Tartaglia, N.: *Quesiti et inventioni diverse*. Fascimile of 1554 edition. Ed. A. Masotti. Ateneo di Brescia, Brescia.

Thomas, G. B., and Finney, R.: *Calculus and Analytic Geometry*, 9th ed. Addison-Wesley, 1996.

Titchmarsh, E. C.: *The Zeta-Function of Riemann*. Cambridge University Press, 1930.

Titchmarsh, E. C.: *Introduction to the Theory of Fourier Integrals*, 2nd ed., Oxford at the Clarendon Press, 1948.

Totik, V.: On Holder's theorem that $\Gamma(x)$ does not satisfy algebraic differential equation. *Acta Sci. Math.* (Szeged), **57**, 1993, 495–496.

Turnwald, G.: Letter to the editor. *Amer. Math. Monthly*, **95**, 1988, 331.

Tweddle, I.: Approximating $n!$. Historical origins and error analysis. *Amer. J. Phys.*, **52**, 1984, 487–488.

Tweedle, I.: *James Stirling: 'This about series and such things'*. Scottish Academic Press, Edinburgh, 1988.

Tyler, D., and Chernhoff, P.: An old sum reappears. *Amer. Math. Monthly*, **92**, Elementary Problem 3103, 1985, 507.

Umemura, H.: Resolution of algebraic equations by theta constants. Appendix to *Tata Lectures on Theta II*, by D. Mumford. Progress in Mathematics, Birkhauser, 1983.

Underwood, R. S.: An expression for the summation $\sum_{m=1}^{n} m^p$. *Amer. Math. Monthly*, **35**, 1928, 424–428.

Underwood, R. S.: Some results involving π. *Amer. Math. Monthly*, **31**, 1924, 392–394.

Vacca, G.: A new series for the Eulerian constant $\gamma = .577 \cdots$. *Quart. J. Pure Appl. Math.*, **41**, 1910, 363–368.

van Poorten, A.: A proof that Euler missed $\cdots$ Apery's proof of the irrationality of $\zeta(3)$. An informal report. *Math. Intelligencer* **1**, 1979, 195–203.

van Poorten, A.: Some wonderful formulae $\cdots$ footnotes to Apery's proof of the irrationality of $\zeta(3)$. *Seminaire Delange-Pisot-Poitou* (Theorie des Nombres), **29**, 1978–1979, 7pp.

van Poorten, A.: Some wonderful formulae $\cdots$ an introduction to Polylogarithms. Proceedings of Number Theory Conference, Kingston, Ontario. *Queen's Papers in Pure Appl. Math.*, **54**, 1979, 269–286.

van Poorten, A.: p-adic methods in the study of Taylor coefficients of rational functions. *Bull. Austral. Math. Soc.*, **29**, 1984, 109–117.

van Poorten, A.: Some facts that should be better known; especially about rational functions. In *Number Theory and Applications*, ed. Richard Mollin (NATO-Advanced Study Institute, Banff, 1988). Kluwer Academic Publishers, Dordrecht, 1989, pages 497–528.

Vardi, I.: Integrals, an introduction to analytic number theory. *Amer. Math. Monthly* **95**, 1988, 308–315.

Venkatachaliengar, K.: Elementary proofs of the infinite product for $\sin z$ and allied formulae. *Amer. Math. Monthly*, **69**, 1962, 541–545.

Verma, D. P.: A note on Euler's constant. *Math. Student*, **29**, 1961, 140–141.

Verma, D. P., and Kaur, A.: Summation of some series involving Riemann zeta function. *Indian J. Math.*, **25**, 1983, 181–184.

Vieta, F.: Variorum de Rebus Mathematicis Reponsorum Liber VII, 1593. In *Opera Mathematica*, Georg Olms Verlag, Hildesheim, New York, 1970, pp. 398–400 and 436–446.

Wadhwa, A. D.: An interesting subseries of the harmonic series. *Amer. Math. Monthly*, **82**, 1975, 931–933.

Wagon, S.: Is π normal? *The Math. Intelligencer*, **7**, 1985, 65–67.

Waldschmidt, M.: On the numbers e^e, e^{e^2} and e^{π^2}. *The Hardy-Ramanujan Journal*, **21**, 1998, 1–7.

Wallis, J. : *Arithmetica Infinitorum*. Oxford, 1656.

Wallis, J.: Computation of π by Successive Interpolations, 1655. In *A Source Book in Mathematics, 1200–1800*, (D. J. Struik, ed.), Harvard University Press, Cambridge, MA, 1969, 244–253.

Walter, W.: Old and new approaches to Euler's trigonometric expansions. *Amer. Math. Monthly*, **89**, 1982, 225–230.

Weinstock, R.: Elementary evaluations of $\int_0^\infty e^{-x^2} dx$, $\int_0^\infty \cos x^2\, dx$, and $\int_0^\infty \sin x^2\, dx$. *Amer. Math. Monthly*, **97**, 39–42.

Weisstein, E. W.: *CRC Concise Encyclopedia of Mathematics*. Chapman & Hall/CRC, 1999.

Whittaker, E. T., and Watson, G. N.: *A Course in Modern Analysis*, 4th ed. Cambridge University Press, 1961.

Wilf, H. S.: *generatingfunctionology*. Academic Press, 1990.

Wilf, H. S.: The asymptotic behavior of the Stirling numbers of the first kind. *Jour. Comb. Theory*, Ser. A, **64**, 1993, 344–349.

Williams, G. T.: A new method for evaluating $\zeta(2n)$. *Amer. Math. Monthly*, **60**, 1953, 19–25.

Williams, K. S.: On $\sum_{n=1}^\infty 1/n^{2k}$. *Math. Mag.*, **44**, 1971, 273–276.

Wilton, J. R.: A proof of Burnside's formula for log $\Gamma(x+1)$ and certian allied properties of Riemann's ζ-function. *Messenger Math.*, **52**, 1922–1923, 90–93.

Wilton, J. R.: A note on the coefficients in the expansion of $\zeta(s, x)$ in power of $s - 1$. *Quart. J. Pure Appl. Math.*, **50**, 1927, 329–332.

Wimp, J.: *Sequence Transformations and Their Applications*. Academic Press, New York, 1981.

Wimp, J.: Review of *Tables and Integrals, Series and Products: CD-ROM Version 1.0* by I. S. Gradshteyn and I. M. Ryzhik, ed. by Alan Jeffrey. Academic Press, 1996. *Amer. Math. Monthly*, **104**, 373–376.

Wolfram, S.: *Mathematica - A System for Doing Mathematics by Computer*. Addison-Wesley, 1998.

Woon, S. C.: Generalization of a relation between the Riemann zeta function and Bernoulli numbers. Preprint available at http://arXiv.org/PS_cache/math/pdf/9812/9812143.pdf [24 Dec 1998]

Yang, B., and Debnath, L.: Some inequalities involving the constant e, and an application to Carleman's inequality. *Jour. Math. Anal. Appl.*, **223**, 347–353.

Young, R. M.: Euler's constant. *Math. Gazette* **75**, 1991, 187–190.

Yue, Z. N., and Williams, K. S.: Some series representations of $\zeta(2n + 1)$. *Rocky Mount. Journal*, **23**, 1993, 1581–1592.

Yue, Z. N., and Williams, K. S.: Some results on the generalized Stieltjes constants. *Analysis*, **14**, 1994, 147–162.

Yue, Z. N., and Williams, K. S.: Values of the Riemann zeta function and integrals involving $\log(2\sinh\theta/2)$ and $\log(2\sin\theta/2)$. *Pac. Jour. Math.*, **168**, 1995, 271–289.

Yzeren, J. Van: Moivre's and Fresnel integrals by simple integration. *Amer. Math. Monthly*, **86**, 1979, 691–693.

Zeilberger, D.: Theorems for a price: tomorrow's semi-rigorous mathematical culture. *Notices Amer. Math. Soc.*, **40**, 1993, 978–981. Repr. in *Math. Intelligencer*, **16**, 1994, 11–14.

Zeitlin, D.: On a class of definite integrals. *Amer. Math. Monthly*, **75**, 1968, 878–879.

Zerr, G. B. M.: Summation of series. *Amer. Math. Monthly*, **5**, 1898, 128–135.

Zhang, N., and Williams, K.: Some series representations of $\zeta(2n + 1)$. *Rocky Mount. J. Math.*, **23**, 1993, 1581–1592.

Zhang, N., and Williams, K.: Values of the Riemann zeta function and integrals involving $\log(2\sinh\theta/2)$ and $\log(2\sin\theta/2)$. *Pacific J. Math.*, **168**, 1995, 271–289.

Zucker, I. J.: On the series $\sum_{k=1}^{\infty} \binom{2k}{k}^{-1} k^{-n}$ and related sums. *Jour. Number Theory*, **20**, 1985, 92–102.

Zucker, I. J., Joyce, G. S., and Delves, R. T.: On the evaluation of the integral $\int_0^{\pi/4} \ln(\cos^{m/n}\theta \pm \sin^{m/n}\theta)\,d\theta$. *The Ramanujan Jour.*, **2**, 1998, 317–326.

Zudilin, W.: One of the numbers $\zeta(5)$, $\zeta(7)$, $\zeta(9)$, $\zeta(11)$ is irrational. *Russian Math. Surveys* **56**, 2001, 774–776.

Zudilin, W.: An elementary proof of Apery's theorem. Available at http://arXiv.org/PS_cache/math/pdf/0202/0202159.pdf (17 Feb. 2002)

Zudilin, W.: A third-order Apery-like recursion for $\zeta(5)$. *Mat. Zametki* [*Math. Notes*] **72** (2002). Available at http://arXiv.org/PS_cache/math/pdf/0206/0206178.pdf

Index

Printed in the United States
By Bookmasters